Animal Diversity, Natural History and Conservation

— Vol. 3 —

The Editors

Dr. Vijay Kumar Gupta, Ph.D., FLS, London (born 1953-) Chief Scientist, CSIR- Indian Institute of Integrative Medicine, Jammu, India. He did his M.Sc. (1975) and Ph.D. (1979) in Zoology both from University of Jammu, Jammu-India. His research capabilities are substantiated by his excellent work on histopathology, ecology and reproductive biology of fishes, turtles, birds and mammals, which has already got recognition in India and abroad.

Dr. Gupta has to his credit more than 100 scientific publications and review articles which have appeared in internationally recognized Indian and foreign journals. Founder fellow, life member and office bearer of many national societies, academies and associations. He has successfully completed a number of research/consultancy projects funded by government, private and multinational agencies. His current areas of interest are histopathology, toxicology, pre-clinical safety pharmacology, reproductive efficacy studies of laboratory animals and biodiversity.

He is the Series Editor of the recently published multi-volume set of books, "**Comprehensive Bioactive Natural Products (Vols. 1-8)"**, published by M/S Studium Press, LLC, USA. He is also Editor-in-Chief of the books, "**Utilisation and Management of Medicinal Plants"**, "**Medicinal Plants: Phytochemistry, Pharmacology and Therapeutics (Vols.1&2)"**, **"Traditional and Folk Herbal Medicine"**, **"Natural Products: Research Reviews"**, **"Bioactive Phytochemicals: Perspectives for Modern Medicine"**, **"Perspectives in Animal Ecology and Reproduction (Vols.1-8)**" and **"Animal Diversity, Natural History & Conservation"**. The Editor-in-chief of the American Biographical Institute, USA, has appointed him as *Consulting Editor* of *The Contemporary Who's Who*. Dr. Gupta also appointed as Nominee for the *Committee for the Purpose of Control and Supervision of Experiments on Animals* (CPCSEA, Govt. of India). The *Linnaean Society of London, U.K.* has awarded fellowship to him in November 2009 in recognition of his contribution towards the cultivation of knowledge in Science of Natural History. Recently, Modern Scientific Press, USA has nominated Dr. Gupta as the Editor of the *International Journal of Traditional and Natural Medicine*.

Dr. Anil K. Verma, Ph.D., M.N.A.Sc., FLS, London (born 1963-) Prof. & Head, Department of Zoology, Govt. (P.G.) Degree College Rajouri, J&K State, did his M.Sc. in Zoology (1986) from University of Jammu, Jammu. He has undergone his M.Phil. (1988) and awarded first rank and Ph.D.(1993) in the field of animal reproduction at the same University. Dr. Verma has published about 60 research papers and review articles in reputed journals and books. He is also a member Editorial Board of the book series "Advances in Fish and Wildlife: Ecology and Biology" a Daya Publishing House, New Delhi. In recognition of his standing in greater scientific community, the Board of Directors of the American Association for the advancement of science (AAAS) New York, Washington, has awarded membership to him. The *Linnaean Society of London, U.K.* has awarded fellowship to him in October 2006 in recognition of his contribution towards the cultivation of knowledge of Science of Natural History.

Dr.Verma was also nominated as Nominee, CPCSEA by Ministry of Environment & Forests, Govt. of India, and also awarded "The Rashtriya Gourav Award" and "The Best Citizen of India" award by the India International Friendship Society in 2009-2010. He is also a nominated member of the Global Academy of Science, India.

Animal Diversity, Natural History and Conservation

— Vol. 3 —

— *Editor-in-Chief* —

Dr. V.K. Gupta

Chief Scientist

CSIR-Indian Institute of Integrative Medicine,
Canal Road, Jammu – 180 001, India

Editor

Dr. Anil K. Verma

Professor and Head

Department of Zoology, Government (P.G.) Degree College,
Rajouri, Jammu – 180 001, India

2013

Daya Publishing House®

A Division of

Astral International Pvt. Ltd.

New Delhi – 110 002

ISBN 9789351301165

Published by : **Daya Publishing House®**
A Division of
Astral International Pvt. Ltd.
4760-61/23, Ansari Road, Darya Ganj
New Delhi-110 002
Phone: 011-4354 9197, 2327 8134
Fax: +91-11-2324 3060
E-mail: info@astralint.com
Website: www.astralint.com

Laser Typesetting : **Classic Computer Services**
Delhi - 110 035

Printed at : **Salasar Imaging Systems**
Delhi - 110 035

PRINTED IN INDIA

Editorial Board

Prof. Cosmas Nathanailides

Department of Aquaculture and Fisheries
TEI of EPIRUS, Igoumenitsa, GREECE
E-mail: cosmasfax@yahoo.com

Prof. A.K. Choubey

Nematology Laboratory,
Department of Zoology, Chaudhary Charan Singh University,
Meerut – 250 005, INDIA
E-mail: kumar_ac2001@rediffmail.com

Prof. D.N. Sahi

Department of Zoology,
University of Jammu ,
Jammu – 180 005, INDIA
E-mail: dnsahi@gmail.com

Prof. B.L. Koul

Ex-Principal
Govt. S.P.M.R.College of Commerce & Management,
Jammu – 180 001, INDIA
E-mail: proban@sancharnet.in

பாரதியார் பல்கலைக்கழகம்
BHARATHIAR UNIVERSITY
COIMBATORE - 641 046
Silver Jubilee Year 1982 - 2007
Accredited with 'A' Grade by NAAC
Ph : 091- 422-2422203, Fax : 091-422-2422387, 2425706, Website : www.b-u.ac.in E-mail : regr@b-u.ac.in Grams : 'UNIBHARATHIAR'

Foreword

Animal Ecology and Reproduction has a long standing tradition of long-term studies on entire animal kingdom, in combination with elegant laboratory and field experiments. The book links the novel research experiments and review studies on various groups and individuals of animal kingdom including behaviour, physiology, biochemical, population dynamics and evolutionary changes in response to environmental variation. It also includes Diversity, Ecology and Season behaviour of Dipterans, Entomopathogenic Nematodes, Bacteria, Termites, Acridoidea, Bugs, Molluscs, Turtles and Tortoise all over the world. Findings in this book will help an underpinning of nature conservation efforts worldwide. Knowledge on Biodiversity and conservation are the basis for environmental health and human being because lives of all human being are dependent on plants, animals and other environmental factors. In a contribution to this volume, a consortium of scientists from the various Universities and Institutions from India described the concerned factor and their management. In particular the present volume will enlighten the readers with knowledge on Biodiversity, Animal Ecology and Conservation.

Dr. K. Murugan, Ph.D., D.Sc.,
Professor
Department of Zoology,
School of Life Sciences
Bharathiar University,
Coimbatore – 641 046,
Tamil Nadu, India

Preface

Planet Earth is endowed with a rich variety of life forms and the teaming millions of these living organisms have been well knit by the laws of nature. The interdependence of the various life forms is a unique feature of the planet earth. Biodiversity is very basic of human survival and economic well being and encompasses all life forms, ecological processes and ecosystems acknowledging the hierarchy at genetic taxon and ecosystem level. Biodiversity is responsible for the essential ecosystems services, including regulation of the atmospheric gaseous composition, climate disturbances and waste assimilation; recreation and biological control. An environment rich in biological diversity offers the broadest array of options for sustainable economic activity, for sustaining human welfare and for adapting to change. Loss of biodiversity has serious economic and social costs for any country.

The past experience has shown that as industrialization and economic development takes place, patterns of consumption, production and needs, change the strain, alter and even destroy the ecosystem. Major threats to the bio-diversity are habitat destruction, extension of agriculture, filling up of the wetlands, conversion of rich bio-diversity site for human settlements industrial development, destruction of costal areas and uncontrolled commercial exploitation etc. The erosion of biodiversity is largely due to habitat loss caused by the expansion of various development projects leading to the fragmentation of the continuous stretches of land and consequently fragmenting the wildlife populations inhabiting their. Apart from the primary loss of the habitats, there are numerous other problems contributing to the loss and endangered status of several plant and animals species. So conservation of biodiversity is essential for human survival.

It was in this backdrop that a book series entitled, "Animal Diversity, Natural History and Conservation" (Volume 1 of which was published in 2011) was imitated by us so as to share and compile the data/ideas of various academicians and researchers from far and wide places of the globe on a common platform for its further dissemination to distant quarters. We have succeeded in getting an overwhelming response from our esteemed contributors who have been instrumental in the speedy and successful compilation of the two more volumes (Volume 2 & 3) reflecting a greater concern by the scientific community on these vital aspects of life and nature, our surroundings.

In this volume, papers/articles on bacterial diversity, diversity of nematodes, population dynamics of butterfly, turtle & tortoises diversity, natural history of *Herpestes palustris* in West Bengal, termite diversity in Western Ghats, animal diversity in Kuldiha Wildlife Sanctuary, Odhisa, Review on reverine fin-fish diversity, ecology of spotted owlet in Tamil Nadu, diversity of spittle bugs and orthoptera in Southern India, distribution and conservation status of syngnathids fish, human-elephant conflicts in Karanjia, ecology and status of Ganga dolphin- the national aquatic animal of India; an account on the great Indian one-horned rhinos are the fascinating chapters to quote a few of them. Other notable contributions included in this volume are: Status, diversity and distribution of freshwater mollusca of India; A glimpse of the avifaunal diversity of Jammu Himalayas; and Exotic ornamental fishes in India: Perspectives and problems.

This volume like the previous one shall definitely prove very useful to the researchers who are interested in seeking knowledge/information pertaining to the diversity, natural history and conservation of the animal. We are thankful to the publisher M/S Astral International, New Delhi for publishing the book well in time without compromising the quality of the contents and the printing. Contributions from our esteemed contributors are also greatly acknowledged and errors if any are owned by the editors.

Dr. V.K. Gupta, *FLS*

Contents

2013, Animal Diversity, Natural History and Conservation, Vol. 3 *Pages* **1–17**
Editors: **V.K. Gupta and Anil K. Verma**
Published by: **DAYA PUBLISHING HOUSE, NEW DELHI**

Chapter 1

Bacterial Diversity in Faecal Microbiota of Badgers *Meles meles* Linnaeus (1758), in Portugal

Luís Miguel Rosalino[1,2], Mafalda Pinto Basto[1], Teresa Sales-Luís[1], Nuno Miguel Pedroso[1], Luís Tavares[3], Cristina Lobo Vilela[3] and Manuela Oliveira*[3]

[1]*Universidade de Lisboa, Centro de Biologia Ambiental, Faculdade de Ciências de Lisboa, Ed. C2, Campo Grande, 1749-016 Lisboa, Portugal*
[2]*Laboratório de Ecologia Isotópica/CENA, Universidade de São Paulo, Caixa Postal 96, Piracicaba, SP 13400-970, Brasil*
[3]*CIISA/Laboratório de Microbiologia e Imunologia, Faculdade de Medicina Veterinária, Universidade Técnica de Lisboa, Avenida da Universidade Técnica, 1300-477 Lisboa, Portugal*

ABSTRACT

The Eurasian badger *Meles meles* Linnaeus (1758) is a medium-sized carnivore, widely distributed in Europe, including Portugal. Badgers can be affected by viral, bacterial, mycotic and parasitic diseases. However, most studies regarding bacterial diseases are focused on *Mycobacterium*, and little is known about the occurrence of other bacterial genera.

This work represents a first approach to characterize the badgers faecal microbiota in samples collected in "Serra de Grândola", a mountainous area located in the coastal south western region of Portugal. Animals were captured and handled following the recommendations of the Animal Welfare Protocol of the European Union. Faecal samples were collected after anaesthesia. The microbiota studied included aerobic and sporulating anaerobic bacteria. Bacterial isolates were obtained using conventional microbiological methods and identified based on morphology and metabolic pathways. It was possible to isolate and identify several gram-negative and gram-positive bacterial species, some of which may have a potential pathogenic role for badger populations.

This study revealed that the faecal microbiota of *Meles meles* presents a high level of bacterial diversity, including potentially pathogenic bacteria, such as *Staphylococcus aureus, Staphylococcus* sp., *Streptococcus mitis* 1, *Bacillus cereus, Escherichia coli, Proteus mirabilis, Salmonella* spp., *Enterobacter aerogenes* and *Klebsiella oxytoca.*

This evidence is an important contribution for understanding the relevance of pathogenic bacteria in badger population health, and also their epidemiologic role as vehicles for human and animal infections or outbreaks.

Keywords: *Eurasian badger, Faecal microbiota, Pathogenic bacteria, Southwest Portugal.*

Introduction

Iberian Badgers

The genus *Meles* originated in the temperate forests of Asia and dispersed west as far as Norway, the British Isles and Portugal, giving rise to its current distribution range. The species *Meles meles* distribution has been somewhat controversial, due to the taxonomic interpretation of its population morphological variation. Some authors argue that it is just one species throughout the entire range, that comprises almost all of Eurasia, including the islands of Crete and Rhodes, ranging from the British Isles to Japan (Neal and Cheeseman, 1996; Mitchell-Jones *et al.,* 1999), with the distribution being delimited by the taiga or boreal forest, in the north, and by desert environment, in the south (Palomo and Gisbert, 2002). However, recent DNA-based studies (Marmi *et al.,* 2006) claim the existence of four distinct groups of this species: an European, a Southwest Asian (considered both *Meles meles* by Abramov and Puzachenko, 2005 in a morphological study), a North and East Asian (identified as *Meles leucurus* according to Abramov and Puzachenko, 2005) and a Japanese group (identified as *Meles anakuma* by Abramov and Puzachenko, 2005).

The Eurasian badger *Meles meles* Linnaeus (1758) is a medium-sized carnivore, whose weight and body size range from 6.7 kg to 18.0 kg and from 45 cm to 88 cm, respectively (Roper, 2010). In the Iberian Peninsula individuals are usually lighter and smaller than their northern and central European counterparts (Mangas *et al.,* 2003), ranging from 4.8 kg to 9.3 kg, while the length may vary from 58 cm to 75 cm (Revilla *et al.,* 1999). In this South Western European region, badgers seem to be widely distributed, reaching from the Valencia Region, on the Spanish Mediterranean east coast, to the western Portuguese seashore, and from Galicia, Asturias and Basque Country, in the north, to Algarve (Portugal) and Andalucía (Spain) in the south (Palomo and Gisbert, 2002; Santos-Reis *et al.,* 2005).

The mentioned above variability in morphometric characteristics is also expressed in ecological adaptations to the Iberian environmental conditions. Although maintaining a gregarious behaviour, they live in social groups as a result of the philopatric behaviour of juveniles, being observed that approximately 80 per cent of the young badgers never leave their natal group (Macdonald, 2001). However, Iberian groups are smaller, being formed by 6 to 8 animals (Rosalino *et al.,* 2004), whereas in the UK groups may reach 35 animals (Neal and Cheeseman, 1996). As a result of this social organization of the Iberian population and of the different use of the availability of trophic resources, territory sizes vary from 0.73 km^2 in North Eastern Spain (Molina-Vacas *et al.,* 2009) to 7.8 km^2 in South Western Spain (Revilla and Palomares, 2002), reaching intermediate values (4.46km^2) in Portugal (Rosalino *et al.,* 2004).

The social organization of each group has the communal sett as the pillar around which the social and spatial activities occur. This structure consists of a burrow system that, as badgers are mostly nocturnal, provides shelter for the entire group during the day and is also used for breeding (Roper, 2010). The system may reach a length of 879 m with 129 entrances (Macdonald, 2001), although a typical Iberian sett would have 3 to 10 large entrances leading to an intricate system of interconnected tunnels and chambers (Neal and Cheeseman, 1996).

The social structure of populations is mainly sustained by maintaining a net of latrines (pits dug near setts, territory boundaries or trophic patches used regularly to defecate) which are used by group members or neighbouring groups as scent marking sites (Kruuk, 1989; Macdonald, 2001) and territories landmarks.

Although Iberian badgers behave more as food gatherers, being mainly adapted for consuming fruits (*e.g.*, olives) and insects (Rosalino *et al.,* 2005a), they may also show some carnivorous behaviour, predating upon wild rabbits (*Oryctolagus cuniculus*) during their nocturnal movements (Rosalino *et al.,* 2005b). This trophic adaptation shows quite a drastic change in the species most common food behaviour, which is mostly focused on earthworms, *Lumbricus terrestris,* in the northern and eastern areas of Europe (Kruuk, 1989; Brøseth *et al.,* 1997; Goszczynski *et al.,* 2000; Roper 2010).

Badgers are one of the European carnivores that better tolerate men's presence, being able to live even within highly populated cities, such as Birmingham in the UK, Brno in the Czech Republic or Vilnius in Lithuania (Roper, 2010). However, in Iberia no records of such proximity to towns were reported, although ecological data reveal that they use resources highly associated with human activities, such as fruit yards and plantations, consuming production fruits such as olives (Rosalino *et al.,* 2005a). Nevertheless, conflicts resulting from this Human linked resources use are common to all European populations of this carnivore, and usually associated with crop damage, lawns and pasture damages, predation on livestock (*e.g.*, lambs) or game species (*e.g.*, wild rabbits). As the species is classified as Least Concern in the IUCN Red List of Threatened Species (http://www.iucnredlist.org/, assessed on 25/01/2012) and is protected by national and European laws, namely by The Protection of Badger Act, 1992 in the UK and the EU Habitat Directive (Cabral *et al.,* 2005; Roper, 2010), this conflict is often difficult to overcome. For this reason, many of the main threats to badger survival in Europe, and particularly in Iberia, are Human related and include road casualties, illegal persecution due to economic damages caused to agriculture and pastoral production, and habitat loss due to deforestation, establishment of new plantations or extensive agricultural fields.

The ability to inhabit environments with human influence also pose some constrains to this species. These aptitudes together with other ecological characteristics (*e.g.*, social behaviour) can promote the dispersion and maintenance of some pathologies. Several studies have pointed out badgers as reservoirs of some zoonotic diseases, which could affect domestic animals, other wildlife or man, such as rabies, canine distemper encephalitis, viral or bacterial pneumonia, salmonellosis, leptospirosis, enteritis or borovskiis disease (Hancox, 1980). In Iberia, badgers have been identified as hosts for several ectoparasites and other infectious agents, namely fleas (*e.g., Pulex* sp.), ticks (*e.g., Ixodes* sp.), lice (*e.g., Trichodectes* sp.), helminths (*e.g., Mastophorus* sp.), bacteria (*e.g., Mycobacterium*) and protozoans (*e.g., Toxoplama*) (Rosalino, 2008).

Iberian Badgers Microbiota

The number of studies concerning the characterization of mammalian wildlife faecal microbiota has exponentially increased over the last few decades (Vulfson *et al.,* 2003). Nonetheless, to our knowledge, in Portugal valid microbiota characterization studies on badgers are missing and little is known about the clinical significance of isolates and environmental contaminant-related diseases.

In Iberia, the badgers' population is widespread, and these animals are frequently in contact with cattle, free-range pigs, sheep and goats, which occupy large areas of extensive farming in this region (Rosalino *et al.,* 2004). The areas surrounding badger's faecal deposits are commonly used by grazing herbivores, and represent high risk transmission of infectious and parasitic diseases via faecal-oral (Scantlebury *et al.,* 2004).

Badgers can be affected by viral, bacterial, mycotic and parasitic diseases. Studies on infectious agents diseases carried by badgers, focusing the characterization of specific animal populations, were already performed (Wilson *et al.,* 2003; Courtenay *et al.,* 2006; Dietz *et al.,* 2006; McCarthy *et al.,* 2009), and include essentially data on tuberculosis, which represents a serious public health problem.

These studies are extremely important to fully understand the epidemiology of such diseases in specific regions. In fact, the international interest in these issues has increased, particularly in the United Kingdom, due to their impact upon livestock economy. Bovine tuberculosis, for example, is one of the more complex veterinary public health issues that farms face in this country (Mill *et al.,* 2012; Reynolds, 2006; Spencer, 2011).

There are several studies that describe a link between infections by *Mycobacterium bovis* in badgers and tuberculosis outbreaks in cattle, showing that badgers may represent a potential obstacle for the control of bovine tuberculosis (Courtenay *et al.,* 2006). Badgers and other wildlife animals are reservoirs of *M. bovis* (Quinn *et al.,* 1994) and the transmission to cattle and other ruminants occurs through the pasture contaminated by bacilli excreted through urine, stool and possibly by pus from abscesses (Scantlebury *et al.*, 2004). The badgers form social groups (Rosalino *et al.*, 2004) that use common latrines whose conditions facilitate the transmission of *M. bovis*, constituting a potential focus of environmental contamination (Courtenay *et al.*, 2006).

M. bovis is the causative agent of bovine tuberculosis (Courtenay *et al.*, 2006), which can also be transmitted to humans (Quinn *et al.*, 1994). This bacterium is responsible for the necrosis of the lymph nodes and for the gaseous erosion of bronchi, blood vessels and viscera (Quinn *et al.*, 1994). While its prevalence is underestimated in many parts of the globe, this zoonosis is widespread worldwide. Tuberculosis infections are observed in almost all species of domestic and wildlife mammals. Although the *M. bovis* main reservoirs are cattle, wildlife and humans, other bacteria belonging to the *Mycobacterium* genus may be present in a very diverse range of hosts, including cats, dogs, deer, birds, fish and turtles, among others (Quinn *et al.*, 1994).

Studies regarding the importance of badgers as potential reservoirs for the transmission of pathogenic bacteria to humans and other animals, including domestic, farming and wild animals, are extremely important for the establishment of effective control programs. Some possible measures may include: (*i*) the identification and control of major animal carriers; (*ii*) vaccination; (*iii*) the sealing of latrines and other potentially contaminated sites; (*iv*) decontamination of sites; (*v*) drainage of soil and pastureland; (*vi*) the use of protective clothing and footwear, and personal hygiene measures, in high-risk situations or activities; and (*vii*) inform the population regarding badgers as potential bacterial carriers, through education and health programs, aiming risk groups in particular.

As above mentioned, the badgers' health status in Portugal is unknown. This work represents a first approach to characterize the viable faecal bacterial microbiota of a wild Eurasian badger Portuguese population, and to evaluate the presence of potential pathogenic microorganisms.

Material and Methods

Study Area

The study derived from biological samples collected in a wild Eurasian badger population inhabiting "Serra de Grândola", a mountainous area, with an elevation range of 150 to 270m a.s.l., located in the coastal south western region of Portugal (38°07′N, 8°36′W) (Figure 1.1). This area is characterized by a heterogeneous landscape where native (*e.g.*, riparian vegetation) and non-native (*e.g.*, pastures, olive groves, orchards and vegetable gardens and eucalyptus plantations) patches are imbibed within a cork oak (*Quercus suber*) woodland matrix (Figure 1.1). The oak forest understory structure ranges from pastures to highly dense scrubs (*e.g.*, *Cistus ladanifer*, *Cistus salvifolius*). Human settlements are sparse and activities are generally restricted to cork extraction occurring in the summer, also cattle, Iberian pigs, and sheep production with small herds in an extensive regime, traditional agriculture and hunting, mainly Red-legged Partridge, *Alectoris rufa*, and European Turtle Dove, *Streptopelia turtur* (Correia and Santos-Reis 1999).

Figure 1.1: Study Area Location in Iberian Peninsula and Cork Oak Woodland Landscape Matrix

The climate in this area is Mediterranean with Atlantic influences, and seasonality is highly marked (Blondel and Aronson, 1999). Summer is the hottest and driest season with temperatures ranging from -0.6°C in the winter and 37.0°C in summer. Rain fall is concentrated in the winter, where it reaches 87.9 mm, while the summer is characterized by droughts and scarce precipitation, averaging 20.2 mm (Loureiro, 2008).

Captures, Handling and Sample Collection

During the implementation of an ecology and conservation study focused on environmental adaptation of badgers to Mediterranean landscapes, developed in 2007-2008, several of these mustelids were live-trapped under an appropriate license from the Institute for Nature Conservation, using box traps (model 608, Tomahawk Live Trap Co., Wisconsin, United States of America) (Figure 1.2).

In order to optimise the number of captures, the species ecological social behaviour was taken into consideration for trap placement. Thus, all traps were set near known badger setts, specifically in the proximity of the most used entrances (Figure 1.3). Traps were bated with a mixture of oatmeal and peach jam, and to minimize the animals' stress, they were checked twice a day, early in the morning, when badger activity has ceased, and early in the night, when animals started their activity (Rosalino *et al.*, 2005b).

All captured animals were handled following the recommendations of the Animal Welfare Protocol of the European Union, and anesthetized with HCl Ketamine (Imalgene 1000, Rhône Mérieux Lab., Lyon, France), at a dose of 0.1 ml per kg. Then, a faecal sample from the rectum of each captured individual was collected using an AMIES swab (FL medical) (Figure 1.4). After recovering from sedation, animals were released at the site of capture.

Bacteria Isolation and Phenotypic Identification

Swabs kept under refrigeration at 4°C until transportation to the Laboratory of Bacteriology of the Faculty of Veterinary Medicine of the Technical University of

Figure 1.2: Eurasian Badger Captured in a Box Trap in "Serra de Grândola"

Lisbon. In the laboratory, swabs were processed for the isolation of aerobic and anaerobic bacteria.

The isolation of aerobic bacteria was performed using Columbia agar supplemented with 5 per cent sheep blood (bioMérieux) and MacConkey agar (Oxoid), incubated at 37°C for 24 to 48 h.

For strict anaerobic bacteria isolation, swabs were plated onto Schaedler agar supplemented with 5 per cent sheep blood (bioMérieux), and incubated for 48 h at 37°C in the absence of oxygen. After incubation, isolates were identified according to their macro and microscopic morphology, staining characteristics and biochemical profiles using API (bioMérieux) and BBL Crystal (Becton Dickinson) systems, according to manufacturer's instructions.

Results

The swab analysis using conventional microbiology procedures allowed for the detection of aerobic and anaerobic bacteria.

It was possible to identify several gram-positive bacterial species. Some of these are not considered pathogens, such as *Enterococcus faecium*, *Enterococcus faecalis*, and *Bacillus circulans*. However, the majority of isolates belong to bacterial species that are recognized as potential pathogens; it is the case of *Staphylococcus aureus* and *Staphylococcus* sp., *Streptococcus mitis* 1, and *Bacillus cereus* (Quinn *et al.*, 1994; McCarthy *et al.*, 2009) (Table 1.1).

Figure 1.3: Badger Sett (A) and Sett Entrances (B), where some of the Animals were Captured in the Frame of this Study

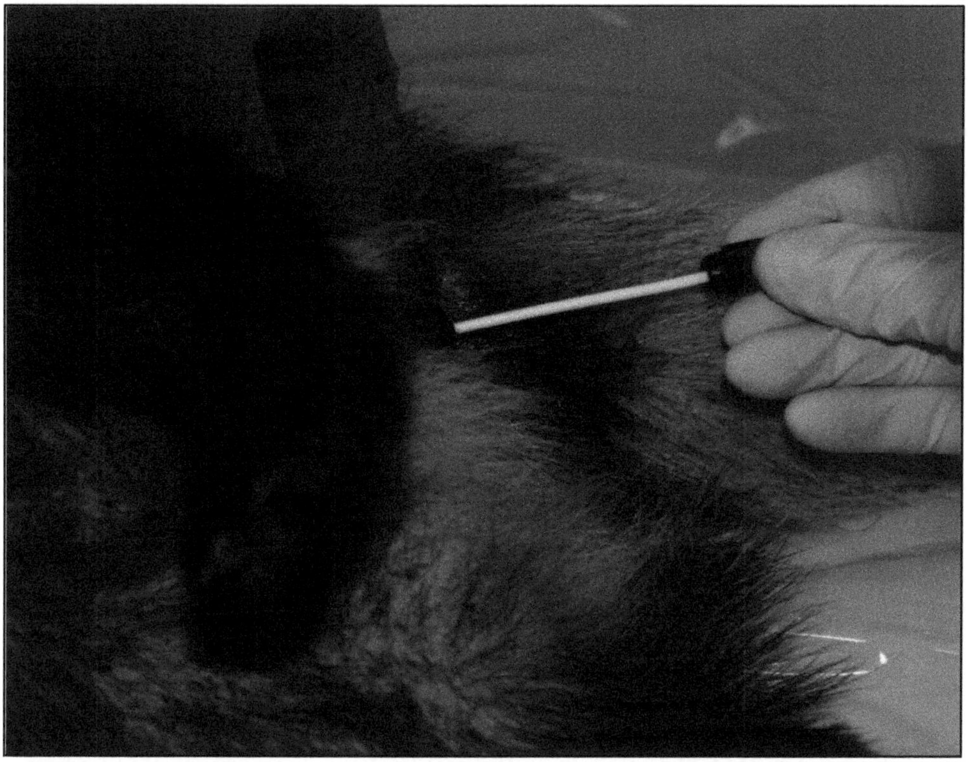

Figure 1.4: Collection of the Bacteriological Samples from the Badger's Rectum

It was also possible to identify several gram-negative species, such as *Escherichia coli*, *Proteus mirabilis*, *Citrobacter* sp. and *Citrobacter youngae*, *Salmonella* spp., *Enterobacter aerogenes* and *Klebsiella oxytoca*. Most of these species have also been already described as potentially pathogenic, namely *E. coli*, *P. mirabilis*, *Salmonella*, *E. aerogenes* and *K. oxytoca* (Quinn *et al.*, 1994; Wilson *et al*, 2003; Dietz *et al.*, 2006; McCarthy *et al.*, 2009) (Table 1.1).

The anaerobic processing of swabs allowed for the identification of *Clostridium barati* and *Clostridium innocuum*, *Bacteroides fragilis* and *Gardnerella vaginalis*. All of these bacteria species have been already described as potentially pathogenic (Table 1.1).

Discussion

In the last decade there has been an increasing concern about wildlife diseases and causes of wild animal mortality. This growing interest has been revealed by the general public that are concerned about animals' well-being, by the physicians that have a special interest in zoonoses, by the environmentalists due to the potential role of these animals as environmental pollution monitors, and by the veterinarians, due to the wildlife potential role as reservoirs of pathogenic microorganisms and as potential disease sources for farm animals or pets (Simpson, 2000).

Table 1.1: Gram-Positive and Gram-Negative Bacteria Isolated from Badgers Anal Swabs

	Genus	*Species*
Gram-positive bacteria	*Enterococcus*	*E. faecium, E. faecalis*
	Staphylococcus	*S. aureus**, *S.* sp.*
	Streptococcus	*S. mitis* 1*
	Bacillus	*B. circulans, B. cereus**
	Clostridium#	*C. barati, C. innocuum**
Gram-negative bacteria	*Escherichia*	*E. coli**
	Proteus	*P. mirabilis**
	Citrobacter	*C.* sp., *C. youngae*
	Salmonella	*S.* spp.*
	Enterobacter	*E. aerogenes**
	Klebsiella	*K. oxytoca**
	Bacteroides#	*B. fragilis**
Gram-variable bacteria	*Gardnerella*#	*G. vaginalis**

*: Potential pathogens; #: Strict anaerobic bacteria.

For these reasons, it is crucial to gather baseline data on wildlife microbiota. The preliminary characterization of the faecal microbiota of a *Meles meles* population inhabiting "Serra de Grândola", in southwest Portugal may contribute to a better understanding of the microbial status of this wild population and potential threats for animal and public health posed by their bacterial intestinal microbiota.

In this study it was possible to identify several potential gram-positive pathogens, belonging to genera *Staphylococcus, Streptococcus, Bacillus* and *Clostridium.*

Staphylococci are ubiquitous bacteria widespread in the environment, frequently found in association with mammals (Foster *et al.,* 1997). They have been already associated with diseases in wild animals, including abscess formation and suppuration, pruritus and mastitis (Ryan *et al.,* 1979; Timm, 1988; Quinn *et al.,* 1994). Staphylococcal infections, in particular due to *S. aureus,* have been described in ferrets (Timm, 1988), otters (Foster *et al.,* 1997) and minks (Ryan *et al.,* 1979), but to our knowledge this is the first description for the identification of *S. aureus* in a sample obtained from badgers. No reference on staphylococcal infections in badgers was found in the literature; but, since the staphylococci isolated in this study were haemolytic, their ability to disrupt erythrocytes may indicate a potential pathogenic role that should be taken into consideration.

It was also possible to identify *Streptococcus mitis* 1. The members of this bacterial species may be responsible for lung, kidney and spleen infections, genitor-urinary tract infections, hepatic hematomas and subcutaneous abscesses, already described in other wildlife animals, such as otters (Hoover and Tyler, 1986; Kimber and Kollias, 2000; Imai *et al.,* 2009). McCarthy *et al.* (2009) have already showed the presence of *Streptococcus* in badgers' bronchoalveolar lavages.

Bacillus cereus is an ubiquitous bacteria frequently found in humans and cattle (Quinn *et al.*, 1994). This bacteria is usually related with severe foodborne diseases (Quinn *et al.*, 1994; Oh *et al.*, 2012) and mastitis (Quinn *et al.*, 1994), but to our knowledge has not yet been related with illnesses in wild animals.

From the gram-positive sporulated anaerobic bacteria isolates, belonging to the *Clostridium* genus obtained in this study, some isolates were further identified as *C. barati* and *C. innocuum*. Clostridia have a worldwide distribution, being frequently found in the environment, namely in soil, freshwater and marine sediments. *C. innocuum* and other clostridia are also commensal bacteria present in the intestinal tract of animals (Mory *et al.*, 1998). In some occasions, these bacteria become opportunistic pathogens and may be responsible for serious infections (Mory *et al.*, 1998). On the other hand, *C. barati* is able to produce toxins that may be responsible for disease similar to botulism (Turton *et al.*, 2002). Although clostridia may be responsible for severe infections in wildlife (Quinn *et al.*, 1994; Kimber and Kollias, 2000), we found no references in the literature regarding the isolation of these bacteria from badgers.

It was also possible to obtain several gram-negative bacterial species, including potential pathogenic isolates. Since gram-negative bacteria survival can be easily impaired by inadequate collection and transport conditions, the isolation of a large number of gram-negative isolates reassures that the sampling methodology used in this study did not affect isolation success. It is also important to point out that a large diversity of intestinal tract bacteria was found in this work, which included bacteria such as *Escherichia coli*, *Klebsiella oxytoca*, *Salmonella* spp., *Proteus mirabilis*, *Enterobacter aerogenes* and *Citrobacter youngae*.

E. coli pathogenic strains have been previously isolated from wild animals (Mora *et al.*, 2012) and were related to severe diseases in wildlife (Hoover and Tyler, 1986). McCarthy *et al.* (2009) have already described the presence of *E. coli* in badgers' bronchoalveolar lavages.

It was also possible to identify some isolates as *Klebsiella oxytoca*. These bacteria may cause infections in wild animals, namely in crocodilians (Novak and Seigel, 1986), in snakes (Goldstein *et al.*, 1981) and in a pygmy sperm whale (Harms *et al.*, 2004), but to our knowledge it has not yet been related with illness cases in badgers.

Salmonella spp. are zoonotic bacteria that represent a serious concern for public health, leading to severe economic losses. *Salmonella* has already being isolated from a wide range of wildlife species, including marine mammals and birds, raptors and snakes (Smith *et al.*, 2002; Téllez *et al.*, 2002). The symptoms of salmonelosis may range from diarrhoea or fever to septicaemia, and animals are frequently in an asymptomatic carriers, but intermittent shedders (Quinn *et al.*, 1994; Wilson *et al.*, 2003).

Badgers are omnivore animals with a wide geographic distribution range. These animals have a diverse diet, which is characterised by a wide variety of preys and plants (Rosalino *et al.*, 2005a), being noticed that sometimes they may explore areas located near dumps (Wilson *et al.*, 2003). Therefore, these animals are frequently exposed to several potential sources of *Salmonella*. They may have an important role as reservoirs for these bacteria, and may transmit it to other animal species, including cattle (Wilson *et al.*, 2003), which may represent a serious problem, since the mortality rate in cattle due to *Salmonella* is high (Dietz *et al.*, 2006).

Some serovars belonging to *Salmonella enterica* are specific to their animal host (Dietz *et al.*, 2006). Previous studies on the occurrence of *Salmonella* infections in badgers showed that these animals can be colonised by several *Salmonella* serovars, especially by the serovars Agama, Bramante, Enteritidis, Coeln, Dublin, Anatum, Bonn, Nagama, Durham and Indiana (Wilson *et al.*, 2003). Thus, *Salmonella* infection diagnosis in badgers should be performed by conventional bacteriological methods, which include the use of selective media and serotyping (Quinn *et al.*, 1994) of isolates obtained from stool samples.

In this study, *Salmonella* isolates were identified based on their biochemical traits, as no serology testing was performed. This isolation suggests that wildlife may play an important role as reservoir animals for the dissemination of *Salmonella* strains (Kimber and Kollias, 2000; Smith *et al.*, 2002).

Regarding *Proteus mirabilis*, this bacterium has already been related with disease cases in wild animals, namely in minks (Martino and Martino, 1996), otters (Hoover and Tyler, 1986) and ferrets (Olsen *et al.*, 1985), but to our knowledge has not yet been related with illnesses in badgers. *P. mirabilis* can be found in mammals' faeces and also in the environment, being responsible for urinary tract infections, otitis externa and diarrhoea (Quinn *et al.*, 1994).

Enterobacter aerogenes can also be found in the environment, namely in water, soil and sewage, but also in animal faeces (Quin *et al.*, 1994). This bacterium is usually associated with diseases in cattle, mares and sows (Quinn *et al.*, 1994), but we found no references in the literature regarding its presence in badgers. Literature is also lacking in information on *Bacteroides fragilis*, a non-sporulating strict anaerobe that may cause diarrhoea, mastitis and abscesses in ruminants and swine (Quinn *et al.*, 1994), and on *Gardnerella vaginalis*, a potentially pathogenic gram-variable anaerobic bacterium, which has already been isolated from mares (Salmon *et al.*, 1990).

The above mentioned characterization was performed using cultured-based conventional microbiological methods, which some authors consider having several limitations when used for faecal bacteriological examination (Leser *et al.*, 2002). In fact, only microorganisms present in the rectal area swabbed at the time of sampling and able to growth in the *in vitro* conditions used could be evidenced. Nonetheless, it should be kept in mind that the main purpose of this study was to perform a preliminary characterization of these animals faecal microbiota, and to evaluate the occurrence of excretion of potential pathogenic microorganisms within the badgers' faeces. More complex methods, like methods based on molecular biology techniques, are often advocated, but they have been considered inadequate for the full characterisation of faecal microbiota (Harmsen *et al.*, 1999), and provide no data on microorganisms' viability.

Regardless of the methods limitations, isolation and identification of pathogenic bacteria potentially responsible for morbidity and mortality in wild animals, including badgers, is an important contribution for understanding the importance of pathogenic bacteria in badgers population health, and also their epidemiologic role as sources for human and animal infection cases or outbreaks. These studies may provide key information for the development of adequate management plans that would allow minimizing environmental contaminations, and prevent population decline.

Ackowledgments

This study was conducted with the financial support of the "Centro de Investigação Interdisciplinar em Sanidade Animal" (CIISA/FMV) from the Faculty of Veterinary Medicine, Technical University of Lisbon. M. Oliveira is a Foundation for Science and Technology funded scientist from program "Ciência 2007". Luís M. Rosalino was funded by the Fundação para a Ciência e a Tecnologia and Fundo Social Europeu (III Quadro Comunitário de Apoio) (Postdoctoral fellowship: SFRH/ BPD/14435/2003 and SFRH/BPD/35842/2007 and SFRH/BPD/35842/2007) and FAPESP (Proc. Ref.: 2011/004084) and Special thanks are due to Luís Neves for the help in the trapping campaign.

References

Abramov, A.V. and Puzachenko, A.Y., 2005. Sexual dimorphism of craniological characters in Eurasian badgers, *Meles* spp. (Carnivora, Mustelidae). *Zoologischer Anzeiger*, 244: 11–29.

Blondel, J. and Aronson, J., 1999. *Biology and Wildlife of the Mediterranean Region*. Oxford University Press, Oxford.

Brøseth, H., Knutsen, B. and Bevanger, K., 1997. Spatial organization and habitat utilization of badgers *Meles meles*: effects of food patch dispersion in the boreal forest of central Norway. *Zeitschrift für Säugetierkunde*, 62: 12–22.

Cabral, M.J., Almeida, J., Almeida, P.R., Dellinger, T., Ferrand de Almeida, N., Oliveira, M.E., Palmeirim, J.M., Queiróz, A. I., Rogado, L. and Santos-Reis, M., 2005. *Livro vermelho dos vertebrados de Portugal*. Instituto da Conservação da Natureza, Lisboa.

Correia, A.I. and Santos-Reis, M., 1999. Área de estudo. *In:* Caracterização da flora e da fauna do montado da Herdade da Ribeira Abaixo (Grândola – Baixo Alentejo). M. Santos-Reis and A.I. Correia (eds). Centro de Biologia Ambiental, Lisboa, Portugal. pp. 5–8.

Courtenay, O., Reilly, L.A., Sweeney, F.P., Hibberd, V., Bryan, S., Ul-Hassan, A., Newman, C., MacDonald, D.W., Delahay, R.J., Wilson, G.J. and Wellington, E.M.H., 2006. Is *Mycobacterium bovis* in the environment important for the persistence of bovine tuberculosis? *Biology Letters*, 22: 460–462.

Dietz, H.H., Chriél, M., Andersen, T.H., Jørgensen, J.C., Torpdahl, M., Pedersen, H. and Pedersen, K., 2006. Outbreak of *Salmonella* Dublin–associated abortion in Danish fur farms. *Canadian Veterinary Journal*, 47: 1201–1205.

Foster, G., Ross, H.M., Hutson, R.A. and Collins, M.D., 1997. *Staphylococcus lutrae* sp. nov., a new coagulase-positive species isolated from otters. *International Journal of Systematic and Evolutionary Microbiology*, 47: 724–726.

Goldstein, E.J., Agyare, E.O., Vagvolgyi, A.E. and Halpern, M., 1981. Aerobic bacterial oral flora of garter snakes: development of normal flora and pathogenic potential for snakes and humans. *Journal of Clinical Microbiology*, 13: 954–956.

Goszczynski, J., Jedrzejewska, B. and Jedrzejewski, W., 2000. Diet composition of badgers (*Meles meles*) in a pristine forest and rural habitats of Poland compared to other European populations. *Journal of Zoology*, 250: 495–505.

Hancox, M., 1980. Parasites and infectious diseases of the eurasian badger (*Meles meles* L.): A review. *Mammal Review*, 10: 151–162.

Harms, C.A., Piccolo, R.L., Rotstein, D.S. and Hohn, A.A., 2004. Struvite penile urethrolithiasis in a pygmy sperm whale (*Kogia breviceps*). *Journal of Wildlife Diseases*, 40: 588–593.

Harmsen, H.J.M., Gibson, G.R., Elfferich, P., Raangs, G.C., Wildeboer-Veloo, A.C.M., Argaiz, A., Roberfroid, M.B. and Welling, G.W., 1999. Comparison of viable cell counts and fluorescence *in situ* hybridization using specific rRNA-based probes for the quantification of human fecal bacteria. *FEMS Microbiology Letters*, 183: 125–129.

Imai, D., Jang, S., Miller, M. and Conrad, P.A., 2009. Characterization of beta-hemolytic streptococci isolated from southern sea otters (*Enhydra lutris nereis*) stranded along the California coast. *Veterinary Microbiology*, 136: 378–381.

Kimber, K.R. and Kollias II, G.V., 2000. Infectious and parasitic diseases and contaminant-related problems of North American river otters (*Lontra canadensis*): a review. *The Journal of Zoo and Wildlife Medicine*, 31: 452–472.

Kruuk, H., 1989. *The Social Badger*. Oxford University Press, Oxford, United Kingdom.

Leser, T.D., Amenuvor, J.Z., Jensen, T.K., Lindecrona, R.H., Boye, M. and Møller, K., 2002. Culture-independent analysis of gut bacteria: the pig gastrointestinal tract microbiota revisited. *Applied and Environmental Microbiology*, 68: 673–690.

Loureiro, C.F.M., 2008. Time and space use of key resources by the Eurasian badger (*Meles meles*) in a Mediterranean cork oak woodland: conservation implications. Universidade de Lisboa, Lisboa.

Macdonald, D.W., 2001. *The New Encyclopedia of Mammals*. Oxford University Press, Oxford.

Mangas, J.G., Kowalczyk, R., Virgós, E., Trua, A. and De Marinis, A., 2003. Body mass and body size changes in badgers (*Meles meles*) across its European distribution range: a test of Bergman rule and niche hypothesis. *In*: *European Mammalogy 2003. 4th European Congress of Mammalogy*. M. Macholán, J. Bryja and J. Zima (eds.). Institute of Vertebrate Biology, Brno, Czech Republic. p. 161.

Marmi, J., López-Giráldez, F., Macdonald, D.W., Calafell, F., Zholnerovskaya, E. and Domingo-Roura, X., 2006. Mitochondrial DNA reveals a strong phylogeographic structure in the badger across Eurasia. *Molecular Ecology*, 15: 1007–1020.

Martino, P.E. and Martino, J.J., 1996. Prevalence of microorganisms in dead mink kits from Aleutian-disease-infected and non-infected farms. *Veterinary Research*, 27: 607–612.

McCarthy, G., Shiel, R., O'Rourke, L., Murphy, D., Corner, L., Costello, E. and Gormley, E., 2009. Bronchoalveolar lavage cytology from captive badgers. *Veterinary Clinical Pathology*, 38: 381–387.

Mill, A.C., Rushton, S.P., Shirley, M.D., Murray, A.W., Smith, G.C., Delahay, R.J. and McDonald, R.A., 2012. Farm-scale risk factors for bovine tuberculosis incidence in cattle herds during the Randomized Badger Culling Trial. *Epidemiology and Infection*, 140: 219–230.

Mitchell-Jones, A.J., Amori, G., Bogdanowicz, W., Krystufek, B., Reijnders, P.J.H., Spitzenberger, F., Stubbe, M., Thissen, J.B.M., Vohralik, V. and Zima, J., 1999. *The Atlas of European Mammals*. T and AD Poyser, London.

Molina-Vacas, G., Bonet-Arboli, V., Rafart-Plaza, E. and Rodríguez-Teijeiro, J.D., 2009. Spatial ecology of European badgers (*Meles meles* L.) in Mediterranean habitats of the north-eastern Iberian peninsula. I: home range size, spatial distribution and social organization. *Vie et Milieu–Life and Environment*, 59: 227–236.

Mora, A., López, C., Dhabi, G., López-Beceiro, A.M., Fidalgo, L.E., Díaz, E.A., Martínez-Carrasco, C., Mamani, R., Herrera, A., Blanco, J.E., Blanco, M. and Blanco, J., 2012. Seropathotypes, phylogroups, Stx subtypes and intimin types of Shiga toxin–producing *Escherichia coli* strains from wildlife animals with the same characteristics as human pathogenic isolates. *Applied and Environmental Microbiology*. [Epub ahead of print].

Mory, F., Lozniewski, A., David, V., Carlier, J.P., Dubreuil, L. and Leclercq, R., 1998. Low-level vancomycin resistance in *Clostridium innocuum*. *Journal of Clinical Microbiology*, 36: 1767–1768.

Neal, E. and Cheeseman, C., 1996. *Badgers*. T and A Poyser Ltd, London, United Kingdom.

Novak, S.S. and Seigel, R.A., 1986. Gram-negative septicemia in American alligators (*Alligator mississippiensis*). *Journal of Wildlife Diseaes*, 22: 484–487.

Olsen, G.H., Turk, M.A. and Foil, C.S., 1985. Disseminated cutaneous squamous cell carcinoma in a ferret. *Journal of the American Veterinary Medical Association*, 186: 702–703.

Palomo, L.J. and Gisbert, J., 2002. Atlas de los mamíferos terrestres de España. Dirección General de Conservación de la Naturaleza–SECEM–SECEMU, Madrid, España.

Oh, M.H., Ham, J.S. and Cox, J.M., 2012. Diversity and toxigenicity among members of the *Bacillus cereus* group. *International Journal of Food Microbiology*, 152: 1–8.

Quinn, P.J., Carter, M.E., Markey, B. and Carter, G.R., 1994. *Clinical Veterinary Microbiology*. Mosby (ed.). Edinburg, p. 648.

Revilla, E., Delibes, M. Traivani, A. and Palomares, F., 1999. Physical and population parameters of Eurasian badgers (*Meles meles* L.) from Mediterranean Spain. *Zeitschrift für Säugetierkunde*, 64: 269–276.

Revilla, E. and Palomares, F., 2002. Spatial organization, group living and ecological correlates in low-density populations of Eurasian Badgers, *Meles meles*. *Journal of Animal Ecology*, 71: 497–512.

Reynolds, D., 2006. A review of tuberculose science and policy in Great Britain. *Veterinary Microbiology*, 112: 119–126.

Roper, T., 2010. *Badger*. HarperCollins Publisher, London.

Rosalino, L.M., 2008. Revisão dos parasitas e outros agentes infecciosos que afectam as populações ibéricas de um carnívoro social-o texugo euroasiático (*Meles meles*, Linnaeus 1758). *Revista Lusófona de Ciência e Medicina Veterinária*, 2: 9–16.

Rosalino, L.M., Loureiro, F., Macdonald, D.W. and Santos-Reis, M., 2005a. Dietary shifts of the badger (*Meles meles*) in Mediterranean woodlands: an opportunistic forager with seasonal specialisms. *Mammalian Biology*, 70: 12–23.

Rosalino, L.M., Macdonald, D.W. and Santos-Reis, M., 2004. Spatial structure and land–cover use in a low–density Mediterranean population of Eurasian badgers. *Canadian Journal of Zoology*, 82: 1493–1502.

Rosalino, L.M. and Macdonald, D.W., Santos-Reis, M., 2005b. Activity rhythms, movements and patterns of sett use by badgers, *Meles meles*, in a Mediterranean woodland. *Mammalia*, 69: 395–408.

Ryan, M.J., O'Connor, D.J. and Nielsen, S.W., 1979. *Staphylococcus aureus* mastitis in nursing mink affected with aleutian disease. *Journal of Wildlife Diseases*, 15: 533–535.

Salmon, S.A., Walker, R.D., Carleton, C.L. and Robinson, B.E., 1990. Isolation of *Gardnerella vaginalis* from the reproductive tract of four mares. *The Journal of Veterinary Diagnostic Investigation*, 2: 167–170.

Santos-Reis, M., Rosalino, L.M., Loureiro, F. and Santos, M.J., 2005. Los tejones en Portugal: distribución, estatus y conservación. *In*: *Ecología, distribución y estatus de conservación del tejón Ibérico*. E. Virgós, J.G. Mangas, E. Revilla and X.-D. Roura (eds). SECEM, Malaga, pp. 241–250

Scantlebury, M., Hutchings, M.R., Allcroft, D.J. and Harris, S., 2004. Risk of disease from wildlife reservoirs: badgers, cattle, and bovine tuberculosis. *Journal of Dairy Science*, 87: 330–339.

Simpson, V. R., 2000. Veterinary advances in the investigation of wildlife diseases in Britain. *Research in Veterinary Science*, 69: 11–16.

Smith, W.A., Mazet, J.A.K. and Hirsh, D.C., 2002. *Salmonella* in California wildlife species: prevalence in rehabilitation centers and characterization of isolates. *Journal of Zoo and Wildlife Medicine*, 33: 228–235.

Spencer, A., 2011. One body of evidence, three different policies: bovine tuberculosis policy in Britain. *Politics*, 31: 91–99.

Téllez, S., Briones, V., González, S., García-Pena, F.J., Altimira, A., Vela, A.I., Blanco, M.M., Ballesteros, C., Fernández-Garayzábal, J.F. and Goyache, J., 2002. *Salmonella* septicaemia in a beauty snake (*Elaphe taeniura taeniura*): a case report. *The Veterinary Record*, 151: 28–29.

Timm, K.I., 1988. Pruritus in rabbits, rodents, and ferrets. *Veterinary Clinics of North America: Small Animal Practice*, 18: 1077–1091.

Turton, K., Chaddock, J.A. and Acharya, K.R., 2002. Botulinum and tetanus neurotoxins: structure, function and therapeutic utility. *Trends in Biochemical Sciences*, 27: 552–558.

Vulfson, L., Pedersen, K., Chriel, M., Holmen Andersen, T. and Dietz, H.H., 2003. Assessment of the aerobic faecal microflora in mink (*Mustela vison* Schreiber) with emphasis on *Escherichia coli* and *Staphylococcus intermedius. Veterinary Microbiology*, 93: 235–245.

Wilson, J.S., Hazel, S.M., Williams, N.J., Phiri, A., French, N.P. and Hart, C.A., 2003. Nontyphoidal Salmonellae in United Kingdom badgers: prevalence and spatial distribution. *Applied and Environmental Microbiology*, 69: 4312–4315.

2013, Animal Diversity, Natural History and Conservation, Vol. 3 *Pages* **19–27**
Editors: **V.K. Gupta and Anil K. Verma**
Published by: **DAYA PUBLISHING HOUSE, NEW DELHI**

Chapter 2

Diversity of Entomopathogenic Nematodes in India: An Overview

Viswa Venkat Gantait*
Zoological Survey of India,
M-Block, New Alipur, Kolkata – 700 053, West Bengal

ABSTRACT

Those Nematodes who are lethal insect parasites are called entomopathogenic nematodes. The nematodes belonging to the families Steinernematidae and Heterorhabditidae are termed as entomopathogenic nematodes because of their capability to kill the insect hosts very quickly. They can act as potential bio-control agents as they cause mortality of the insect hosts within 24-48 hours after the invasion of the later. Due to their high bio-control potential and virtually no environmental hazards, they perform a major role in integrated pest management.

Family Steinernematidae possesses two genera: *Steinernema* and *Neosteinernema*. There is only one genus, *Heterorhabditis* under the family Heterorhabditidae. Globally, *Steinernema* possesses 68 species, *Neosteinernema* with only one species and 18 species are under *Heterorhabditis* till-to-date. Of these, only 5 species have been described from India. Poinar *et al.* (1992) described *H. indicus* from Coimbatore, Tamil Nadu. *S. thermophilum* (Ganguly and Singh, 2000) was described from IARI farm, New Delhi. Ali *et al.* (2005) described *S. mashhoodi* and *S. seemae* from Kanpur, Uttar Pradesh. Very recently, *S. meghalayensis* (Ganguly *et al.*, 2011) was described from Upper Shillong, Meghalaya. Besides these, *S. abbasi*, *S. bicornutum*, *S. carpocapse*, *S. feltiae*, *S. glaseri*, *S. riobrave*, *S. siamkayai*, *S. tami* and *H. bacteriophora* are reported from different states of India (Cabanillas *et al.*, 1994;

* *Corresponding author.* E-mail: v.gantait@rediffmail.com

Elawad *et al.*, 1997; Filipjev, 1934; Luc *et al.*, 2000; Poinar, 1976; Steiner, 1929; Stock *et al.*, 1998; Tallosi *et al.*, 1995; Weiser, 1955; Wouts *et al.*, 1982). Studies on EPNs were done only in the last two decades. Extensive works have to be done on these potential bio-control agents.

Keywords: *Entomopathogenic nematode, Diversity, India.*

Introduction

Pathogenic nematodes associated with insects are referred to as entomophilic, entomogenous or entomopathogenic. More than 30 nematode families are known to parasitize host taxa or are associated with insects (Nickle, 1972; Poinar, 1975, 1990; Maggenti, 1981; Kaya and Stock, 1997). Nematodes belonging to the families Steinernematidae and Heterorhabditidae are commonly termed as entomopathogenic nematodes (EPNs) because of their capability to kill the insect hosts very quickly. These are used inundative, augmentative or inoculative for the biological control of insects (Gaugler and Kaya, 1990; Kay and Gaugler, 1993). They also act as excellent bio-control agents of soil-dwelling stages of different insect pests (Gaugler, 2002; Ansari *et al.*, 2007). EPNs provide an environmentally-safe and economical alternative for the control of a wide range of arthropod pests (Grewal and Georgis, 1998; Grewal *et al.*, 2005). Due to their high bio-control potential and without causing any environmental hazards, they perform a major role in integrated pest management.

Entomopathogenic nematodes have been known since 1923, but interest in the use of them as biological control agents waned until the 1970s and 1980s. During that period, increasing environmental concerns about the use of chemical pesticides and their reduced availability reawakened interest in EPNs to control insect pests (Nguyen *et al.*, 2007). The first entomopathogenic nematode *Steinernema krausei* (*Aplectana krausei*) was described by Gotthold Steiner from Czechoslovakia in 1923. In 1929, Rudolf William Glaser discovered the second EPN *S. glaseri* (syn. *Neoaplectana glaseri*) parasitizing the Japanese beetle (*Popillia japonica*) in New Jersey. He was the first to culture the nematodes axenically, on solid culture, and became the first to conduct field tests with cultured nematodes against the Japanese beetle. Since then, entomopathogenic nematology has made tremendous strides worldwide (Grewal, 2002).

Interest in EPNs has increased rapidly in recent years and researches with these beneficial organisms are being conducted in approximately 100 laboratories representing 60 countries worldwide (Stock, 2005). Numerous surveys provide evidence of the omnipresence of these entomopathogens in natural and agricultural soils (Hominick, 2002) and new species and isolates are described every year. More importantly, many of these indigenous species appear to possess unique traits that enable their survival in unique environments, and these traits can be exploited in novel pest management strategies (Lopez-Nunez *et al.*, 2007). The most up-to-date bio-geographical accounts indicate that these nematodes have been isolated from all continents (except Antarctica) and all regions of the world (Hominick, 2002; Adams *et al.*, 2006; Nguyen and Hunt, 2007). Nguyen (2006) reported 56 species of EPNs in the family Steinernematidae and 12 species in Heterorhabditidae worldwide. Ganguly *et al.* (2011) reported 65 species under the family Steinernematidae and 15 species belonging to Heterorhabditidae, are known from all over the world.

During the last few decades, taxonomy of entomopathogenic nematodes have undergone several remarkable developments with respect to the use of novel molecular approaches and description of several new species of EPNs and their bacterial symbionts. Though, some taxonomic reviews of EPNs including the diagnostic key to the species are already available (Poinar, 1990; Hominick *et al.*, 1997; Rahman *et al.*, 2000; Adams and Nguyen, 2002; Ganguly, 2003), due to lack of adequate taxonomic expertise pertaining to the extraction of adults from the insects, processing techniques and non-availability of sufficient literature, many more species are yet to be identified. In the present article, an effort has been made to highlight the species of EPNs, described and recorded so far from India.

Systematics of EPNs

Phylum: Nematoda Rudolphi, 1808

Class: Secernentia von Linstow, 1905

Order: Rhabditida (Orley, 1880) Chitwood, 1933

Suborder: Rhabditina (Orley, 1880) Chitwood, 1933

Superfamily: Rhabditoidea (Orley, 1880) Travassos, 1920

Family: Steinernematidae Chitwood and Chitwood, 1937

Syn. Neoaplectanidae Sobolev, 1953

Type genus: *Steinernema* Travassos, 1927

Syn. *Steineria* Travassos, 1927 nec *Steineria* Micoletzky, 1922

Neoaplectana Steiner, 1929

Type species: *Steinernema kraussei* (Steiner, 1923) Travassos, 1927

Syn. *Aplectana kraussei* (Steiner, 1923) Travassos, 1927

Other genus: *Neosteinernema* Nguyen and Smart, 1994

Type and the only species: *Neosteinernema longicurvicauda* Nguyen and Smart, 1994

Family: Heterorhabditidae Poinar, 1976

Type and the only genus: *Heterorhabditis* Poinar, 1976

Syn. *Chromonema* Khan, Brooks and Hirschmann, 1976

Type species: *Heterorhabditis bacteriophora* Poinar, 1976

Syn. *Chromonema heliothidis* Khan, Brooks and Hirschmann, 1976

= *Heterorhabditis heliothidis* (Khan, Brooks and Hirschmann, 1976) Poinar, Thomas and Hess, 1977

= *H. argentinensis* Stock, 1993

EPNs in India

In India, much of the earlier work was confined to the use of exotic strains of *Steinernema carpocapsae, S. glaseri, S. feltiae,* and *Heterorhabditis bacteriophora,* which had been reviewed by Rahman *et al.* (2000). Some efforts were also made to isolate the indigenous populations of EPNs, of which five have been described as new species *viz. Heterorhabditis indica* by Poinar *et al.* (1992) from Coimbatore, Tamil Nadu, *Steinernema thermophilum* Ganguly and Singh, 2000 from IARI farm, New Delhi, *S. mashhoodi* and *S. seemae* from Kanpur, Uttar Pradesh by Ali *et al.* (2005). Very recently, Ganguly *et al.* (2011) described *S. meghalayensis* from Upper Shillong, Meghalaya. Besides these, *S. abbasi* Elawad *et al.,* 1997, *S. bicornutum* Tallosi *et al.,* 1995, *S. carpocapse* (Weiser, 1955) Wouts *et al.,* 1982, *S. feltiae* (Filipjev, 1934) Wouts *et al.,* 1982, *S. glaseri* (Steiner, 1929) Wouts *et al.,* 1982, *S. riobrave* Cabanillas *et al.,* 1994, *S. siamkayai* Stock *et al.,* 1998, *S. tami* Luc *et al.,* 2000 and *H. bacteriophora* Poinar, 1976 and few others have been reported from different states of India (Table 2.1). Ganguly *et al.* (2006) reported a list of EPNs, described and recorded so far from India. In India, research on EPNs are being carried out at different centers which have led to the isolation of several EPN strains and only few have been identified, while several strains are yet to be identified (Ganguly, 2003).

Table 2.1. Described/reported EPNs in India

Nematode Species	*Locality/State*	*Crop/Soil*	*References*
		Described species	
Heterorhabditis indica	Coimbatore, Tamil Nadu	Soil	Poinar *et al.,* 1992
S. mashhoodi	Kanpur, Uttar Pradesh	Pigeonpea fields	Ali *et al.,* 2005
S. meghalayensis	Upper Shillong, Meghalaya	Soil	Ganguly *et al.,* 2011
S. seemae	Kanpur, Uttar Pradesh	Pigeonpea fields	Ali *et al.,* 2005
S. thermophilum	IARI farm, New Delhi	Greengram	Ganguly and Singh, 2000
		Reported species	
S. abbasi	Aligarh, Moradabad, Uttar Pradesh	Mango	Khan and Haque, 2010
S. bicornutum	IARI farm, New Delhi; Karnataka	Mango	Hussaini *et al.,* 2001
S. carpocapsae	Andhra Pradesh, Tamil Nadu	Tomato, Okra, Banana, Cowpea, Field bean	
	Meghalaya, Arunachal Pradesh	Soil	Ganguly and Rathour, 2004
	Banaskantha, North Gujarat	Soil	Shinde *et al.,* 2008
	Aligarh, Moradabad, Uttar Pradesh	Mango	Khan and Haque, 2010
S. feltiae	Banglore, Karnataka	Maize	Hussaini *et al.,* 2001
	ICRISAT, Andhra Pradesh	Soil	Singh *et al.,* 1992
	CPCRI, Kerala	Soil	Ganguly *et al.,* 2006

Contd...

Table 2.1–*Contd...*

Nematode Species	*Locality/State*	*Crop/Soil*	*References*
S. glaseri	Bangalore, Karnataka	Okra	Hussaini *et al.*, 2001
S. riobrave	Kanpur, Uttar Pradesh	Field bean	Hussaini *et al.*, 2001
	Anand, Gujarat	Soil	Ganguly *et al.*, 2002
S. siamkayai	Champawat, Uttaranchal	Soil	Ganguly *et al.*, 2005
	Tamil Nadu	Soil	Maheswari *et al.*, 2005
	Aligarh, Moradabad, Uttar Pradesh	Mango	Khan and Haque, 2010
S. tami	Jorhat, Assam	Mustard	Hussaini *et al.*, 2001
	Aligarh, Moradabad, Uttar Pradesh	Mango	Khan and Haque, 2010
Heterorhabditis bacteriophora	Coonoor, Tamil Nadu	Citrus	Hussaini *et al.*, 2001
	Assam, Meghalaya	Soil	Ganguly and Rathour, 2004
	Aligarh, Moradabad, Uttar Pradesh	Mango	Khan and Haque, 2010
H. indica	Kerala	Soil	Banu *et al.*, 1998
	Bangalore, Karnataka	Maize	Hussaini *et al.*, 2001
	Karnataka, Tamil Nadu	Sugarcane, Sapoda, Rose	Hussaini *et al.*, 2001
	Manipur, Meghalaya	Soil	Ganguly and Rathour, 2004
	Guntur, Andhra Pradesh	Soil	Subbanna *et al.*, 2008
	Aligarh, Moradabad, Uttar Pradesh	Mango	Khan and Haque, 2010

Catalogue of Described EPNs in India

1. *Heterorhabditis indica* Poinar *et al.*, 1992

Type host: *Scirpophaga excerptalis* (Pyralidae: Lepidoptera) used as a trap insect in soil.

Type locality: Coimbatore, Tamil Nadu, India.

Type collector: George O. Poinar, Jr.

Type materials: Holotype (male) and allotype (hermaphroditic female) deposited in the Nematology Collection at the University of California, Davis, California. Paratypes (one male, one female) deposited in the Laboratoire des Vers, museum at national d' Histoire Naturelle, Paris.

Distribution: India: Tamil Nadu (Coimbator), Karnataka (Bangalore), Manipur, Meghalaya and Kerala.

2. *Steinernema masoodi* Ali *et al.*, 2005

Type host: Unknown, but likely to be *Helicoverpa armigera* (Hubner).

Type locality: Sandy soil in pigeonpea (*Cajanus cajan* L.) fields in Bithoor village, Kanpur district, Uttar Pradesh, India.

Type collector: Ali *et al.*

Type materials: Holotype male, 8 paratype males, 8 paratype females and 10 paratype juveniles deposited at Nematology Unit, Indian Institute of Pulses Research, Kanpur; 2 each paratype females, males and juveniles deposited at CAB Bioscience (UK) Centre, Egham, U.K.

Distribution: India: Uttar Pradesh (Kanpur).

3. *Steinernema meghalayansis* Ganguly *et al.*, 2011

Type host: Natural insect host unknown. Nematode specimens were isolated by baiting the soil samples with larvae of *Galleria mellonella.*

Type locality: Soil in the farm area of Upper shilling, Meghalaya, India.

Type collector: K.S. Rathour and Sushil Kumar.

Type materials: Holotype male (first generation), allotype female (first generation) and paratypes comprising 13 first generation males, 18 first generation females, 13 second generation males and 21 second generation females and 20 infective juveniles were deposited in the National Nematode Collection of India at Division of Nematology, IARI, New Delhi (NNCT/Type/2304 -2363). Permanent mounts of 2 paratypes each of first generation males and females, second generation males and females, 10 infective juveniles were deposited in the University of California Nematode Collection, Davis, California, USA.

Distribution: India: Meghalaya.

4. *Steinernema seemae* Ali *et al.*, 2005

Type host: Unknown, but likely to be *Helicoverpa armigera* (Hubner).

Type locality: Sandy soil in pigeonpea (*Cajanus cajan* L.) fields in Hamirpur, Uttar Pradesh, India.

Type collector: Ali *et al.*

Type materials: Holotype male, 5 paratype males, 3 paratype females and 8 paratype juveniles deposited at Nematology Unit, Indian Institute of Pulses Research, Kanpur; 2 each paratype females, males and juveniles deposited at CAB Bioscience (UK) Centre, Egham, U.K.

Distribution: India: Uttar Pradesh (Hamirpur).

5. *Steinernema thermophilum* Ganguly and Singh, 2000

Type host: The nematode specimens were isolated by baiting the soil samples with larvae of *Corcyra cephalonica.*

Type locality: Soil around roots of green gram (*Vigna aureus*) growing at farm area of Indian Agricultural Research Institute, New Delhi, India.

Type collector: Collected in August, 1999 by the senior author, *i.e.*, Sudershan Ganguly.

Type materials: Holotype first generation male, allotype first generation female, 8 paratypes of first generation females, 15 paratypes of second generation females, 18 paratypes of first generation males, 16 paratypes of second generation males and 20 paratypes of third stage juveniles deposited in the Nation Nematode Collection of India, IARI, New Delhi. One paratype each of males and females of both the generations and 5 paratypes of third stage infective juveniles deposited with CABI Biosciences, U.K. Centre, Egham, U.K.

Distribution: India: New Delhi (IARI farm).

Conclusion

Keeping in view the vast and diverse agro-climatic conditions and the insect biodiversity in India, it is expected that there must be several new strains or species of entomopathogenic nematodes existing in this country, which need to be explored and identified for harvesting their bio-control potential. There is an urgent need to build a strong biosystematics base in the country in order to identify and characterize the existing EPN biodiversity.

Lack of adequate taxonomic expertise and non-availability of sufficient literature on various described species have been major constraints to identify the species of entomopathogenic nematodes. Therefore, this is the high time to highlight the recent taxonomic developments, approaches and techniques used in EPN taxonomy, identifying the diagnostic characters at familial, generic and species levels, which may serve as an aid for species identification as well as strengthening taxonomic research on steinernematid and heterorhabditid nematodes in the country (Ganguly, 2001). Development of proper and advanced isolation techniques and culture methods of different EPN species are very much essential for taxonomic study of EPNs.

References

Adams, B.J. and Nguyen, K.B., 2002. Taxonomy and systematics. In: *Entomopathogenic Nematology,* (Ed.) R. Gaugler. CABI Publishing, CAB International, UK, pp. 1–33.

Adams, B.J., Fodor, A., Klein, M.G., Smith, H.L., Stackebrandt, E. and Stock, S.P., 2006. Biodiversity and systematics of nematode bacterium entomopathogens. *Biological Control*, 37: 32–49.

Ali, S.S., Shaheen, A., Pervez, R. and Hussain, M.A., 2005. *Steinernema masoodi* sp. n. and *S. seemae* sp. n. (Nematoda: Rhabditida: Steinernematidae) from India. *International Journal of Nematology*, 15(1): 89–99.

Ansari, M.A., Waeyenberge, L. and Moens, M., 2007. Natural occurrence of *Steinernema carpocapse* Weiser, 1955 (Rhabditida: Steinernematidae) in Belgian turf and its virulence to *Spodoptera exigua* (Lepidoptera: Noctuidae). *Russian Journal of Nematology*, 15: 21–24.

Ganguly, S. and Singh, L.K., 2000. *Steinernema thermophilum* sp. n. (Rhabditida: Steinernematidae) from India. *International Journal of Nematology*, 10 (2): 183–191.

Ganguly, S., 2001. Taxonomy of entomopathogenic nematodes. In: *Abstract: National congress on centenary of nematology in India, appraisal and future plans.* Division of Nematology, IARI, New Delhi, pp. 20–21.

Ganguly, S., 2003. In: *Current Status of Research on Entomopathogenic Nematodes in India* (Eds.) S.S. Hussaini, R.J. Rabindra and M. Nagesh. Project Directorate of Biological Control (ICAR), Bangalore, pp. 69–108.

Ganguly, S., Rathour, K.S., Kumar, S. and Singh, M., 2011. *Steinernema meghalayensis* sp. n. (Rhabditida: Steinernematidae) from hilly region of India. *Indian Journal of Nematology*, 41 (1): 83–97.

Ganguly, S., Rathour, K.S., Pandey, J. and Sosamma, V.K., 2006. Current status of entomopathogenic nematodes occurring in India along with new record of *Steinernema feltiae* (Filipjev) Wouts *et al.,* 1982 from Kerala. *Indian Journal of Nematology*, 36 (2): 283–285.

Gaugler, R. and Kaya, H.K., 1990. *Entomopathogenic nematodes in biological control.* CRC Press, Boca Raton, FL, USA, pp. 365.

Gaugler, R., 2002. *Entomopathogenic Nematology.* CABI Publishing, CAB International, New York, USA, pp. 388.

Grewal, P.S. and Georgis, R., 1998. Entomopathogenic nematodes. In: *Biopesticides: Use and Delivary,* (Eds.) F.R. Hall and J.J. Menn. Humana Press, Totowa, NJ, USA, pp. 271–299.

Grewal, P.S., 2002. Formulation and application technology. In: *Entomopathogenic Nematology,* (Ed.) R. Gaugler. CABI Publishing, CAB International, UK, pp. 265–287.

Grewal, P.S., Ehlers, R.U. and Shapiro-Ilan, D., 2005. *Nematodes as Biocontrol Agents.* CABI Publishing, CAB International, Wallingford, UK, pp. 505.

Homonick, W.M., 2002. Biogeography. In: *Entomopathogenic Nematology,* (Ed.) R. Gaugler. CABI Publishing, CAB International, Wallingford, UK, pp. 115–144.

Homonick, W.M., Briscoe, B.R., Del Pino, F.G., Heng, J.A., Hunt, D.J., Kozodoy, E., Mracek, Z., Nguyen, K.B., Reid, A.P., Spiridonov, S.E., Stock, S.P., Sturhan, D., Waturu, C. and Yoshida, M., 1997. Biosystematics of entomopathogenic nematodes: current status, protocols and definitions. *Journal of Helminthology,* 71: 271–298.

Kaya, H.K. and Gaugler, R., 1993. Entomopathogenic nematodes. *Annual Review of Entomology,* 38: 181–206.

Kaya, H.K. and Stock, S.P., 1997. Techniques in insect nematology. In: *Manal of Techniques in Insect Pathology,* (Ed.) L.A. Lacey. Academic Press, San Diego, CA, USA, pp. 281–324.

Lopez-Nunez, J.C., Cano, L., Carmenza, E., Gongora, B. and Stock, S.P., 2007. Diversity and evolutionary relationship of entomopathogenic nematodes (Steinernematidae and Heterorhabditidae) from the Central Andean region of Colombia. *Nematology,* 9(3): 333–341.

Maggenti, A.R., 1982. Nemata. In: *Synopsis and Classification of Living Organisms*, (Ed.) S.P. Parker. McGraw-Hill, New York, USA, pp. 878–923.

Nguyen, K.B., 2006. *Morphology and Taxonomy of Entomopathogenic Nematodes*. Available from http: / / kbn.ifas.ufl.edu / kbnstein.htm (accessed 1st October, 2006).

Nguyen, K.B. and Hunt, D.J., 2007. *Entomopathogenic Nematodes: Systematics, Phylogeny and Bacterial Symbiots.* E. J. Brill, Leiden, pp. 816.

Nguyen, K.B., Stuart, R.J., Andalo, V., Gozel, U. and Rogers, M.E., 2007. *Steinernema texanum* n. sp. (Rhabditida: Steinernematidae), a new entomopathogenic nematode from Texas, USA. *Nematology*, 9 (3): 379–396.

Nickle, W.R., 1972. A contribution to our knowledge of the Mermithidae (Nematoda). *Journal of Nematology*, 4: 113–146.

Poinar, G.O. Jr., 1975. *Entomogenous Nematodes: A Manal and Host List of Insect-Nematode Associations.* E.J. Brill, Leiden, Netherlands, pp. 317.

Poinar, G.O. Jr., 1990. Taxonomy and biology of Steinernematidae and Heterorhabditidae. In: *Entomopathogenic Nematodes in Biological Control*, (Eds.) R. Gaugler, and H.K. Kaya. CRC Press, Boca Raton, FL. pp. 23–60.

Poinar, G.O. Jr., Karunakar, G.K. and David, H., 1992. *Heterorhabditis indicus* n. sp. (Rhabditida: Nematoda) from India: separation of Heterorhabditis spp. by infective juveniles. *Fundamental and Applied Nematology*, 15: 467– 472.

Rahman, P.F., Sharma, S.B. and Wightman, J.F., 2000. A review of insect-parasitic nematodes research in India: 1927–1997. *International Journal of Pest Management*, 1: 19–28.

Steiner, G., 1923. *Aplectana kraussei* n. sp. eine in der Blattwespe *Lyda* sp. parasitierende Nematodenform, nebst Bemerkungen Uber das Seitenorgan der parasitischen Nematoden. *Zentralblatt fur bakteriologie, Parasitenkunde, Infektionskrankheiten und Hygiene, Abteilung II*, 59: 14–18.

Stock, S.P., 2005. Insect-parasitic nematodes: more than model organisms. *Journal of Invertebrate Pathology*, 89: 57–66.

2013, Animal Diversity, Natural History and Conservation, Vol. 3 *Pages* **29–35**
Editors: **V.K. Gupta and Anil K. Verma**
Published by: **DAYA PUBLISHING HOUSE, NEW DELHI**

Chapter 3

Investigation of Termite Biodiversity in the Foot Hills of Maruthamalai, Western Ghats Region, Coimbatore, Tamil Nadu

N. Sathishkumar,[1] B. Kalaivani[2]* and G. Vanithakumari[1]
[1]*Division of Endocrinology, Department of Zoology, Bharathiar University, Coimbatore – 641 046, Tamil Nadu, India*
[2]*Department of Zoology, Sri Sarada College for Women (Autonomous), Salem – 636 016, Tamil Nadu, India*

ABSTRACT

Termites are social insects widely distributed in Western Ghats of Coimbatore District of Tamil Nadu and generally feed on dead and dying trees. They also cause damage to standing trees hence biodiversity of termites have been studied. The present study was aimed to investigate, identify and record the diversity of termites in the Bharathiar University campus located in the foot hills of Maruthmalai, Coimbatore, Tamil Nadu. Three speceis of termites belonging to two genera falling under the family termitidae identified in the area, included *Odontotermes obesus, Odontotermes feae* and *Eremotermis fletcheri.* Among the three recorded termite species, the dominant was *Odontotermes obesus,* whose diversity was very high when compared to others.

Keywords: *Biodiversity, Termites, Western Ghats, Odontotermes obesus, Odontotermes feae, Eremotermis fletcheri.*

---

* *Corresponding author.* E-mail: kalai03win@yahoo.co.in

Introduction

Planet earth is endowed with a rich variety of life forms and the teeming millions of these living organisms have been well knit by the laws of nature. The inter dependence of the various life forms starting from the unicellular primary producers to the complexity built higher plants and animals is a unique feature of this green planet. Biodiversity is very basis of human survival and economic well being and encompasses all life forms, ecosystems and ecological processes, acknowledging the hierarchy at genetic taxon and ecosystem levels (McNeely, 1990). Biodiversity is responsible for the essential ecosystem services, including regulation of the atmospheric gaseous composition, climate disturbance and waste assimilation, pollination, biological control, recreation etc. (May, 1988).

The Western Ghats region is considered as one of the most important biogeographic zones of India, as it is one of the richest centres of endemism. Due to varied topography and micro-climatic regimes, some areas within the region are considered to be active zones of speciation. The region has 490 arborescent taxa, of which as many as 308 are endemics. This endemism of tree species shows a distinct trend, being the highest (43 per cent) in 8N-10°30′N location and declining to 11 per cent in 16N–16°30′N location.

Termites are dominant soil animals in tropical terrestrial ecosystems. The biomass of termites is high, for example approaching 10 gm^{-2} in hermid forests and they are, therefore, one of the most abundant animal groups on earth. In the present investigation, we studied the diversity of termites in different habitats such as Barren grass land, human altered habitat and also in scrub jungle in the foot hills of Maruthamalai, Coimbatore District, Tamil Nadu.

Materials and Methods

The present study sites fall broadly into two categories reflecting different levels of disturbances, namely the moderately disturbed area (MDA) and disturbed area or Human altered area (HAA). The moderately disturbed area was frequently grazed by cattle and also had some human interference. The disturbed area or the human interference or altered habitat consists of administrative buildings and the residential areas.

Quadrate Method

The study areas selected included human altered area, grass land and thorny scrub jungle. A colony survey involves using a repeatable sampling method to estimate total number of colonies of a termites species in the study area. Study area was divided into sampling segments and the number of colonies of different termites species in each segment counted. The number of termites colonies was counted in a series of 20 m × 20 m quadrates through different study sites of Bharathiar University Campus. The average number of termites colonies in each of 6 such quadrates were counted.

The values are expressed as number of colonies of each species of different study sites/400 m^2. There are several circumstances in which termites may be collected. For convenience a camel hair brush wetted with alcohol was applied on the termites. Termites adhered to the brush and was finally transferred to the vials. The softer bodied insects were preserved in fluid. The best fluid in which to store the collection is ethyl alcohol 75-80 per cent.

Later the collected specimens were identified by consulting appropriate literature cited in bibliography. Survey of termites has been made by spot observation by visiting above places fortnightly.

Results and Discussion

An environment rich in biological diversity offers the broadest array of options for sustainable economic activity, for sustaining human welfare and for adapting to change. Loss of biodiversity has serious economic and social costs for any country. The experience of the past few decades has shown that as industrialization and economic development in the classical sense takes place, patterns of consumption, production and needs, change, straining, altering and even destroying ecosystems. Termites are dominant soil animals in tropical terrstrial ecosystems. The biomasses of termites is high, for example approching 10 gm^2 in hermid forests and they are, therefore, are of the most abundant animal groups on earth.

Plate 3.1 denotes different selected sites in Bharathiar University, while, Table 3.1 shows the termite diversity in different selected sites in Bharathiar University. In the present study, area-1 was dominated by *Odontotermes obesus*, the distribution of mean was 27.50 followed by *Eremotermis fletcheri* with the distribution mean of 27.16, *Odontotermes feae*, distribution mean was very less (9.33) when compared to other species. In area-2, *Odontotermes feae* distribution mean was only 11.33, whereas, the distribution of other 2 species namely *Eremotermis fletcheri* was 23.66, *O. obesus* was 24.00 respectively and both of these 2 species were equally dominant.

Table 3.1: Termite Diversity in Different Sites of Bharathiar University Campus (Quadrate Method)

No. of Quadrate	*Trivervitermes fletcheri*			*Odontotermes obesus*			*Odontotermes feae*		
	A1	*A2*	*A3*	*A1*	*A2*	*A3*	*A1*	*A2*	*A3*
1	22	20	16	21	18	14	8	12	6
2	30	25	14	28	22	19	10	13	7
3	26	21	17	32	27	13	6	7	4
4	28	23	15	26	30	20	9	11	5
5	32	26	18	30	28	17	11	16	13
6	25	27	16	28	19	18	12	9	6
Mean	27.16	23.66	16.00	27.50	24.00	16.83	9.33	11.33	6.83

A1: Barren Grassland; A2: Thorny scrub jungle; A3: Human altered habitat.

Major biodiversity threats are habitat destruction, extension of agriculture, filling up of wetlands, conversion of rich bio-diversity site for human settlement and industrial development, destruction of coastal areas, uncontrolled commercial exploitation. This erosion of biodiversity is largely due to habitat loss caused by the expansion of various development projects such as mines, dams, roads and canal construction. Habitat loss leads to the fragmentation of continuous stretches of land and consequently fragments the wildlife populations inhabiting them. These small populations are increasingly vulnerable to inbreeding depression, high infant mortality, susceptibility to environmental stochasticity, and in the long run, possibly to extinction. Apart from the primary loss of habitats, there are numerous other problems contributing to the loss and endangered status of several plant and animal species.

India occupies only 2.4 per cent of the world's land area but its contribution to the world's biodiversity is approximately 8 per cent of the total number of species (Khoshoo, 1996), which is estimated to be 1.75 million (As per Global Biodiversity Assessment of UNEP of 1995, described number of species so far is 1.75 million). The survey carried out in Area-3 shows the mean distribution of the three species namely *Odontotemes obsus*, *Eremotermis fletcheri* and *Odontotermes feae* as 16.83, 16.00 and 6.83 respectively. Like in other two areas the area is also primarily dominated by *Odontotermes obesus* followed by *Eremotermis fletcheri* and by *Odontotermes feae*.

Table 3.2: Showing Ecology of Three different Termite Species

Sl.No.	*Name of the Species*	*Habitat*	*Feeding Habitat*	*Predators*
1.	*Trinervitermes flecheri*	Subterranean dwellers	Feed on dry grasses and dead leaves	Ants lizards some birds
2.	*Odontotermes obesus*	Mound dwellers (and) subterranean dwellers	Feed the dead wood, papers, cow dung and attack the trees.	Mostly Ants, birds.
3.	*Odontotermes feae* dwellers	Subterranean	Feed the dead wood	Ants

Table 3.2 show the ecology of the three different termite species. Table 3.2 shows the termite species and its specific host plants. In the University Campus *Odontotermes obesus* were seen to attack the trees and cause severe damage to these host trees. Mostly bark of the mature tree and the entire plants of younger ones were completely infested. Many of such infested trees were economically and medicinally important *e.g.*, *Eucalyptus terticornis*, *Azardirachta indica*, *A. jursaracta indica*, *Mangifera indica* and *Tectona grandis* (Plate 3.2).

Table 3.2 shows the predators of the 3 different species of termites. Ants, lizards and birds act as predators of *Eremotermis fletcheri*. However, *Odontotermes obesus* appears to be attacked mostly by ants and birds, where as *Odontotermes feae* is purely preyed by ants. Among the three termite species *Eremotermis fletcheri* never formed a mound but *Odontotermes obesus* has foraging gallaries just below or sometimes on the soil surface (Munthali *et al.*, 1999).

Harris (1971) reported that termites cause damage to a wide range of crops in eastern and Southern Africa; including crops such as rubber, coconut, cotton,

Barren Grassland

Thorny Scrub Jungle

Human Altered Habitat

Plate 3.1

Ficus reliosa

Cocus nucifera

Mangifera indica

Azadirachta indica A. juss

Plate 3.2

sunflower, and food crops. In the present study *Odontotermes obesus* was also observed to cause significant damage to *Acacia arabica, Ficus religiosa, Prosophis liniera* etc. in Bharathiar University Campus.

Biogeographically, India is situated at the trijunction of three realms namely Afrotropical, Indo-Malayan and Paleo-Arctic realms and, therefore, has characteristic elements from each of them. This assemblage of three distinct realms probably is a fact which is believed to partly account for its rich and unique biological diversity.

Now-a-day's most of researchers focus on major pests and their control. Due to pollution, habitat destruction, deforestation, over exploitation results in the loss of biodiversity. This will lead to mass extinction of many organisms. So conservation of biodiversity is essential for human survival. Based on the data collected it is concluded that the reduction in their number of species was observed to be increased in the human altered habitat when compared to natural habitats. The present study was intended to identify and record the diversity of termites in the Bharathiar University Campus, to study the termite – host plants interaction and to understand about the termite predators.

References

Harris, W.V., 1971. *Termites: Their Recognition and Control,* Second edition. Longmans, London.

Khoshoo, T.N., 1996. 'Biodiversity in the Indian Himalayas: conservation and utilization'. In: *Banking on Biodiversity,* (Ed.) Sheggi P. Kathmandu. International Centre for Integrated Mountain Development.

May, R.M., 1988. *J. Science,* 241: 1441–1449.

McNeely, G.A., Miller, K.R. Reid, W.V. Miltermeier, R.A. and Werner, T.R., 1990. Conserving the World's Biological Diversity, IUCN.

Munthali, D.C., Logan, J.W.M., Wood, T.G. and N. Yirenda, G.K.C., 1999. Termite distribution and damage to crop on small holdes farms in southern Malawi. *Insect Sci. Applic.,* 19(1): 43–49.

2013, Animal Diversity, Natural History and Conservation, Vol. 3 *Pages* **37–54**
Editors: **V.K. Gupta and Anil K. Verma**
Published by: **DAYA PUBLISHING HOUSE, NEW DELHI**

Chapter 4

Seasonal Population Dynamics and Migration of Butterflies in Coastal Plains of Kalpakkam, Southern India

T. Ramesh[1], K. Jahir Hussain[2]*, K.K. Satpathy[2] and M. Selvanayagam[1]

[1]Loyola Institute of Frontier Energy (LIFE), Loyola College, Chennai – 600 034, T.N., India
[2]Environmental and Safety Division, Indira Gandhi Centre for Atomic Research, Kalpakkam – 603 102, India

ABSTRACT

We assessed seasonal population trends of butterflies of the Department of Atomic Energy (DAE) campus at Kalpakkam in southern India from June 2008 to May 2009. We set up five permanent, 300 m-long line transects and recorded all the species within a five meter distance around the observer. In all, 2,223 individuals of 66 butterfly species were recorded. Species richness and relative abundance of butterflies were highest during October and January, coinciding with the beginning and end of the North-East Monsoon (NEM), and remained high during the two intervening months as compared to the rest of the year. Patterns of variation in climatic variables and butterfly populations showed that butterfly richness and abundance were negatively correlated with temperature.

We also report a large bidirectional migratory flight of three species, *Catopsilia pyranthe*, *Catopsilia pomona* and *Papilio demoleus* at Kalpakkam during October (2009)

* *Corresponding author.* E-mail: zakir781@yahoo.co.in

and July (2010). The number of butterflies observed in this swarm was 9697 individuals during October and 1727 individuals during July. The most abundant migratory species were *Catopsilia pyranthe* and *Catopsilia pomona*, together comprising 58.5 per cent of the migratory swarms. The activity of migratory swarms increased during the peak day hours (11.30 h to 13.00 h). Further in-depth studies are required to find out, whether the migration is undertaken to avoid unfavorable weather or searching of adult food source/larval host.

Keywords: *Butterfly, Seasonality, Population Dynamics, Migration, DAE campus, Kalpakkam.*

Introduction

Fluctuation in insect population in relation to season is a common phenomenon and this has been reported from all parts of the world, including the tropics (Wolda, 1978, 1988; Denlinger, 1986). Compared to other insects, the butterfly populations are monitored extensively worldwide. The world contains roughly 18,000-20,000 species of butterflies and there is a clear latitudinal gradient in butterfly species diversity, with numbers highest in the tropics (Shields, 1989; Larsen, 2005). In addition, roughly 90 per cent butterflies species live in the tropical areas (Bonebrake *et al.*, 2010; Larsen, 2005). In India, approximately 1,800 species and subspecies (Kunte *et al.*, 2011) of butterflies were documented of which peninsular India hosts 350 (Kunte, 2000-01).

Butterflies are often good candidates to study population and community ecology (Pollard, 1991) because they are good indicators of ecosystem health (Thomas, 2005) and many of the species are strictly seasonal (Kremen, 1992; Kunte, 1997, 2005). Butterflies are suited for indirect measure of environmental changes, because they are very sensitive to changes in microclimate and habitat (Kremen, 1992; Daily and Ehrlich, 1995; Hill, 1995; Brown and Freitas, 2000; Bobo *et al.*, 2006; Akite, 2008; Bonebrake and Sorto, 2009). Eventhough much has been studied in terms of butterfly seasonal diversity and abundance, especially at the Western Ghats, little is known from the dryer coastal plains of south India. Migration is one of the main driving mechanisms of insect population dynamics and cannot be viewed separately from the seasonal dynamicity of meta-populations. The tropical butterfly species are well-known for their migratory behavior, but the physiological mechanisms of orientation are unknown (Oliveira *et al.*, 1998) and the pattern of butterfly migration at these localities remains unclear. Hence, the purpose of present study is to determine the temporal trends in diversity, abundance and the effects of abiotic factors on butterfly community at coastal plain areas, and also inquire into their migration pattern.

Materials and Methods

Study Area

The Department of Atomic Energy (DAE) campus at Kalpakkam on the east coast of southern India (12° 33.7′N and 80° 10.5′E; ~2,500 acres) encompasses seashore and a large open area. The main natural vegetation observed at DAE campus is dry evergreen and scrub comprising of members predominantly belonging to the families

Poaceae, Fabaceae, Cyperaceae, Asteraceae, Euphorbiaceae, Verbenaceae, Solanaceae, Rubiaceae, Convolvulaceae and Amaranthaceae respectively (Gajendiran and Ragupathy, 2002). Some of the dominant tree species are: *Acacia auriculiformis, Acacia polyacantha, Albizia lebbeck, Albizia saman, Cassia siamea, Casuarina equisetifolia, Ficus benghalensis, Prosophis juliflora* and *Terminalia arjuna*. Predominant herbs and shrubs are: *Achyranthes aspera, Anacardium occidentale, Anisomelous* sp., *Calotropis gigantean, Canthium dicoccum, Capparis* sp., *Catharanthus roseus, Clitoria ternatea, Crotalaria verucosa, Croton bonplandianum, Euphorbia antiquorum, Gomphrena serrata, Heliotropium indicum, Lantana camera, Leucas aspera, Martynia annu, Mimosa pudica, Ocimum basilicum, Solanum* sp., *Tephrosia purpurea, Tinospora cordifolia, Toddalia asiatica, Tragia involucrate, Urena lobata, Vernonia cinerea* and *Zizyphus oenoplia*.

Butterfly Census Technique

Butterflies were sampled between June 2008 and May 2009 using a modified version of the line transect method described by Kunte (1997). In our method, five permanent 300 m line transects were setup by using Global Positioning System (GPS) (Garmin, 76CSx). The fixed transects covered all the microhabitats like gardens, scrubs, riparian corridor, sandy area and monoculture (*Casuarina* plantation). All the transects were walked once in a month from 09.30 h to 11.30 h during good weather period. Number of individuals of each species was recorded around a radius of five meter from the observer covering his either sides, above and front. The observed butterfly numbers from all transect were pooled and considered as a collection for the month. This is considered a suitable method for surveying butterflies in a wide range of habitats (Walpole and Sheldon, 1999; Caldas and Robbins, 2003; Koh and Sodhi, 2004).

Seasonal mass migratory movements of butterflies were observed only during October and July, and lasted for a few days. Quantification of swarm was carried out on 16th October 2009 and 14th July 2010. A 10 m wide imaginary strip was chosen as per Kunte (2005), and the numbers of butterflies passing through this strip over a 15 minutes duration were counted. Twenty two similar observations were made between 09.30 h to 15.00 h.

Data Analysis

For data analyses, the whole year was divided into four periods namely South West Monsoon-SWM (June to September), North East Monsoon- NEM (October to December), winter (January to February) and summer (March to May). Data on mean temperature, mean relative humidity, monthly rainfall and number of rainy days were collected from meteorological station, Indira Gandhi Centre for Atomic Research (IGCAR), Kalpakkam. Pearson correlation analysis was carried out to correlate between abiotic factors with richness and abundance of butterfly community. Species richness (sample based rarefaction) at different seasons was calculated using BiodiversityPro software version 2 (Neil Mcaleece *et al.*, 1997).

Results

We recorded 2,223 individuals belonging to 66 butterfly species, 48 genera and 5 families. The Nymphalidae was the dominant family in terms of species richness

(23 species; 29.1 per cent of genera) and abundance, followed by Pieridae (16 species, 23 per cent genera), Lycaenidae (15 species, 31.2 per cent genera) and Papilionidae (8 species, 8.3 per cent genera). Hesperiidae was represented by only four species (Table 4.1). Interestingly 59.24 per cent of the species, and 82 per cent of the individuals collected belonged to two families (Nymphalidae and Pieridae). Eventhough, family Nymphalidae was the most rich, Lycaenidae was more diverse in generic richness (15 species belongs to 15 genera).

Temporal Abundance and Seasonality Profile of Butterflies

Three major abundance peaks were observed in the months of July, October, January to February and the population rapidly declined during the period of March to June (Figure 4.1). Estimate of species richness during different seasons are expressed through sample base rarefaction curve (Figure 4.2). During NEM period highest curvature was noticed indicating that species richness per occurrence of individuals was the highest during this season. The other extreme season was summer during which relatively low species richness was observed. SWM period appeared to have comparatively steeper rarefaction curve, indicating relatively high evenness as compared with most other communities, whereas winter has relatively low evenness. Moreover, significant numbers of species were observed in the month of October (33 species) which comprised 50 per cent of the species and 16 species were represented by a single individual. The results indicated unique species contributed 45 per cent

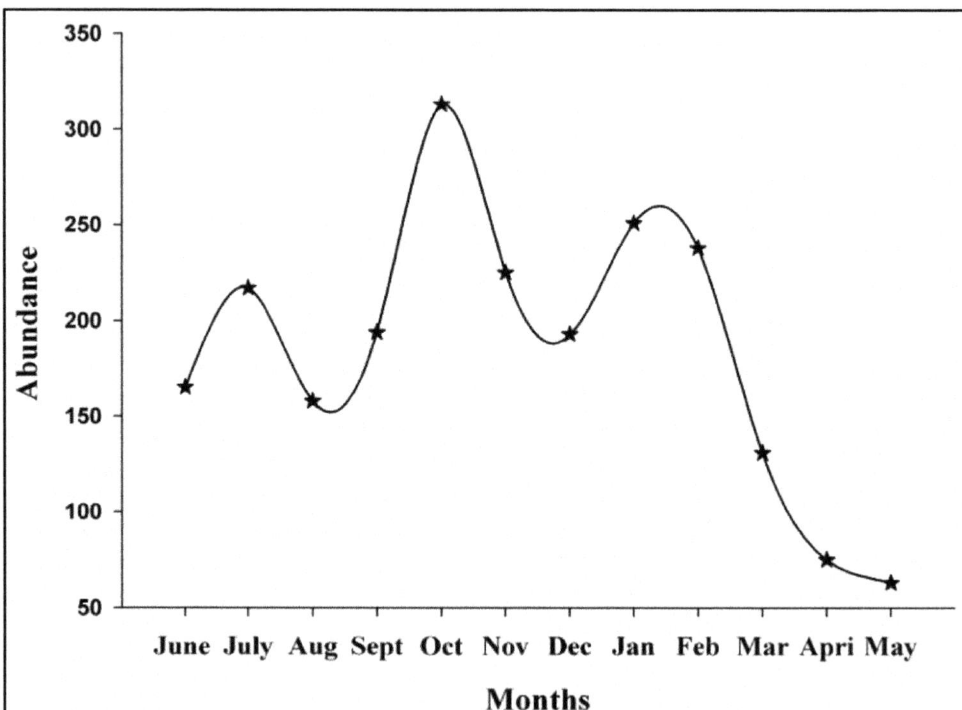

Figure 4.1: Abundance Profile of Butterflies Observed at different Months

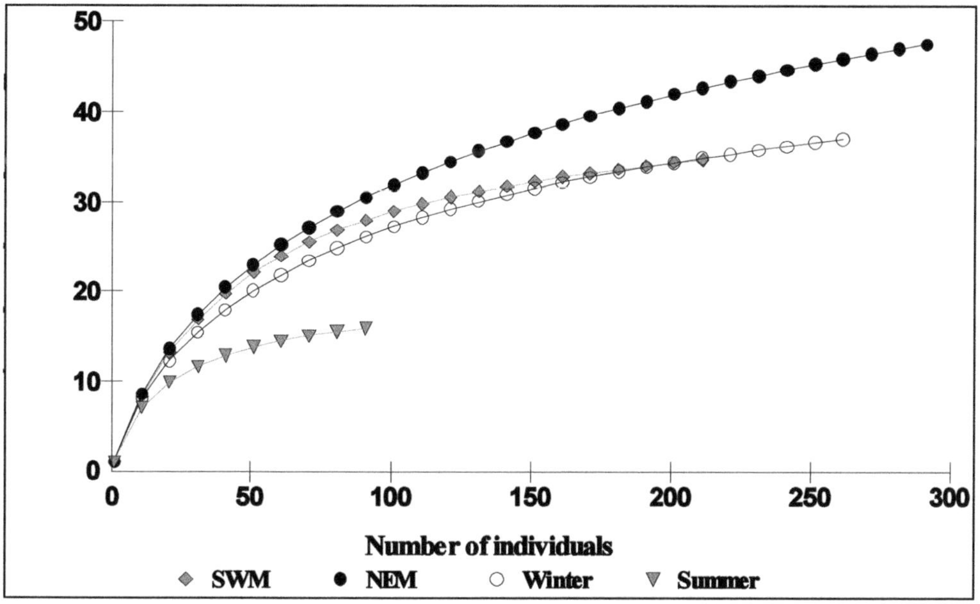

Figure 4.2: Sample Based Rare Faction Curve for the different Seasons

of the total richness during the NEM period (Table 4.2). Species such as *Colotis danae, Colotis eucharis, Delias eucharis, Hebomoia glaucippe, Ixias pyrene, Ariadne ariadne, Polyura athamas, Hypolimnas bolina, Junonia hierta, Mycalesis perseus, Curetis thetis, Chilades laius, Euchrysops cnejus, Zizina otis, Spindasis vulcanus* and *Spialia galba* were observed only during the NEM period; *Graphium sarpedon, Graphium nomius, Papilio polymnestor, Euthalia aconthea, Junonia atlites, Cynthia cardui, Melanitis leda, Azanus ubaldus, Rapala melampus* were observed only during the SWM period. *Colotis etrida, Ixias marianne, Catochrysops strabo, Jamides celeno, Gomalia elma* were observed during winter. Some species, *viz. Danaus chrysippus, Acraea violae, Tirumala septentrionis, Catopsilia pomona, Catalius rosimon, Eurema hecabe* and *Ariadne merione* were observed regularly and were more common at the southern coastal plains of India (Appendix 1).

Table 4.1: Total Number, Percentage of Species and Individuals Collected per Family

Sl.No.	*Family*	*Number of Genera*	*Number of Species*	*Number of Individual*
1	Papilionidae	4 (8.3 per cent)	8 (12.12 per cent)	154 (7 per cent)
2	Pieridae	11 (23 per cent)	16 (24.24 per cent)	636 (28.6 per cent)
3	Nymphalidae	14 (29.1 per cent)	23 (35 per cent)	1188(53.4 per cent)
4	Lycaenidae	15 (31.2 per cent)	15 (22.7 per cent)	232 (10.4 per cent)
5	Hesperiidae	4 (8.3 per cent)	4 (6 per cent)	13 (0.5 per cent)
	Total (5)	48	66	2, 223

Table 4.2: Seasonality of Butterflies in different Seasons at Kalpakkam

No. of Unique Species (Family-wise)	*SWM*	*NEM*	*Winter*	*Summer*
Papilionidae	3	–	–	–
Pieridae	–	5	2	–
Nymphalidae	4	5	–	–
Lycaenidae	2	4	2	–
Hesperiidae	–	2	1	–
Unique (total)	9	16	5	–
Richness (number of species)	35	48	41	16
Abundance (mean)	219	300	263	95

Apart from Nymphalid population, the overall family abundance varied across the seasons (Figure 4.3). Pieridae abundance was more during NEM, whereas, Lycaenidae abundance was more during winter. The observed fluctuation among different families could be due to the abundance of dominant species of the particular family. For example, the *Catalius rosimon* contributed about 50 per cent of the total abundance during winter when compared to rest of the year. Similarly, the emergence of *Catopsilia* spp. at coastal plains of Kalpakkam during NEM also considerably influenced the overall population trend of Pieridae. Eventhough the significant numbers of unique species were observed in different families across different seasons

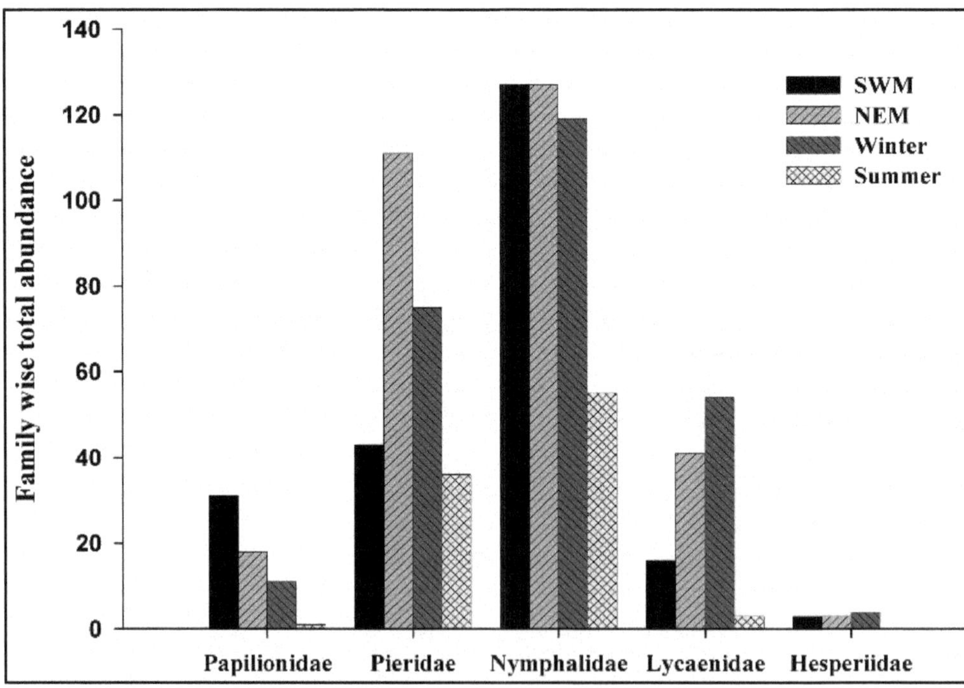

Figure 4.3: Population Trends of Butterfly Families in different Seasons

(Table 4.2), the preponderance of the dominant species governs the overall population trend.

Abundance, Richness and their Correlation with Weather Parameters

The butterfly abundance and richness increased with decreasing temperature and increasing humidity, the abundance drastically decreased at higher temperature during summer months, March to May (Figure 4.4).

The correlation analysis between weather parameter and butterfly diversity and abundance at Kalpakkam is given in Table 4.3. During the present study the butterfly abundance was positively correlated with richness ($R=0.903$ $P= <0.01$ per cent). Temperature was negatively correlated with richness ($R=-0.585$ $P= <0.05$) and abundance ($R=-0.654$ $P= <0.05$). The common species such as *Danaus chrysippus* and *Acraea violae* were observed during all the seasons and significant numbers of these species were observed even during summer. In this context it is worth mentioning that species presence during summer are presumed to be stress-tolerant species, and their occurrence all around the year (in considerable numbers) would perhaps be also attributable to their chemical defense.

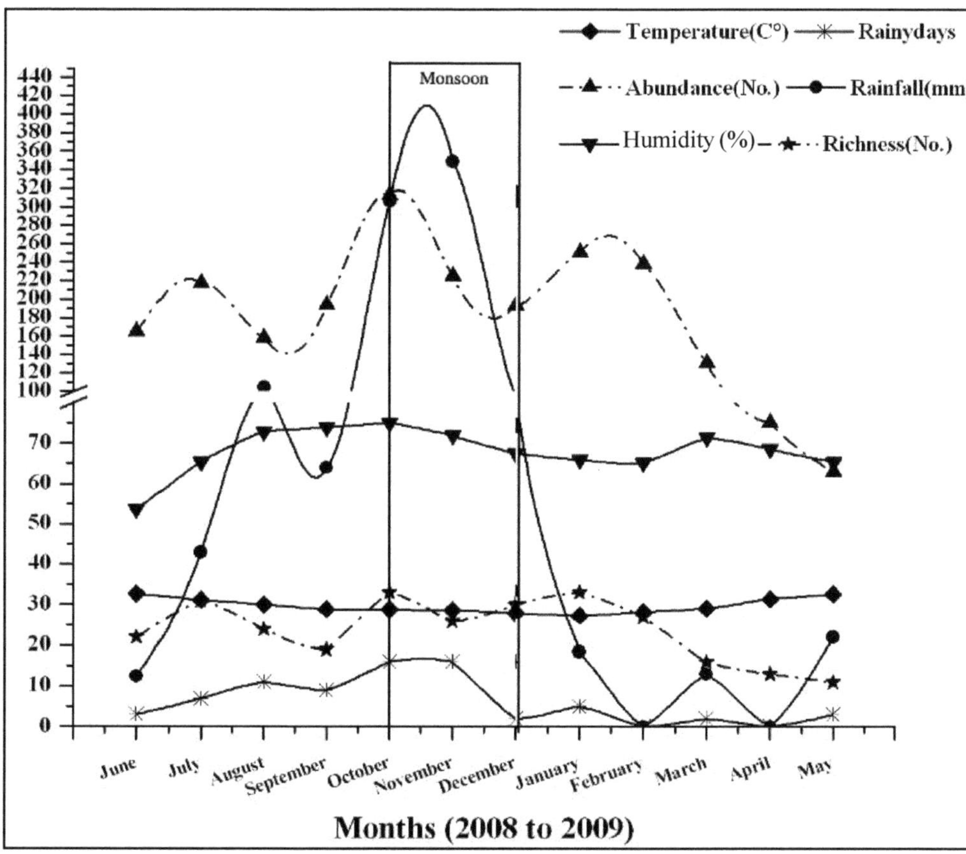

Figure 4.4: Abiotic and Butterfly Variation in different Months

Table 4.3: Correlation between Weather Parameter and Butterfly Population in DAE Campus Kalpakkam

Factors	*Temperature*	*Humidity*	*Rainfall*	*Rainy day*	*Richness*	*Abundance*
Temperature	1					
Humidity	–0.509*	1				
Rainfall	–0.316	0.524*	1			
Rainy day	–0.234	0.561*	0.889**	1		
Richness	–0.585*	0.044	0.399	0.406	1	
Abundance	–0.654*	0.206	0.535*	0.550*	0.903**	1

*: Significant at <0.05; **: Significant at <0.01.

Seasonal Migration

We quantified butterfly migratory swarms at Kalpakkam in October 2009 and July 2010. The swarm comprised three main species *viz., Catopsilia pyranthe, Catopsilia pomona,* and *Papilio demoleus* along with some common species, *viz. Euploea core, Tirumala septentrionis, Graphium nomius, Graphium sarpedon, Papilio polytes* and *Pachliopta hector.* The direction of the flight of the swarm was from south to north during October and north to south during July. The naturally migrating butterflies maintain a constant direction throughout the day. A total of 9697 individuals were counted on 16th October, 2009. *Catopsilia pyranthe* was dominant in the swarm (5792 individuals, 60 per cent of the swarm) followed by *Papilio demoleus* (3356 individuals, 34.6 per cent of the swarm) and *Catopsilia pomona* (549 individuals, 5.6 per cent of the swarm). Whereas, 1727 individuals were observed on 14th July 2010. *C. pomona* was dominant during July (896 individuals), which comprised 52 per cent of the total count swarm followed by *P. demoleus* (475 individuals, 27.5 per cent of the swarm) and *C. pyranthe* (356 individuals, 20.6 per cent of the swarm) (Table 4.4). This reduction in density could be due to the fact that, the southerly migration starts immediately after the unfavorable hot summer. The October migrants were freshly eclosed, whereas the July migrants were older adults.

Table 4.4: Total Number and Percentage of Migrant Species Observed during Different Seasons

Scientific Name	*No. of Individuals (October)*	*Per cent*	*No. of Individuals (July)*	*Per cent*
Catopsilia pyranthe	5792	60	356	20.6
Papilio demoleus	3356	34.6	475	27.5
Catopsilia pomona	549	5.6	896	52
Total	9697	100	1727	100

Periodicity of Migrant Swarms

The active flight of butterflies started at around 9.30h and was over by 15 h. During October the swarm showed continuous increasing trend between 11.15 h to

13.30 h and started to decline after that (Figure 4.5). Whereas, the peak density was observed at 10.15 h during July and later it started to decline with two peaks at 11.30 h and 12.30 h (Figure 4.6). During October the population trend of *C. pyranthe* perfectly resembled the overall periodicity of migratory swarm. Similarly *C. pomona* decided the overall trend during July.

Discussion

Nymphalid dominance in terms of species (Nimbalkar *et al.*, 2011) and Lycaenid in terms of genus (Tiple and Khurad, 2009) seems to be the general trend in Kalpakkam and else where in India. In tropical region with distinct wet and dry seasons, many insect species attain maximum adult abundance during the wet seasons (Didham and Springate, 2003; Tiple and Khurad, 2009). Moreover, Spitzer *et al.* (1993) states that the butterfly seasonal abundance fluctuation are closely associated with the monsoon climate. In agreement with above observation, the present study also revealed that the butterfly abundance and species diversity were more during wet season (NEM) than the other period. Earlier Nayak *et al.* (2004) had observed peak butterfly activity from October to January/February at different heterogeneous habitats of Western Ghats, India. Similar observations were made by Kunte (1997, 2000-01) at tropical, moist deciduous and scrub habitats in Northern Western Ghats. Since, the

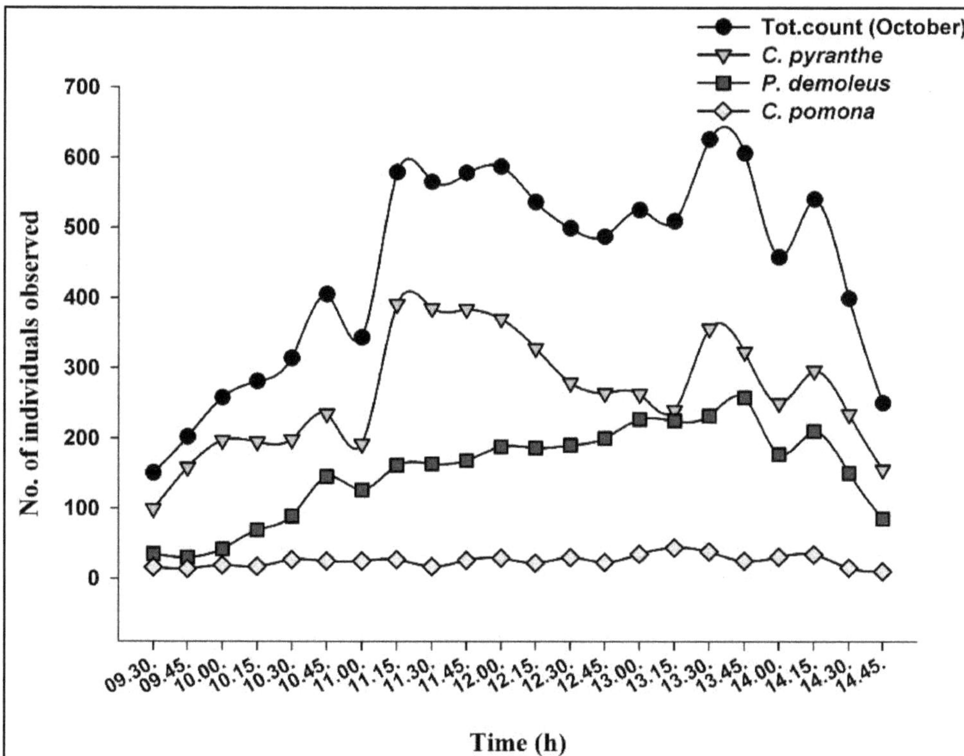

Figure 4.5: Number of Butterflies Passing through a 10m Strip in 15 Minutes Duration on 16th October 2009

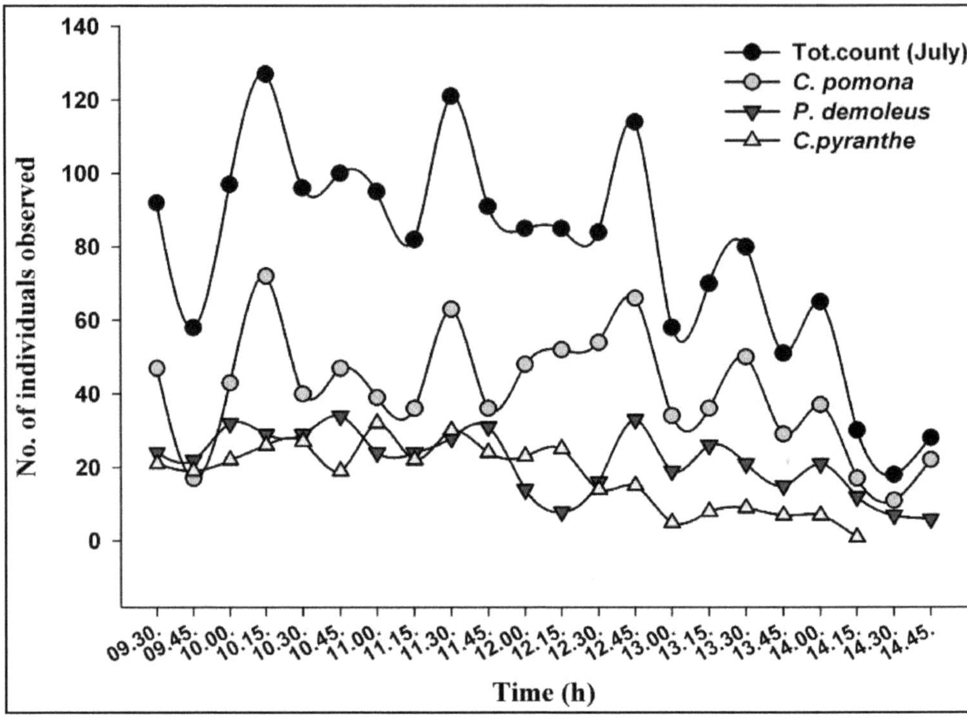

Figure 4.6: Number of Butterflies Passing through a 10m Strip in 15 Minutes duration on 14th July 2010

present study area receives sufficient rain (Mean-744.5 mm) and with the prevalence of conducive temperature (28° C) during NEM (Figure 4.5), this period appears to be the ideal breeding season for most of the butterflies in this region, and such favorable conditions continue till winter.

Many researchers have reported that butterflies are good indicators of environmental changes since they prefer a suitable climatic condition feeding and reproduction, respond reasonably to even subtle change in climate, which has been attributed to the fact that their entire life directly depend on temperature and monsoons (Kunte, 1997; Arun, 2002; Mathew and Binoy, 2002; Borkar and Komarpant, 2004; Kunte, 2005; Padhye *et al.*, 2006; Tiple *et al.*, 2006; Joshi, 2007; Mathew and Anto, 2007). Mathew and Anto (2007) have reported that, temperature ranges between 27°C to 29°C and humidity ranging between 60 per cent to 80 per cent are the most favorable for butterfly growth.

In the present study it was observed that a temperature range of 27-29°C and a relative humidity between 80-85 per cent were most suitable climatic conditions for butterfly assemblage of the coastal plains. Moreover, the period September to February (NEM and winter) was conducive climate during which temperature was less. This dominant factor facilitated the increase in species richness and population. Earlier studies (Kunte, 2000-01; Padhye *et al.*, 2006; Tiple and Khurad, 2009) also suggest that temperature and precipitation are two vital factors which influence butterflies richness and population directly. The butterfly population of the study area was correlated negatively with temperature and positively with relative humidity. It is known that elevated atmospheric temperature affects adversely the butterfly abundance, life cycle (Roy *et al.,* 2001). In the present study, butterfly population rapidly declined during the summer (Kunte, 1997; 2000-01; Barua *et al.,* 2010). Swaay (1990) suggests that butterflies, like any other insects are very vulnerable to changes in their environment because of their specialized life cycle. Moreover, factors such as lacking of larval host, nectar plants and dry vegetation, results in less butterfly abundance and lower survival ability of most species in the south-east coastal plains of India. Food habits among species (Gilbert and Singer 1975; Kitahara *et al.,* 2000; Vane-Wright, 2003) also influence the relationships between climate and butterfly diversity and abundance (Southwood, 1975). Availability and phenology of some predominant nectar plants *viz., Lantana camera, Lucas aspera, Tridax procumbens, Mimosa pudica, Gomphrena serrata, Vernonia cinerea, Tephrosia purpurea, Canthium dicoccum, Euphorbia antliquaram, Crotalaria verucosa, Heliotropium indicum, Calotropis gigantean* have appeared to play major role on diversity and abundance patterns of butterfly communities at Kalpakkam.

Like birds, butterflies also migrate to avoid unfavorable conditions of environmentally benign region. They are well known as insect migrants because many species make conspicuous unidirectional flights close to the ground maintaining the same general compass direction (Williams, 1930). In India, about 52 species are known to migrate of which 27 species are also known to migrate in Sri Lanka (Williams, 1938). Their migrations patterns in southern India have been documented since the 19th century (Prall, 1898; Aitken, 1897, 1898, 1900). Moreover, occasional butterfly migration that take place in central India, northern Western Ghats (Aitken, 1897, 1898, 1900; Home, 1935; Williams, 1938; Chaturvedi and Satheesan. 1979; Chaturvedi, 1992; Bharosa, 2000; Palot *et al.,* 2002), and Anamalai hills, Palani and Nilgiri hills have also been well documented (Evershed, 1910; Williams, 1938; Wynter-Blyth, 1957; Williams, 1949; Briscoe, 1952; Larsen, 1978, 1987; Kunte, 2005). *Danaus genutia, Tirumala septentrionis, Euploea sylvester, Euploea core, Phalanta phalantha, Cynthia cardui, Catopsilia pomona* and *Catopsilia pyranthe* are few common migratory species from south India. Moreover, the following species appear to be frequent, conspicuous migrants in other region: *Catopsilia florella* (Africa), *Catopsilia sennae* (America), *Catopsilia statira* (South America), *Catopsilia pomona* (Java), *Ascia monuste* (America), *Danaus plexippus* (circum continental), *Vanessa cardui* (worldwide). It is perhaps of significance that these species belong to the Pieridae and Nymphalidae, asset of families that are evolutionarily interlinked (Shields, 1974). *Lampides boeticus* found to be the only migrating species in Lycaenidae (Smetacek, 2002). We report a bidirectional seasonal migration of butterflies and also quantified the migratory swarms of butterfly which largely consisted of *Catopsilia pyranthe* and *Catopsilia pomona.* To the best of our knowledge this is the first scientific report on quantification of butterfly migration at costal plains of South India.

In tropics, flights of a species tend to recur in the same direction at the same season in different years (Williams, 1949), on the other hand the migratory path and season of the same species may differ regionally. These differences may attribute to the local climatic changes especially monsoon. Kunte (2005) has observed longitudinal migration of Danaine butterflies between the Western Ghats and plains. He hypothesized that, this movement occurred to avoid upcoming torrential rain (SWM) at Western Ghats and to breed alternatively in the plains. Hence, we believe that, the observed migration of butterflies is mainly driven by the monsoonal rainfall. However the density of migratory swarm is decided by the conducive climate prevailed at northern and southern feeding and breeding grounds at east coast. Unfortunately, the exact patterns of breeding phenology, voltinism and occurrence of this migratory species at southern and northern parts of south-eastern coastal areas remain obscure. To bring more light on this interesting phenomenon and to test specific hypothesis about this seasonal migration, large scale recapture studies and close examination during migration with the help of a network comprising of many individuals and institutes is required.

Acknowledgments

Authors are grateful to Dr. Baldev Raj, former Director, IGCAR for his continuous encouragement and support. They are also indebted to Dr. Krushnamegh Kunte, Post-doctoral Research Fellow, FAS Center for Systems Biology, Harvard University, Cambridge, USA for confirming the species identification and for commenting on the manuscript.

References

Aitken, E.H., 1897. The migration of butterflies. *Journal of Bombay Natural History Society*, 11: 336–337.

Aitken, E.H., 1898. Migration of *Euploea core*. *Journal of Bombay Natural History Society*, 12: 229–230.

Aitken, E.H., 1900. Butterflies as weather prophets. *Journal of Bombay Natural History Society*, 13: 540–541.

Akite, P., 2008. Effects of anthropogenic disturbances on the diversity and composition of the butterfly fauna of sites in the Sango Bay and Iriiri areas, Uganda: implications for conservation. *African Journal of Ecology*, 46: 3–13.

Arun, P.R., 2002. Butterflies of Siruvani forest of Western Ghats, with notes on their seasonality. *Zoo's print Journal*, 18(2): 1003–1006.

Barua, K.K., Slowik, J., Bobo, K.S. and Muehlenberg, M., 2010. Correlations of rainfall and Forest Type with Papilionid Assemblages in Assam in Northeast India. *Psyche 2010*: ID 560396, 1–10.

Bharosa, K.M., 2000. Large scale emergence and migration of the common emigrant butterflies *Catopsilia pomona* (Family– Pieridae). *Journal of Bombay Natural History Society*, 97: 301.

Bobo, K.S., Waltert, M., Fermon, H., Njokagbor, J. and Muhlenberg, M., 2006. From forest to farmland: butterfly diversity and habitat associations along a gradient of forest conversion in Southwestern Cameroon. *Journal of Insect Conservation*, 10: 29–42.

Bonebrake, T.C and Sorto, R., 2009. Butterfly (Papilionoidea and Hesperioidea) rapid assessment of a coastal countryside in El Salvador. *Tropical Conservation Science*, 2: 34–51.

Bonebrake, T.C., Ponisio, L.C., Boggs, C.L. and Ehrlich, P.R., 2010. More than just indicators: A review of tropical butterfly ecology and conservation. *Biological Conservation*, 143: 1831–1841.

Borkar, M.R. and Komarpant, N., 2004. Diversity, abundance and habitat associations of butterfly species in Bondla Wildlife Sanctuary of Goa, India. *Zoo's Print Journal*, 19(10): 1648– 1653.

Briscoe, V., 1952. Butterfly migration in the Nilgiris. *Journal of Bombay Natural History Society*, 50: 417–418.

Brown, K.S. and Freitas, A.V.L., 2000. Atlantic forest butterflies: indicators for landscape conservation. *Biotropica*, 32: 934–956.

Caldas, A. and Robbins, R.K., 2003. Modified Pollard transects for assessing tropical butterfly abundance and diversity. *Biological Conservation*, 110: 211–219.

Chaturvedi, N. and Satheesan, S.M., 1979. Southward migration of *Euploea core core* Cr. at Khandala, Western Ghats. *Journal of Bombay Natural History Society*, 76: 534.

Chaturvedi, N., 1992. Northward migration of the Common Indian Crow butterfly *Euploea core* (Cramer) in and around Bombay. *Journal of Bombay Natural History Society*, 90: 115–116.

Daily, G.C. and Ehrlich, P.R., 1995. Preservation of biodiversity in small rainforest patches: rapid evaluation using butterfly trapping. *Biodiversity and Conservation*, 4: 35–55.

Denlinger, D.L., 1986. Dormancy in tropical insects. *Annual Review of Entomology*, 31: 239–264.

Didham, R.K. and Springate, N.D., 2003. Determinants of temporal variation in community structure. In: *Arthropods of Tropical Forests: Spatial-Temporal Dynamics and Resource Use in the Canopy*, (Eds.) Y. Basset, V. Novotny, S.E. Miller and R.L. Kitching. Cambridge University Press, Cambridge.

Evershed, J., 1910. Note on migration. *Journal of Bombay Natural History Society*, 20: 390–391.

Gajendiran, N. and Ragupathy, S., 2002. *The Macroflora of East Coast at Kalpakkam*. Published by IGCAR, Kalpakkam.

Gilbert, L.E. and Singer, M.C., 1975. Butterfly ecology. *Annual Review of Ecology, Evolution and Systematics*, 6: 365–397.

Hill, C.J., 1995. Linear strips of rain forest vegetation as potential dispersal corridors for rain forest insects. *Conservation Biology*, 9: 1559–1566.

Home, L.W.M., 1935. Notes on Coorg butterflies. *Journal of Bombay Natural History Society*, 37: 669–674.

Joshi, P.C., 2007. Community structure and habitat selection of butterflies in Rajaji National Park, a moist deciduous forest in Uttaranchal, India. *Tropical Ecology*, 48(1): 119–123.

Kitahara, M., Sei, K. and Fujii, K., 2000. Patterns in the structure of grassland butterfly communities along a gradient of human disturbance: Further analysis based on the generalist/specialist concept. *Population Ecology*, 42: 135–144.

Koh, L. and Sodhi, N.S., 2004. Importance of reserves, fragments and parks for butterfly conservation in a tropical urban landscape. *Ecological Applications*, 14: 1695–1708.

Kremen, C., 1992. Assessing the indicator properties of species assemblages for natural areas monitoring. *Ecological Applications,* 2: 203–217.

Kunte, K., 1997. Seasonal patterns in butterfly abundance and species diversity in four tropical habitats in the northern Western Ghats. *Journal of Bioscience,* 22: 593–603.

Kunte, K., 2000-01. Butterfly diversity of Pune city along the human impact gradient. *Journal of Ecological Society*, 13 and 14: 40–45.

Kunte, K., 2005. Species composition, sex-ratios and movement patterns in Danaine butterfly migrations in southern India. *Journal of Bombay Natural History Society*, 102 (3): 280–286.

Kunte, K., Kalesh, S. and Kodandaramaiah, U. (eds.), 2011. *Butterflies of India. v. 1.04.* Indian Foundation for Butterflies. URL at http: / / ifoundbutterflies.org/ Accessed on 1 December 2011.

Larsen, T.B., 1978. Butterfly migrations in the Nilgiri Hills of South India. *Journal of Bombay Natural History Society,* 74: 546–549.

Larsen, T.B., 1987. The butterflies of the Nilgiri Mountains of Southern India (Lepidoptera: Rhopalocera). *Journal of Bombay Natural History Society*, 84: 291–316.

Larsen, T.B., 2005. *Butterflies of West Africa.* Apollo Books, Stenstrup.

Mathew, G. and Binoy, C.F., 2002. Migration of butterflies (Lepidoptera: Rhopalocera) in the new Amarambalam reserve forest of the Nilgiri biosphere reserve. *Zoo's Print Journal*, 17(8): 844–847.

Mathew, G. and Anto, M., 2007. *In situ* conservation of butterflies through establishment of butterfly gardens: A case study at Peechi, Kerala, India. *Current Science India*, 93(3): 337–347.

Nayak, G., Subramanian, K.A., Gadgil, M., Achar, K.P., Acharya, Padhye, A., Deviprasad, Bhatta, G., Ghate, H., Murugan, Pandit, P., Thomas, S. and Thomas, W., 2004. Patterns of diversity and distribution of butterflies in heterogeneous landscapes of the Western Ghats, India. Envis technical report no.18. Centre for ecological science, Bangalore, pp. 1–33.

Neil McAleece., P.J.D. Lambshead and Paterson, P.L.J., 1997. Biodiversity Pro (version 2). The Natural History Museum, London.

Nimbalkar, R.K., Chandekar, S.K. and Khunte, S.P., 2011. Butterfly diversity in relation to nectar food plants from Bhor Tahsil, Pune District, Maharashtra, India. *Journal of Threatened Taxa*, 3(3): 1601–1609.

Oliveira, E.G., Srygley, R.B. and Dudley, R., 1998. Do Neotropical migrant butterflies navigate using a solar compass? *The Journal of Experimental Biology*, 201: 3317–3331.

Padhye, A.D., Dahanukar, N., Paingankar, M., Deshpande, M. and Deshpande, D., 2006. Season and landscape wise distribution of butterflies in Tamhini, Northern, Western Ghats, India. *Zoo's print Journal*, 21(3): 2175–2181.

Palot, M.J., Radhakrishnan, C., Balakrishnan, V.C. and Kambrath, B., 2002. A report on the migration of butterf, lies in Aralam wildlife sanctuary, Kerala. *Zoo's print Journal*, 17(2): 722

Pollard, Å., 1991. *Monitoring Butterfly Numbers in Monitoring for Conservation and Ecology*. pp. 87–111, (Ed.) F.B. Goldsmith. Chapman and Hall, London.

Prall, S.E., 1898. The migration of butterflies. *Journal of Bombay Natural History Society*, 11: 533.

Roy, D.B., Rothery, P., Moss, D., Pollard, E. and Thomas, J.A., 2001. Butterfly numbers and weather: predicting historical trends in abundance and the future effects of climate change. *Journal of Animal Ecology*, 70: 201–217.

Shields, O., 1974. Toward a theory of butterfly migration. *Journal of research on the Lepidoptera*,13(4): 217–238.

Shields., O., 1989. World numbers of butterflies. *Journal of the Lepidopterists Society*, 43: 178–183.

Smetacek, P., 2002. The study of butterflies (congregations, courtship and migration). *Resonance*, 7: 1–9.

Southwood, T.R.E., 1975. The dynamics of insect populations: In *Insects Science and Society*, (Ed.) D. Pimentel. Academic Press, New York, pp. 151–199.

Spitzer, K., Novotny, V., Tonner, M. and Leps, J., 1993. Habitat preferences, distribution and seasonality of the butterflies (Lepidoptera, Papilionidae) in a montane tropical rain forest. *Vietnam. Journal of Biogeography*, 20: 109–121.

Swaay, C.A.M., 1990. An Assessment of the Changes in Butterfly Abundance in the Netherlands during the 20th Century. *Biological Conservation*, 52: 287–302

Thomas, J.A., 2005. Monitoring change in the abundance and distribution of insects using butterflies and other indicator groups. *Philosophical Transactions of the Royal Society B: Biological Sciences*, 28 360(1454): 339–357.

Tiple, A.D., Deshmukh, V.P. and Dennis, R.L.H., 2006. Factors influencing nectar plant resource visits by butterflies on a university campus: implications for conservation. *Nota lepidopterologica*, 28: 213–224.

Tiple, A.D. and Khurad, A.M., 2009. Butterfly species diversity, habitats and seasonal distribution in and around Nagpur City, Central India. *World Journal of Zoology*, 4(3): 153–162.

Vane-Wright, D., 2003. *Butterflies.* The Natural History Museum/Smithsonian, p. 112.

Walpole, M.J. and Sheldon, I.R., 1999. Sampling butterflies in tropical rainforest: an evaluation of a transect walk method. *Biological Conservation,* 87: 85–91.

Williams, C.B., 1930. *The Migration of Butterflies.* In: Oliver and Boyd, Edinburgh and London.

Williams, C.B., 1938. The migration of butterflies in India. *Journal of Bombay Natural History Society,* 40: 439–457.

Williams, C.B., 1949. Migrant butterflies outside North America. *Lepidoptera news,*3(4–5): 39–40.

Wolda, H., 1978. Fluctuations in abundance of tropical insects. *American Naturalist,* 112: 1017– 1045.

Wolda, H., 1988. Insect seasonality: Why? Annual *Review of Ecology and Systematics,* 19: 1– 18.

Wynter-Blyth, M.A., 1957. Butterflies of the Indian region. *Journal of Bombay Natural History Society,* 20: 523.

Appendix 1

Family/Subfamily	*Scientific Name*	*Common Name*	*SWM*	*NEM*	*Winter*	*Summer*	*Total*
		Papilionidae					
Papilioninae	*Pachliopta aristolochiae*	Common rose	5	–	2	–	7
Papilioninae	*Graphium sarpedon*	Common bluebottle	3	–	–	–	3
Papilioninae	*Graphium agamemnon*	Tailed joy	–	1	1	–	2
Papilioninae	*Graphium nomius*	Spot swordtail	6	–	–	–	6
Papilioninae	*Pachliopta hector*	Crimson rose	9	13	5	–	27
Papilioninae	*Papilio demoleus*	Lime butterfly	5	2	–	–	7
Papilioninae	*Papilio polytes*	Common mormon	2	2	3	1	8
Papilioninae	*Papilio polymnestor*	Blue mormon	1	–	–	–	1
		Pieridae					
Coliadinae	*Catopsilia pyranthe*	Mottled emigrant	–	41	1	–	42
Coliadinae	*Catopsilia pomona*	Common emigrant	6	18	1	2	27
Coliadinae	*Eurema hecabe*	Common grass yellow	11	12	29	7	59
Pierinae	*Anaphaeis aurota*	Pioneer	13	9	11	9	42
Pierinae	*Appias libythea*	Striped albatross	1	2	2	–	5
Pierinae	*Cepora nerissa*	Common gull	5	6	–	–	11
Pierinae	*Colotis amata*	Small salmon arab	4	2	4	2	12
Pierinae	*Colotis danae*	Crimson tip	–	5	–	–	5
Pierinae	*Colotis etrida*	Little orange tip	–	–	1	–	1
Pierinae	*Colotis eucharis*	Plain orange tip	–	1	–	–	1
Pierinae	*Delias eucharis*	Common jezebel	–	3	–	–	3
Pierinae	*Hebomoia glaucippe*	Great orange tip	–	3	–	–	3
Pierinae	*Pareronia valeria*	Common wanderer	–	3	5	3	11
Pierinae	*Leptosia nina*	Psyche	3	5	20	13	41
Pierinae	*Ixias pyrene*	Yellow orange tip	–	1	–	–	1
Pierinae	*Ixias marianne*	White orange tip	–	–	1	–	1
		Nymphalidae					
Biblidinae	*Ariadne merione*	Common castor	5	15	16	9	45
Biblidinae	*Ariadne ariadne*	Angled castor	–	4	–	–	4
Charaxinae	*Polyura athamas*	Common nawab	–	1	–	–	1
Danainae	*Danaus chrysippus*	Plain tiger	42	39	33	19	133
Danainae	*Danaus genutia*	Striped tiger	4	1	8	1	14
Danainae	*Euploea core*	Common crow	7	4	4	4	19
Danainae	*Tirumala limniace*	Blue tiger	8	9	3	–	20

Contd...

Appendix 1–*Contd...*

Family/Subfamily	*Scientific Name*	*Common Name*	*SWM*	*NEM*	*Winter*	*Summer*	*Total*
Danainae	*Tirumala septentrionis*	Dark blue tiger	16	15	5	4	40
Heliconiinae	*Acraea violae*	Tawny coster	21	20	36	14	91
Heliconiinae	*Phalanta phalantha*	Common leopard	5	5	2	–	12
Limenitidinae	*Neptis hylas*	Common sailer	1	1	4	3	9
Limenitidinae	*Euthalia aconthea*	Common baron	1	–	–	–	1
Nymphalinae	*Hypolimnas bolina*	Great eggfly	–	4	–	–	4
Nymphalinae	*Hypolimnas misippus*	Danaid eggfly	–	4	1	–	5
Nymphalinae	*Junonia orithya*	Blue pansy	2	–	1	–	3
Nymphalinae	*Junonia iphita*	Chocolate pansy	–	2	1	–	3
Nymphalinae	*Junonia atlites*	Grey pansy	5	–	–	–	5
Nymphalinae	*Junonia lemonias*	Lemon pansy	6	1	2	1	10
Nymphalinae	*Junonia almana*	Peacock pansy	1	–	3	–	4
Nymphalinae	*Junonia hierta*	Yellow pansy	–	1	–	–	1
Nymphalinae	*Cynthia cardui*	Painted lady	2	–	–	–	2
Satyrinae	*Melanitis leda*	Common evening brown	1	–	–	–	1
Satyrinae	*Mycalesis perseus*	Common bush brown	–	1	–	–	1
		Lycaenidae					
Curetinae	*Curetis thetis*	Indian sunbeam	–	3	–	–	3
Polyommatinae	*Azanus ubaldus*	Bright babul blue	2	–	–	–	2
Polyommatinae	*Castalius rosimon*	Common pierrot	9	12	26	3	50
Polyommatinae	*Catochrysops strabo*	Forget me not	–	–	1	–	1
Polyommatinae	*Chilades laius*	Lime blue	–	4	–	–	4
Polyommatinae	*Euchrysops cnejus*	Gram blue	–	1	–	–	1
Polyommatinae	*Everes lacturnus*	Indian cupid	3	1	1	–	5
Polyommatinae	*Jamides celeno*	Common cerulean	–	–	7	–	7
Polyommatinae	*Lampides boeticus*	Pea blue	–	2	5	–	7
Polyommatinae	*Leptotes plinius*	Zebra blue	1	–	5	–	6
Polyommatinae	*Pseudozizeeria maha*	Pale grass blue	–	13	6	–	19
Polyommatinae	*Zizina otis*	Lesser grass blue	–	2	–	–	2
Theclinae	*Arhopala amantes*	Large oakblue	–	2	3	–	5
Theclinae	*Spindasis vulcanus*	Common silverline	–	1	–	–	1
Theclinae	*Rapala melampus*	Indian red flash	1	–	–	–	1
		Hesperiidae					
Hesperiinae	*Parnara guttata*	Common straight swift	–	1	3	–	4
Hesperiinae	*Suastus gremius*	Indian palm bob	3	1	–	–	4
Pyrginae	*Gomalia elma*	African mallow skipper	–	–	1	–	1
Pyrginae	*Spialia galba*	Indian grizzled skipper	–	1	–	–	1
	Total	**66 species**	**220**	**300**	**263**	**95**	

2013, Animal Diversity, Natural History and Conservation, Vol. 3 *Pages* **55–62**
Editors: **V.K. Gupta and Anil K. Verma**
Published by: **DAYA PUBLISHING HOUSE, NEW DELHI**

Chapter 5

Male Genitalia as Taxonomic Tool in the Classification of Indian Acridoidea (Orthoptera)

Mohammad Kamil Usmani, Md. Rashid Nayeem* and Md. Humayoon Akhtar
Section of Entomology, Department of Zoology, Aligarh Muslim University, Aligarh – 202 002, U.P., India

ABSTRACT

The present study is based on Indian species representing the families Acrididae, Catantopidae and Pyrgomorphidae belonging to the Superfamily Acridoidea. The significance of male genitalic structures in the classification of Indian Acridoidea is shown. For better understanding of the significance of morphological structures, comparative study has been done on male genitalia with reference to subgenital plate, supra-anal plate and cerci, epiphallus and aedeagus with an aim to discover their significance in order to make the identification of genera and species together with other conventional morphological characters more perfect and convenient.

Keywords: *Significance, Male genitalia, Indian species, Acridoidea, Orthoptera.*

Introduction

Superfamily Acridoidea includes locusts and grasshoppers. They are also called short-horned grasshoppers in contrast to Ensifera (Tettigonoidea and Grylloidea) or

* *Corresponding author.* E-mail: rashidnayeem48@gmail.com

long horned grasshoppers which constitute one of the other suborders of Orthoptera. Suborder Caelifera is divided into four superfamilies *viz.* Acridoidea, Tridactyloidea, Tetrigoidea and Eumastacoidea.

Acridoidea possess short antennae, usually shorter than the body, a short and stout ovipositor and their tarsi are three segmented. The other superfamilies of Caelifera are easily recognizable at sight and are not frequently encountered. Superfamily Acridoidea has shown maximum diversity and divided into various families of which family Acrididae, Catantopidae and Pyrgomorphidae are widely distributed in India.

In case of Acridoidea, genitalic characters along with already recognized conventional characters have made the identification of families, subfamilies, genera and species more stable and practicable. Bridge-shaped condition of epiphallus; presence or absence of dorso-lateral appendages and oval scelerites on epiphallus; divided and undivided or flexured condition of aedeagus, presence or absence of gonopore process on aedeagus are taken as stable characters for separating various families. Smooth or toothed condition of the apex of male cercus, presence or absence of ancorae on epiphallus, long or short condition of valves of aedeagus are taken as stable characters for separating various subfamilies. Shape and length of basal and apical valves of aedeagus, broad or narrow condition of bridge, mono, bi and trilobite condition of lophi of epiphallus and shape of male supra-anal plate are suggested as useful generic characters. Shape and size of ancorae, shape of subgenital plate and supra-anal plate and cerci are taken as specific characters.

Morphological characters of adult specimens are still used more frequently than any other but they are supplemented to an ever increasing extent by other characters. The new taxonomic characters (*e.g.* proteins, chromosomes, behavior etc.) supplement but do not displace the use of morphological characters. Fortunately it turns out to be true that the new characters usually confirm the classification based on the more classical approaches. It seems that general morphology is a reflection of a large part of genotype and this permits, on the whole, reliable conclusion on the relationships.

The genetalia of many animals, particularly Arthropods, not only show a great deal of structural detail but are also highly species-specific. Genetalic structures are most useful among the arthropods. In many group of insects genetalic structures are more important for species diagnosis than any other character. In most vertebrates they are very soft, but the gonopodium of fishes, the hemipenis of snakes and the baculum of mammals supply good taxonomic information.

It has ever been suggested by Dufour and others that a lock a lock and key relationship exists as regards the copulatory structures of the males and females of those species with sclerotised genetalia. Such appears the case in certain group of insects. In general, genetalic structures must be evaluated just like other characters. In groups where their significance has been proved they are usually very useful, because genetalic structures appear to be among the first to change in the course of speciation. Recently in 2009 Usmani studied the male and female genitalia in some Libyan species of Acrididae. The present study is based on the conventional as well as genitalic characters, for a better understanding of the significance of morphological structures. Comparative study has been done on genitalia with reference to subgenital plate, supra-anal plate and cerci, epiphallus and aedeagus.

Materials and Method

Preparation for genitalic studies: For a detailed study of the various components of genitalia, the apical part of male bodies was cut off and boiled in 10 per cent potassium hydroxide for a variable period till the material became transparent (usually about 10 minutes) to remove unsclerotized and non-chitinous tissues. They were then thoroughly washed in tap water for complete removal of KOH and examined in 70 percent ethyl alcohol on a cavity slide. Later, every specimen was dissected under a binocular microscope with the help of fine needles to separate various components viz., supra-anal plate and cerci, subgenital plate, epiphallus and aedeagus. The normal process of dehydration was adopted and clearing was done in clove oil. The genitalic structures were mounted separately on cavity slides in Canada balsam. A 22mm square cover-glass over the cavity of the slide was normally used when examining the genitalic structures. This was made to prevent them from curling upwards and inwards at the edges. The slides were kept in a slide drier at a temperature of approximately 400°C for about one week to get them completely dry.

The permanent slides were examined under the microscope in order to make a detailed study of the genitalic structures. Drawings were initially made with the help of a camera lucida. Details were filled in by conventional microscope examination.

Results

For a better understanding of the significance of genitalic structures, comparative study has been done on genitalia with reference to supra-anal plate and cerci, subgenital plate, epiphallus and aedeagus of males. A comparative study of these characters may therefore help to trace the interrelationship of the groups more clearly than the conventional external characters.

Male Genitalic Characters

Male Supra-anal Plate (Epiproct) and Cerci (Plate 5.1 Figures A, B, C, D)

Tergite XI is the epiproct (Supra-anal plate); in its simplest form it is triangular in shape, but it may be semielliptical, rectangular, pentagonal etc.; its surface is somewhat uneven, and there may be on it either a furrow or a ridge along the middle as well as some tubercles. It is lying above the half concealed paraprocts (XIth sternite) and with the membrane between them and the epiproct are articulated the cerci. Cerci are tubular and plate like appendages which arise between supra-anal plate and paraprocts. Both assist to grip the female abdomen during the course of copulation. Cerci also function as sense organs. A cercus consists of two sclerotised parts: the main body and a small basal lobe usually concealed under the epiproct, although in the subfamily Euryphyminae it is large and conspicuous. The lobe has a well defined dense patch of hairs called brustia, possibly sensory (Jannone, 1939). The cercus itself is usually conical but there is much variation.

Smooth or toothed condition of the apex of male cercus may be taken as a subfamilial character. The subfamily Calliptaminae may be characterized by the

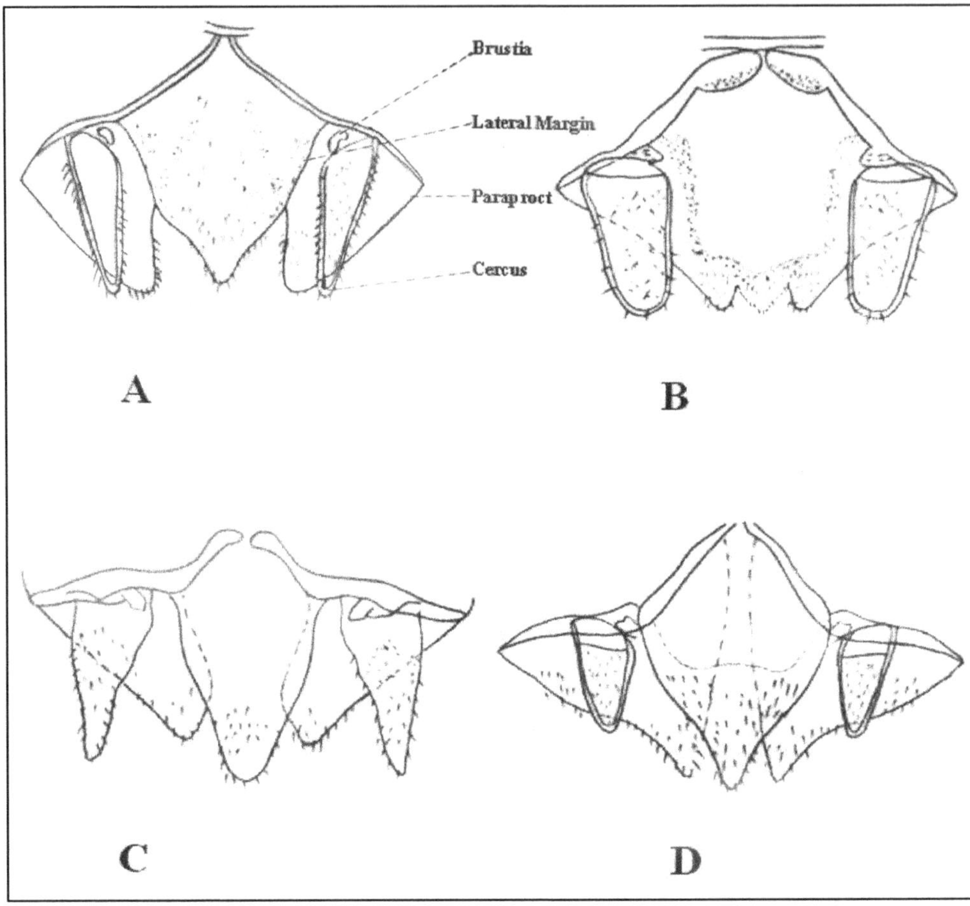

Plate 5.1: Male Supra-Anal Plate and Cerci
A. Acrididae; B. Catantopidae; C. Pyrgomorphidae; D. Pamphagidae

presence of a tooth on apex of cercus. By this character it differs from other subfamilies of Acrididae. Shape and length of supra-anal plate and cerci are regarded as generic as well as specific characters.

Subgenital Plate (Plate 5.2 Figures A, B, C, D)

The greatly enlarged IXth sternite is fully or partially subdivided by transverse furrow into two parts which are more or less movable on each other. When the division is complete, basal portion is enlarged as the true IXth sternite and apical portion as subgenital plate. The plate which envelops the phallic complex is generally conical but there is much variation.

Epiphallus (Plate 5.3 Figures A, B, C, D)

It is a strongly sclerotised structure located on dorsal side of the phallic organ. It severs to grasp the edge of female subgenital plate and to fix the phallus firmly

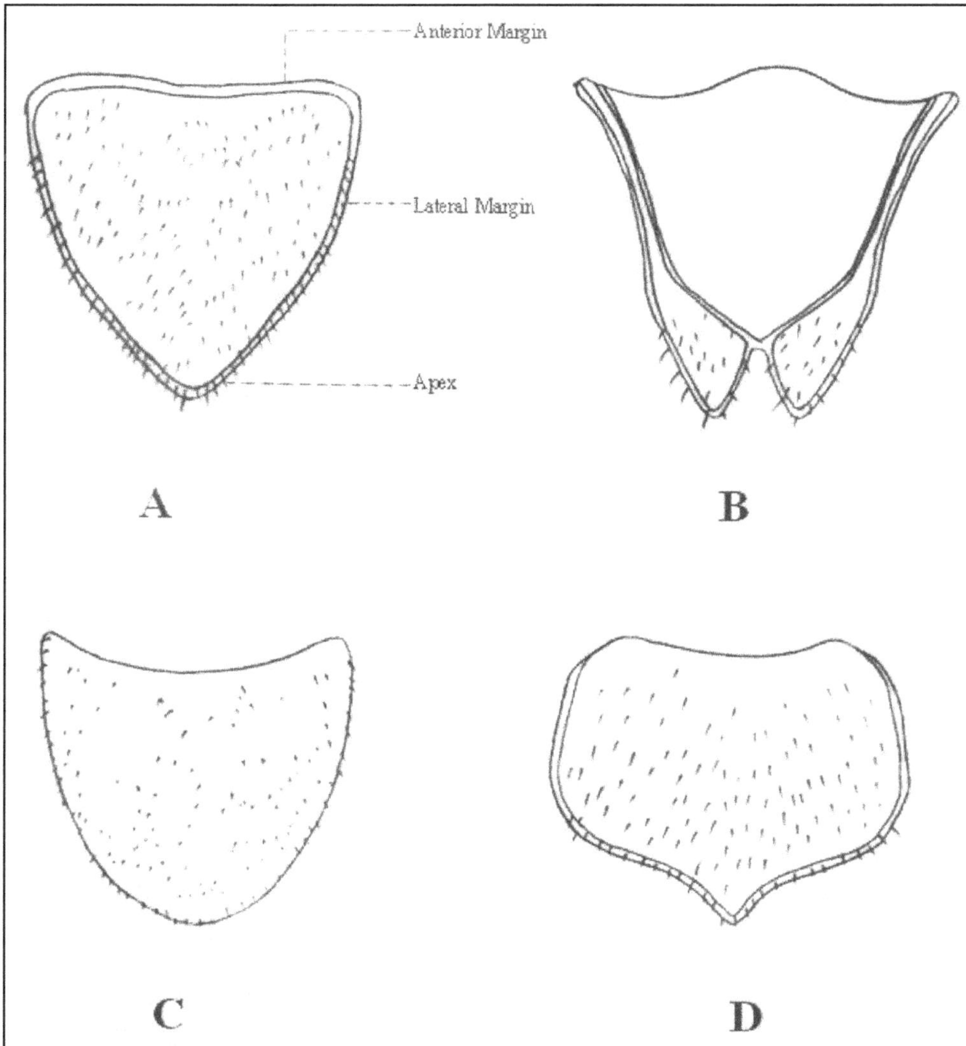

Plate 5.2: Male Subgenital Plate
A. Acrididae; B. Catantopidae; C. Pyrgomorphidae; D. Pamphagidae

during copulation. It provides an excellent taxonomic character which may be used as a single character for differentiating the subfamilies.

Shield or bridge-shaped condition of epiphallus; presence or absence of dorso-lateral appendages, oval sclerites and lophi on epiphallus are taken as stable characters for separating various families.

Presence or absence of ancorae on epiphallus; divided or undivided condition of bridge are taken as subfamilial characters. Plate like Epiphallus in the subfamily Calliptaminae; bridge of epiphallus undivided in Acridinae, Oedipodinae, Truxalinae,

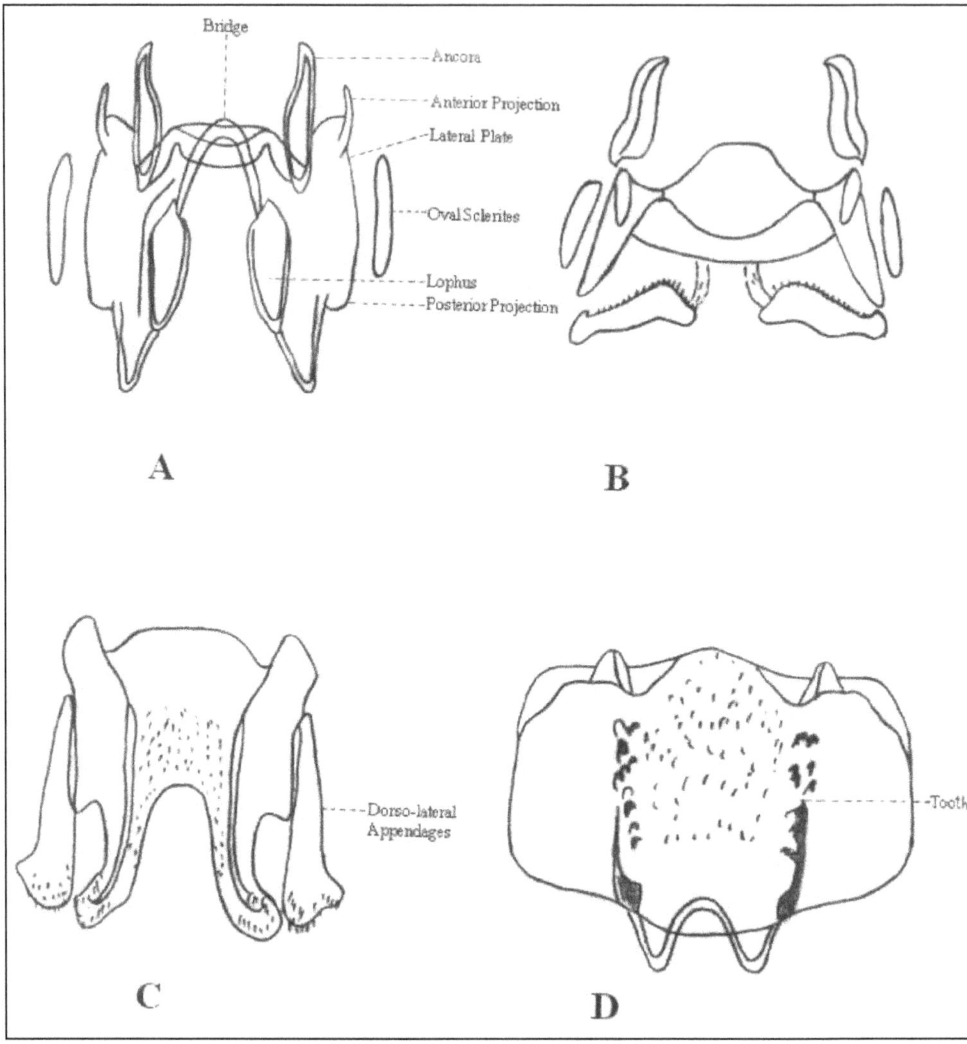

Plate 5.3: Epiphallus
A. Acrididae; B. Catantopidae; C. Pyrgomorphidae; D. Pamphagidae

Gomphocerinae, Hemiacridinae, Catantopinae, Cyrtacanthacridinae, Eyprepocnemidinae, Tropidopolinae and Romaleinae; bridge of epiphallus divided in Oxyinae and Coptacridinae.

Short or long condition of ancorae, broad or narrow condition of bridge, mono, bi or trilobite condition of lophi are taken as generic characters. Shape of lophi and ancorae are considered as specific characters.

Aedeagus (Plate 4 Figures A, B, C, D)

It is a main intromittent organ consisting of a pair of basal and apical valves. During the course of copulation, it is inserted between the ventral ovipositor valves of the female into the vagina and its tip reaches the spermathecal duct.

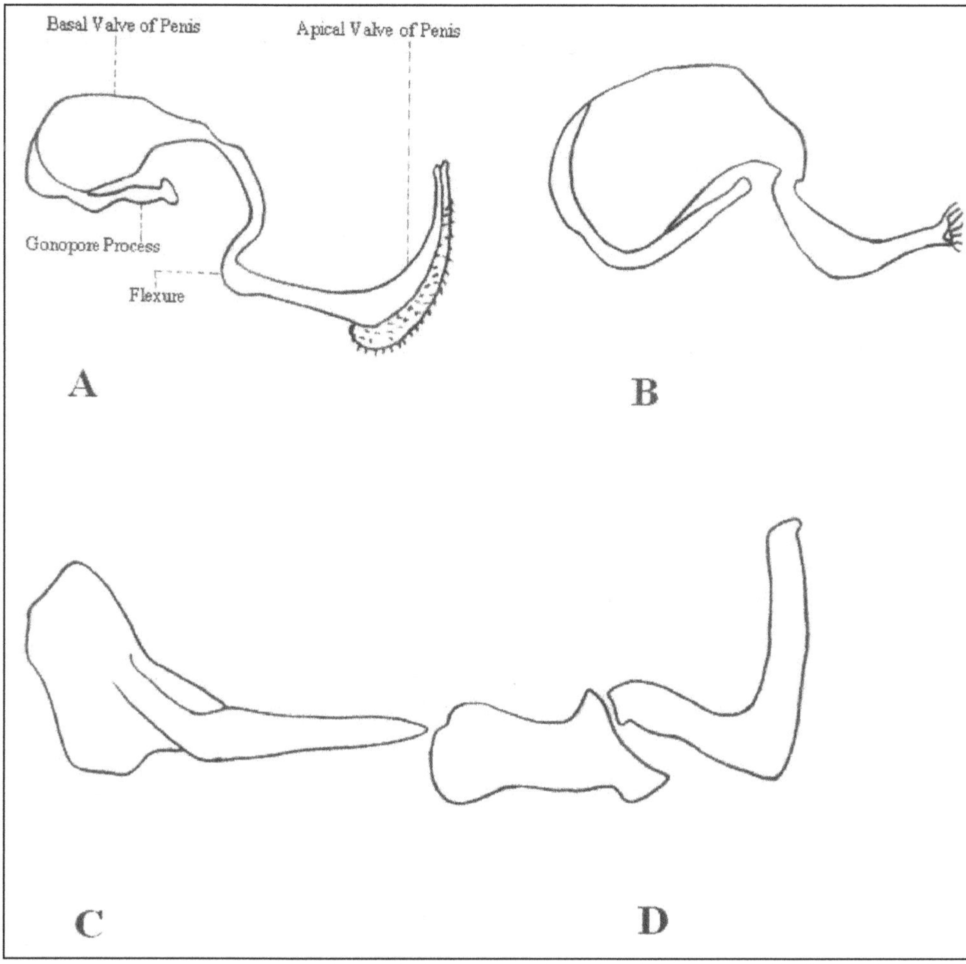

Plate 5.4: Aedeagus
A. Acrididae; B. Catantopidae; C. Pyrgomorphidae; D. Pamphagidae

Divided, undivided or flexure condition of aedeagus, presence or absence of gonopore process on aedeagus are taken as stable characters for separating various families. Flexured condition of aedeagus is a constant character in the families Acrididae (Plate 5.4 Figure A) and Catantopidae (Plate 5.4 Figure B) except Hemiacridinae in which divided and weakly flexure condition is observed.

Discussion

Earlier studies on the systematic of Indian Acridoidea are exclusively based on external visible characters, namely shape, size, coloration, texture, number of antennal segments etc. A revolution in Acridoid systematic was brought by genetalic characters especially of phallic complex. This has resulted into a profound change in the systematic concept of this group. The genetalic structures particularly epiphallus and aedeagus are less affected than the external characters by environmental conditions. A comparative study of these structures makes it possible to put forward some suggestions regarding interrelations of families and subfamilies of Acridoidea more clearly than external characters.

There are two main trends of specialization of epiphallus in this group: (1) the most primitive shield-like epiphallus is found in Pamphagidae (Plate 5.3 Figure D), and (2) Bridge-shaped epiphallus in the families Pyrgomorphidae (Plate 5.3 Figure C), Acrididae (Plate 5.3 Figure A) and Catantopidae (Plate 5.3 Figure B). Therefore, Pamphagidae should be regarded as the most primitive family among this group. Moreover dorso-lateral appendages articulated with bridge of epiphallus in Pyrgomorphidae and oval sclerites of epiphallus in Catantopidae and Acrididae are most characteristic features whereas in Pamphagidae, the epiphallus is devoid of any of these structures.

Beside the primitive epiphallus, Pamphagidae (Plate 5.4 Figure D) show primitiveness in having aedeagal sclerites completely divided while it is contiguous in Pyrgomorphidae (Plate 5.4 Figure C) and flexure in Catantopidae (Plate 5.4 Figure B) and Acrididae (Plate 5.4 Figure A).

References

Jannone, G. (1939). Sulla diffusione della vescicola ghiandolare protoracica negli Ortotteri della subfam. Oedipodinae. *Bollettino di Zoologia,* 10: 1–3.

Usmani, M.K. (2009). Male and female genitalia in some Libyan Acrididae (Orthoptera:Acridoidea). *Entomological Research,* 39: 1–35.

2013, Animal Diversity, Natural History and Conservation, Vol. 3 *Pages* **63–70**
Editors: **V.K. Gupta and Anil K. Verma**
Published by: **DAYA PUBLISHING HOUSE, NEW DELHI**

Chapter 6

Biodiversity of Spittle Bugs in the Foot Hills of Maruthamalai, Western Ghats Region, Coimbatore, Tamil Nadu

P. Malarmathi[1], B. Kalaivani[2]* and G. Vanithakumari[1]
[1]Division of Endocrinology, Department of Zoology,
Bharathiar University, Coimbatore – 641 046, Tamil Nadu, India
[2]Department of Zoology, Sri Sarada College for women (Autonomous)
Salem – 636 016, Tamil Nadu, India

ABSTRACT

The ecology of the scrub jungle is profoundly influenced by prolonged drought conditions and, therefore, presents a remarkable range of microhabitat with a wealth of insects fauna that manifest unlimited diversities of ectopic specialization in their Bio-ecology, Ecophysiology and Ethiology. The spittle bugs are from the cercopidae family of insects in the suborder Homoptera, order Hemiptera. It is found in the western Ghats region especially in the university campus during the month of October to January. They attack the tall grasses *Cymbopogon caesius* and the dwarf grasses Aristida species and also the plant (Shrub)

---

* *Corresponding author.* E-mail: kalai03win@yahoo.co.in

Trichoderma indicum. Therefore, an attempt has been made in the present study to investigate the Biodiversity of cercopid species, namely *Clovia puncta* walker and *Clovia quadridens* (walker) with reference to their Host – preference.

Keywords: *Biodiversity, Cymbopogon caesius, Trichoderma indicum, Clovia puncta, Clovia quaderidens.*

Introduction

India, with its diversified ecosystems ranging from the snow-boreal forests in the Himalayas and tropical evergreen forests along the Western Ghats, to the dry deserts of Rajasthan is considered to be one of the mega diversity countries. Of the various ecosystems, the Eastern Himalayas and Western Ghats are known to harbour rich diversity (Myers *et al.,* 2000). The Western Ghats mountain range comprises of a series of hills running almost parallel to the west coast of Indian sub-continent from Tapti river in southeastern Gujarat to Kanyakumari in Tamil Nadu. The 'ghats' descend steeply to the coastal plains on the west but merge rather gently through a series of hills in the Deccan plateau in the east. The western slopes receive rainfall of about 2000 mm/yr and support juxuriant evergreen forests while the eastern slopes, which come under the rain shadow area harbour moist and dry deciduous types of vegetation. The area covers 1,59,000 km^2 with elevation ranging from sea level to 2.695 m. (Anamudi peak).

The spittle bugs are from the cercopidae family of insects in the sub order Homoptera, order Hemiptera. Spittle bug is found in the Bharathiar University Campus during the month of October to January. They attack the tall grasses *Cymbopogen caesius* and the dwarf grasses *Aristida* sp. and also the plant (shrub) *Tirchoderma indicum*. Therefore, an attempt has been made in the present study to investigate the Biodiversity of the cercopid species, namely *Clovia punkta* (walker) and *Clovia quadridens* (Walker) with reference to their Host-preference inside the Bharathiar University Campus at Coimbatore.

Materials and Methods

Study Sites

The study area is located in Coimbatore district of the State of Tamil Nadu, India. The present study area falls broadly into the categories of Human altered and unaltered (=Natural) habitat. Sites representing this type of area were chosen in the Bharathiar University Campus at the foot hills of Maruthamalai, at an elevation of 410-m altitude above mean sea level (MSL). It is situated geographically in the 78'93°E longitude and 11°N latitude. The local area of our University Campus covers about 1000 acres.

The area was disturbed by human interference or by altered habitat and consisted of administrative buildings residential areas etc. The study was conducted during the months of October, 2002 to January, 2003, when the weather was dry.

C. puncta **Inhabiting** ***Cymbopogon caesius*** **Grass**

C. puncta **Inhabiting** ***Aristida*** **spp. Grass**

C. quadridens **Inhabiting** ***Trichiodesma indicum*** **Plants**

Plate 6.1

Spittle of *C. puncta* on *Cymbopogon caesius*

Spittle of C. quadridens on *Trichiodesma indicum*

Plate 6.2

Collection and Preservation

Collection of samples were made from the respective host plants namely *Trichoderma indicum, Cymbopogen ocesius* and *Aristida* spp. The nymphs, adults and the spittle were collected and kept in the refrigerator for further investigation.

Biochemical Analysis

In the nymph and spittle of the spittle bug species biochemical analysis was also done so as to estimate Mucoproteins (Winzler *et al.*, 1998) Sialic acid (Warren, 1959) and Proteins (Lowry's Method, 1951).

Statistical Analysis

The data were statistically analyzed and expressed as Mean ± Standard error of mean (S.E.M.). Students 't' test was used to calculate the degree of significance.

Results and Discussion

Biodiversity is defined as 'the variability among living organisms from all sources, including terrestrial, marine and other aquatic ecosystems and the ecological complexes of which they are a part; this includes diversity within species, between species and of ecosystems'. Conservation and sustainable use of biodiversity is fundamental to ecologically sustainable development. Biodiversity is part of our daily lives and livelihood, and constitutes resources upon which families, communities, nations and future generations depend. Every country has the responsibility to conserve, restore and sustainably use the biological diversity within its jurisdiction. Biological diversity is fundamental to the fulfillment of human needs.

The study was undertaken inside the University campus due to the richness of grasses and plants. Based on the abundance of the host plants, the population of spittle bug species is also high. Similarly, based on the abundance of the host plants of spittle bug species, inside the Bharathiar University campus different area have been selected.

The present investigation revealed that *C.puncta* occurs on grasses such as *Cymbogon caesius* and *Aristida* spp. the *C. quadridens* are seen on the shrubs namely, *Trichodesma indicum, C. puncta* and *C. quadridens* show relative host specificity (Plate 6.1). Even though the area had plant varieties like shrubs, trees and other types of grasses, the above two species inhabited their corresponding host plants only and were seen to avoid the other plants. Plate 6.2 shows the spittle of *C. puncta* on *Cymbopogon caesius* and spittle of *C. quadridens* on *Trichodesma indicum.*

During the present observations the two spittle bug species can be easily differentiated. Nymphs of *C. puncta* seen on *Cymbopogon caesius* are greenish in colour and smaller in size. The nymphs are found in the internodal regions or in between the nodes. But *C.quadridens* inhabiting *Trichodesma indicum*, the nymphs are somewhat larger in size when compared to *C. puncta* and are light brownish in colour. The dorsal surface of the nymphs and abdomen are stripped with dark lines and the head is black in colour transformed to light brownish shade.

The Western Ghats region is considered as one of the most important biogeographic zones of India, as it is one of the richest centres of endemism. Due to varied topography and micro-climatic regimes, some areas within the region are considered to be active zones of speciation. The region has 490 arborescent taxa, of which as many as 308 are endemics this endemism of tree species shows a distinct trend, being the highest (43 per cent) in 8N-10°30′N location and declining to 11 per cent in 16N–16°30′N location. About 1 500 endemic species of dicotyledonous plants are reported from the Western Ghats. 245 species of orchids belonging to 75 genera are found here, of which 112 species in 10 genera are endemic to the region.

The nymphs of all known species of this family become enveloped in froth. The nymphs of *C. puncta* and *C. quadridens* are noticed singly or in group of twos. At the end of the nymphal period the forth formation stops and by drying up of the froth, a vault appears through which the adults bug emerges out.

It has been noted that *C.puncta* and *C.quadridens* has five nymphal instars. The two species can be easily differentiated by their appearance and behaviour. It is thus evident that there are different in the number of instars among the various species of spit bugs. Chenshire (1886) and Janet (1893) have also observed such dissimilarities among the European species of spit bugs. Burst and Manoharan (1952) have recorded that the other related species in South India, *i.e. C.bisignata* passes through only 8 nymphal instars in both the sex.

The composition of the spittle was determined and it mainly consisted of the micro proteins, sialic acids and proteins. Table 6.1 shows the concentration of mucoproteins present in different spittle bug species inhabiting different host plants. In nymph, the mucoprotein concentration was significantly higher in group I–*C. quadridens* (87.6 µg/100mg wet tissue). When compared to group II-*C. puncta* (67.2 µg/100 mg wet tissue) and Group III- *Aristia* spp. (33.8 µg/100 mg wet tissue). In spittle, the mucoprotein concentration, was significantly very high in Group I (20.4 µg/100 mg wet tissue) when compared to other groups. Group II spittle had mucoprotein concentration of 9.6 µg/100mg wet tissue than in Group III.

Table 6.1: Concentration of Mucoproteins in different Spittlebug Species Inhabiting different Plants

Group	*Species name*	*Nymph*	*Percentage*	*Spittle*	*Percentage*
Group I	*Clovia quadridens*	87.6±4.006***[a]	100	20.4±4.1723	100
Group II	*Clovia puncta*	67.2±2.216***[a]	–23.29	9.6±1.0807	–52.94
Group III	*Clovia puncta*	33.8±3.8512***	–61.42	8.8429±0.4744	–56.65

Values are expressed as mean ± S.E.M of five nymphs + µg mucoprotein/100 mgwet tissue

***: – P < 0.001 Group II Vs Group III; ***: – P < 0.001 Group I Vs Group III; a: – P < 0.01 Group I Vs Group II; b: – P < 0.05 Group I Vs Group III.

Group I: *Clovia quadriadens* in *Trichodesma indium*, Group II: *Clovia puncta* in *Gymbopogon caesius*, Group III: *Clovia puncta* in *Aritada* spp.

Table 6.2 shows the concentration of sialic acid in different spittle bug species. In the present study Group I nymph had higher sialic acid concentration of about 3.2964 µmoles/100 mg wet tissue, which is comparatively higher than the other two groups. In spittle, the concentration was 2.8128 µ moles/100 mg wet tissue, 1.7536 µ moles/100 mg wet tissue and 1.6019 µ moles/100 mg wet tissue in Group I, Group II and Group III respectively.

Table 6.2: Concentration of Sialic Acid+ in different Spittlebug Species Inhabiting different Plants

Group	*Species name*	*Nymph*	*Percentage*	*Spittle*	*Percentage*
Group I	*Clovia quadridens*	3.2964±0.1776***	100	2.8128±0.2326 [a]	100
Group II	*Clovia puncta*	2.1183±0.213***	–35.74	1.7536±0.1615 [a]	–37.66
Group III	*Clovia puncta*	1.9271±0.1315***	–41.54	8.8429±0.4744 [a]	–43.05

Values are expressed as mean ± S.E.M of five nymphs + µ moles sialic acid/100 mg wet tissue.

***: – P < 0.001 Group II Vs Group II and Group III; a: P < 0.001 Group II Vs Group II and Group III

Group I: *Clovia quadriadens* in *Trichodesma indium*, Group II: *Clovia puncta* in *Gymbopogon caesius*, Group III: *Clovia puncta* in *Aritada* spp.

Table 6.3 shows the concentration of proteins present in different spittle bug species inhabiting host plants. In the present study, group I nymph had higher protein concentration of about 80.8 µg/100 mg wet tissue, which is comparatively higher than the other two groups. The concentration of protein in group I, II and III were 56.8 µg/100 mg wet tissue, 43 µg/100 mg wet tissue and 34.2 µg/100 mg wet tissue, respectively and accordingly the percentage decrease between Group I and Group II and Group I and Group III was 24.30 per cent and 39.79 per cent respectively.

Table 6.3: Concentration of Protein+ in different Spittlebug Species Inhabiting different Plants

Group	*Species name*	*Nymph*	*Percentage*	*Spittle*	*Percentage*
Group I	*Clovia quadridens*	80.8±3.3395*** [b]	100	56.8±1.5849***	100
Group II	*Clovia puncta*	71.2±1.610*** [b]	–11.88	43±0.9798*** [a]	–24.30
Group III	*Clovia puncta*	45.2±1.6829***	–44.06	34.2±1.4805	–39.70

Values are expressed as mean ± S.E.M of five nymphs + µg protein/100 mg wet tissue.

***: P < 0.001 Group II Vs Group II and Group III; ***: P < 0.01 Group I Vs Group II and Group III; ***: P < 0.001 Group I Vs Group II and Group III; a: P < 0.01 Group II Vs Group III; b: P < 0.05 Group I Vs Group II.

Group I: *Clovia quadriadens* in *Trichodesma indicum*, Group II: *Clovia puncta* in *Gymbopogon caesius*, Group III: *Clovia puncta* in *Aritada* spp.

The composition of the spittle depends on the stage and host plants. Murray (2000) evaluates that with regards to function, mucines help in lubrication and form a protective physical barrier on epithelial surfaces. According to Wolman (1956) the weakly acidic mucopolysaccharides having a pH range of 4.9 to 5.4 is hyaluranic acid. Hale (1953) observes that hyaluronic acid has water-retention capacity and high viscosity.

In conclusion, there is a definite diversity in the spittlebugs species studied with regard to their nymphal stages, spittle formation and composition as well as the adult behaviour pattern.

References

Myers, N., Metermeier, R.A. Metermeier, C.G., Da Francisco, G.A.B. and Kent. J., 2000. Biodiversity hot spots for conservation priorities. *Nature*, 403: 853–858.

Burst, L. and Manoharan, T., 1952. Some observations on the biology of a south Indian spittle bug *Clovia bisignata. Zool. Anz.*, 108: 267–281.

Chenshire, F.R., 1886. *Biology of Spittle Bugs of United Kingdom*. Scientific and practical publishers, London.

Janet, C., 1893. Etudes Sur les Formis. Les Abcilles, Paris.

Murray, K.R., 2000. *Harper's Biochemistry*. Mc Graw Hill, Health profession Division, New York, pp. 675–659.

Warren, L., 1959. The thiabarbituric acid assay of sialic acid. *J. Biol. Chem.,* 244: 1971.

Winzler, R.J., Devor, A.W., Mehl, J.W and Symth, I.M.C., 1948. Studies on the mucoproteins of human plasma. I. Determination and isolation. *J. C. in Investm.,* 27: 609.

Wolman, M., 1956. Hydrogen-ion concentation of mucopolysaccharides. *Bull. Res Comm. Israel*, 6: 27–29.

2013, Animal Diversity, Natural History and Conservation, Vol. 3 *Pages* **71–93**
Editors: **V.K. Gupta and Anil K. Verma**
Published by: **DAYA PUBLISHING HOUSE, NEW DELHI**

Chapter 7

Ultra Structural Comparison of Spinneret Morphology of Some Orb Weaving Spiders

Elizabeth V. Mathew* and Pothalil Antony Sebastian
Division of Arachnology, Dept. of Zoology, Sacred Heart College, Thevara, Cochin – 682 013, Kerala, India

ABSTRACT

Structural details of spinnerets vary with family and are very closely related to the behaviour of the spider. Hence comparison of ultra structure of spinnerets is an effective method to study not only taxonomic differences, but also behavioural features. This is the pioneering study on spinneret morphology of any spider species in the whole of India. Western Ghats, one of the biodiversity hotspots, is home to a diverse assemblage of spiders. Four araneoid orb weavers *viz. Neoscona mukerjei, Cyrtophora cicatrosa, Gasteracantha geminata* (Araneidae) and *Nephila pilipes* (Nephilidae) were collected from the southern stretch of Western Ghats and their spinneret morphology examined and compared using Scanning Electron Microscope (SEM).

Among araneoids, Neoscona is the most typical. The Posterior Lateral Spinneret (PLS) of female *Neoscona* is most complex bearing many aciniform spigots, two cylindrical spigots, two aggregate spigots and a flagelliform spigot sticking out between the tips of the aggregate spigots. Its Posterior Median Spinneret (PMS) holds a cylindrical spigot and an ampullate spigot apart from the numerous aciniform gland spigots. The Anterior Lateral Spinneret (ALS) has a prominent primary major ampullate spigot along with the major ampullate nubbin and many piriform spigots. This pattern conforms to that observed in other typical

---

* *Corresponding author.* E-mail: lizshz@gmail.com

orb weavers. Gasteracantha is atypical for having a sclerotized ring around the spinnerets. Cyrtophora is atypical for lacking aggregate and flagelliform spigots. *Nephila* is different from the other three in that the number of aciniform spigots on the PMS is very small. It is also characterized by two kinds of cuticular sculpturing on the spinnerets. Despite belonging to the same family, the four genera differ in spinneret structure to some extent.

Keywords: *Spinnerets, Spigot, Araneoid, Orb web, Scanning electron microscope.*

Introduction

Spiders constitute the largest order of arachnids. Unique among all organisms in their modes of silk production and usage and of reproduction, spiders are common predatory arthropods in all terrestrial and many aquatic ecosystems. Among all organisms, spiders form the seventh largest order. Furthermore, this is the most diverse, female-dominated and entirely predatory order in the arthropod world. As such, spiders are key components of all ecosystems in which they live (Noss, 1990). Currently more than 43,678 valid species in 3,898 genera and 112 families have been described (Platnick, 2012). The estimated total extent of world spiders can only be guessed at and many researchers speculate that up to 1,70,000 species could exist. Evidently many species remain to be discovered, especially in the tropics.

Spiders are unique among arachnids as they possess spinnerets, situated near the hind end on the ventral side of the abdomen, which produces silk. Some mites also produce silk, but this is produced by glands in the palp and there are no spinnerets. Spiders use silk for many purposes–to protect their young, catch food, make homes and move around. They are the only animals which use silk in almost every part of their daily lives. The individual spinnerets move independently, yet in a highly coordinated manner to build cocoons or webs.

The Western Ghats, locally known as the Sahyadri Hills, are formed by the Malabar Plains and the chain of mountains running parallel to India's western coast, about 30 to 50 kilometers inland between the latitudes 8° N to 20° N and longitudes 73° E to 77° E. They cover an area of about 160,000 km^2 and stretch for 1,600 kilometers from the country's southern tip to Gujarat in the north, interrupted only by the 30 kilometers wide Palghat Gap. This area is one of the world's ten "Hottest biodiversity hotspots" and at least 325 globally threatened species occur here (Myers *et al.,* 2000). This region exhibits very high level of endemism, which ranges from 75 per cent in amphibians to 40 per cent in insects and hence, the Western Ghats is listed among the 34 biodiversity hotspots of the world (Mittermeier *et al.,* 2005). It possesses a rich assemblage of spider diversity which is largely left undocumented till date. All the more study on spinneret morphology of no species has been attempted in India yet. Hence this study rightly focuses on the spinneret morphology of a few Western Ghat spiders and becomes the pioneering effort in this direction in the country.

Silk Glands

There are seven different glands known (Figure 7.1; Peter, 2008). But each spider possesses only some of these glands and not all together. The glands known as Glandula ampullaceae major and minor produce non-sticky silk used as draglines or to form the frames of their webs (non-sticky foundations). All spiders have this to produce the safety lines they leave behind them as they move about. Glandula piriformes is used for the production of the attaching threads. Glandula aggregata produces the sticky material for the production of the attaching threads. Glandula aciniformes produces threads for the encapsulation of prey. Glandula tubuliformes produces thread for cocoons. Glandula flagelliform is used for the production of the adhesive threads. The corresponding spigots are named Major ampullate (MaA), Minor ampullate (MiA), Piriform (Pi), Aciniform (Ac) and Aggregate (Ag). The large number of spigots produces multiple fibrils, which then are combined into bundles similar to cables.

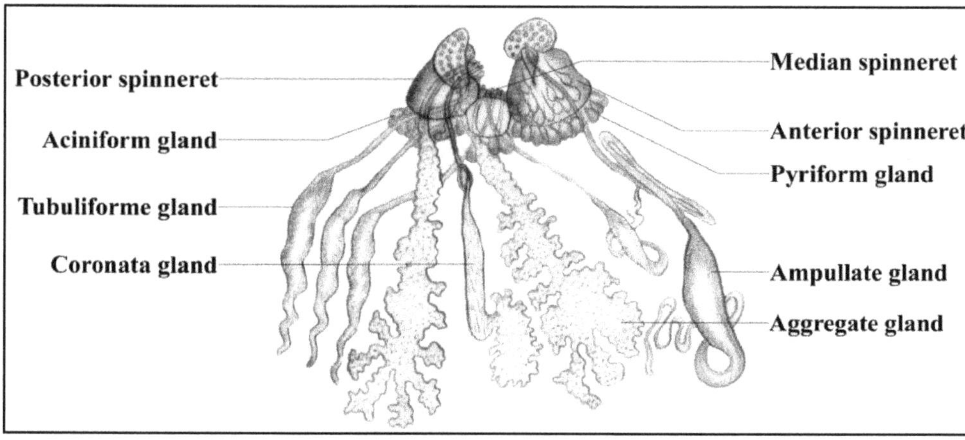

Figure 7.1: Different Types of Silk Glands in Spider Body.

Arrangement of Spinnerets

The spinnerets are usually arranged in three pairs, anterior, median and posterior (named from front to the rear of the spider–see Plate 7.1A).The median pair is often small and usually completely obscured by the two other pairs. The paired spinnerets have one to three segments. The spinnerets are highly manoeuvrable for silk-spinning and may be quite short or relatively long (as long as the abdomen in some mygalomorph spiders). The end segment of each spinneret is provided with hollow, tiny hair-like silk outlets called the spigots connected to particular paired silk glands within the abdomen. A spinneret may produce different types of silk that are derived from different pairs of silk-producing glands located within the abdomen.

Silk

Although silk is basically an aqueous solution, once the silk leaves the spinnerets, it is insoluble in water, dilute acids and bases. In some families of spiders (cribellate spiders), there is an additional special silk-producing organ, the cribellum. This is a narrow plate situated immediately in front of the spinnerets. Spider silk is an elastic and sometimes adhesive material. The dragline silk of the golden orb weaver, *Nephila clavipes* (Nephilidae) is one of the strongest silks and one of the most studied. Silk differ depending on the temperature and speed at which the silk is spun, and the spider's diet. The tensile strength is about the same as that of high tensile engineering steel, but has lower relative density than steel and in weight is much stronger.

Orb Web

It would be appropriate to describe orb web in this context. The principal characteristic feature of an orb web is that the central portion, the part lying within the supporting framework, consists of a series of radiating lines of dry and inelastic silk that support a thread of viscid and elastic silk (Plate 7.5). Orb webs vary in structure, shape and size according to the families and genera of the spiders. Webs of the families Araneidae, Tetragnathidae and Uloboridae make good examples of orb webs.

By judicious use of spigot number, placement, appearance and known ontogenetic patterns, one can often work out spigot homologies without reference to histological data. A basic rule of inference is the difference between morphological singulars and multiples, or homologues and homonoms (Riedl, 1979). In the case of spigots one has both singulars, such as major ampullates, minor ampullates, cylindrical, pseudoflagelliformes, aggregates and flagelliformes, and "multiples", such as paracribellars, piriforms and aciniforms.

Multiples

1. If the spigots are unique to the ALS and are small, numerous and occupy most of the ectal spinning field, they are *piriform* (Pi) *gland spigots.* They are used to attach draglines together or to the substrate. They should occur in all instars and both sexes.
2. If they occur both on the PMS and PLS and are small, but slightly more elongate than the ALS piriforms, they are probably *aciniform* (Ac) *gland spigots.* They have several uses: prey wrapping, retreats and egg sacs. They should occur in all instars and both sexes.
3. If they occur only on the PMS and are thin, very long and with annulate shafts, they are probably *paracribellar spigots.* Their function is unclear, although they contribute a component to the sticky thread. They should occur in all instars and both sexes of all cribellate orb weavers, at least.

Singulars

4. If a set of three spigots occur as two on the araneoid PLS and one on the anterior portion of the araneoid PMS, they are *cylindrical* (Cy) *gland spigots.* They are apparently used to produce the specialized silk for egg sacs. They should only occur in adult females. The situation in cribellate orb weavers is more complex because more cylindricals are present, but their distribution is as in araneoids.

5. If a pair of spigots occurs as one on the mesal margin of the ALS and one on the posterior margin of the PMS, they are *ampullate gland spigots,* usually one major (MaA) and one minor (MiA). They are used for components of the dragline and major structural lines. They should occur in all instars and both sexes.
6. If a pair of large spigots with wide openings is juxtaposed on the anterolateral margin of the PLS, they are *aggregate* (Ag) *gland spigots,* used to make the viscid glue of the sticky line. They should occur in juveniles and adult females, but are supposed to be absent in all adult males (exceptions may occur).
7. If a single large spigot is unique to the PLS anterolateral or apical margin, it is the *flagelliform* (Fl) or *pseudoflagelliform gland spigot,* used to make the base fibers of the sticky line. Flagelliform spigots are usually between the two aggregate spigots, but not always. Like the aggregate gland spigots, flagelliform and pseudoflagelliform spigots should occur in juveniles and adult females, and should be absent in adult males.

Eberhard (1982) described several ways in which orb weavers lay and connect radii; these behaviours are phylogenetically informative (Coddington, 1990; Hormiga *et al.,* 1995). Nephilinae interrupt hub loop construction (itself continuous with nonsticky spiral construction) and move along a pre-existing radius to a vacant spot on the frame, paying out a new line behind. The spider attaches this radius to the frame, moves a few millimeters along the frame and attaches it again, and finally returns to the hub and attaches its dragline, thus having laid two juxtaposed but distinct radii in a single pass from hub to frame and back. This radius construction behavior is unique to nephilines.

Peters (1955) described the spinning apparatus of Nephila as being composed of six different types of glands, of which one is first distinguished as "Glandula coronata" (now "Glandula flagelliform"). He described the distribution of the various glands on the three pairs of spinnerets and the construction of the spools. It was found that the adhesive threads of the web comes from an aggregate of three spools, of which one produces the groundline from the secretion of Glandula coronata while from the other, two the adhesive flows which pours over it.

Definitions

Many spiders can draw silk in the days preceding ecdysis due to the collared openings formed in the new cuticle, surrounding the ducts of those silk glands that are functional during the proecdysial period. After ecdysis, the collared openings remain in the new exoskeleton in a collapsed form. These structures came to be called *tartipores* (Shear *et al.,* 1989; Yu and Coddington, 1990; Townley and Tillinghast, 2003).

In adult spiders only vestiges of some spigots or vestiges along with functional spigots form, for *e.g.,* in adult *Mimetus* and *Araneus,* only vestiges of 2° MaA and 2° MiA spigots are found (Townley and Tillinghast, 2003). Such nonfunctional, partially formed spigots, which are sometimes no more than small cuticular knobs are called *nubbins*. Nubbins may also be present for Ac and Pi spigots.

In general, nubbins occur in adults, being more abundant in males (largely since silk glands used solely or primarily in prey capture tend to regress in adult males), and are ontogenetically vestigial. That is, they are located in adults in positions where functional spigots would have formed if the spider had remained a juvenile after its most recent moult. In addition to the MaA and MiA nubbins present in a variety of adult male and female araneoids , flagelliform and aggregate nubbins form on the PLS of many adult male araneoids (Sekiguchi 1955; Platnick *et al.*, 1991: Fig. 275; Townley *et al.*, 1991: Fig. 16; Townley 1993: Fig. 16; Griswold *et al.*, 1998: Figs. 25D, 39D, 43D). Among the examples of exceptions to the general rule are nubbins occasionally seen in early instar *Cyrtophora* (Araneidae) that suggest phylogenetic vestiges of either aggregate or flagelliform spigots. In an examination of six first instar *Cyrtophora citricola*, on one PLS (Peters, 1993: Figs. 11b, c) observed a single "shaft-like structure" on the vestigial plate of the aggregate-flagelliform triad. 2° MiA spigots and MiA nubbins seemed to switch positions after each ecdysis (Townley *et al.*, 1991).

Materials and Methods

Four species of orb weaving spiders *viz.*, *Neoscona mukerjei, Cyrtophora cicatrosa, Gasteracantha geminata* (Araneidae) and *Nephila pilipes* (Nephilidae) were collected by hand picking from Athirapally region of Western Ghats and preserved in 70 per cent alcohol. Spinnerets were processed for SEM imaging by following Townley and Tillinghast (2003). A circle was cut around the spinnerets and the spinnerets removed from the rest of the abdomen. The rest of the spider's body was preserved in ethanol for species confirmation.

Pre-Processing

As much soft tissue as possible should be teased away from beneath the spinnerets, but in the process care is taken not to let debris from these soft tissues stick to the spinnerets. Spinnerets were washed with double distilled water as needed. If there is little soft tissue beneath the spinnerets, it is more likely that a good, uncollapsed preparation is obtained after drying.

Each preparation was checked for any collapsed condition. If the spinneret was collapsed to some extent, fine insect pins were used to stick in to the spinneret from underneath so that the spinneret is fully extended (expanded). Insect pins of different sizes were stuck in to a wax-covered Petri dish, with their points sticking up, and then spinnerets were pulled on to these points. The pins and spinnerets were immersed in a detergent-containing buffer during the process.

Spinnerets were then passed through the ethanol series – 30 per cent, 50 per cent, 70 per cent, 90 per cent and 100 per cent to dehydrate the tissue. The abdomen was immersed in each solution of the ethanol series for at least a day. From the second 100 per cent ethanol, the spinnerets were subjected to critical point drying by immersing them in Hexamethyldisilazane (HMDS) for at least a day and then letting them air dry. Critical point drying is an established method of dehydrating biological tissue prior to examination in the scanning electron microscope. Larger abdomens can be left in HMDS for 2-3 days. HMDS has started being in use for this purpose only recently. The other option is to use a critical point dryer. The spinnerets were then taken from the HMDS and allowed to air dry. Most of the excess HMDS from the preparation was blotted off with blotting paper, but care was taken not to touch the spinnerets with the paper.

SEM Analysis

The spinnerets were mounted on an Aluminium stub, sputter-coated with a thin layer of gold at 10mA for 120s using JEOL JFC -1600 Autofine coater. Later the spinnerets were examined from different angles in the Japan made Analytical SEM serial number JEOL JSM – 6490LA at 8kV under various magnifications ranging 33x to 5000x. The micrographs obtained were carefully observed and compared with available SEM images to arrive at conclusions.

Results

Plate 7.1: *Neoscona mukerjei*

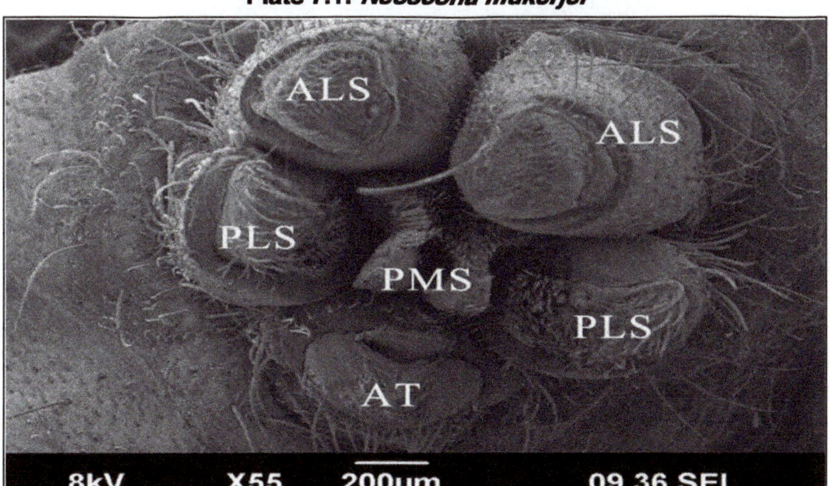

Plate 7.1A: All six spinnerets and the anal tubercle are clearly seen.

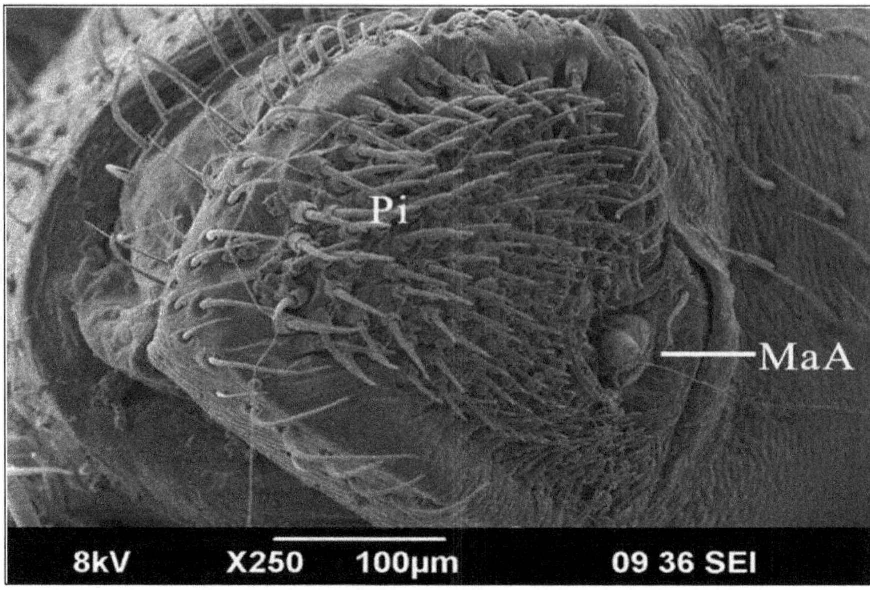

Plate 7.1B (Right ALS): Almost the entire terminal segment of the spinneret containing many piriform spigots and one primary major ampullate spigot is shown. The major ampullate nubbin is partially visible just below the primary major ampullate spigot.

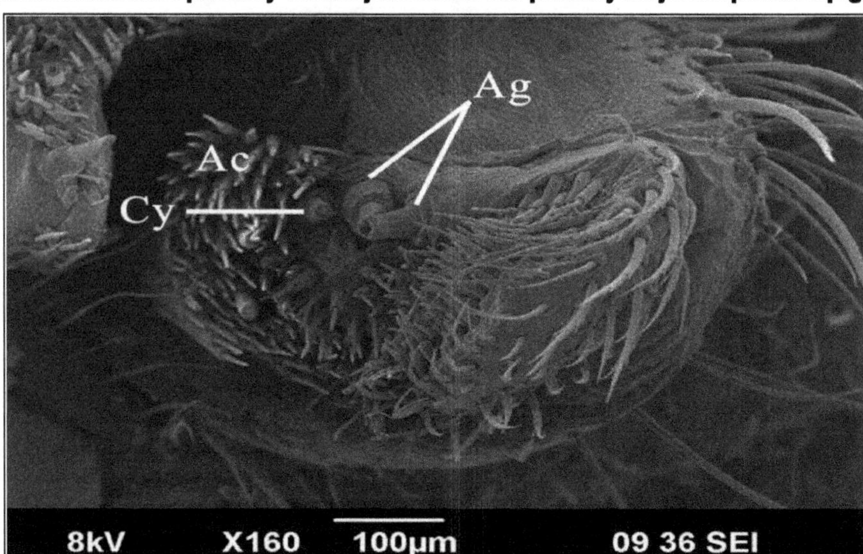

Plate 7.1C (Left PLS): The solitary spigot on the left is a cylindrical spigot (so this must be a female) for making the egg sac. On the left you see two very wide spigots. These are the aggregate spigots. The tip of the flagelliform spigot can be seen sticking out between the tips of the two aggregate spigots. When the spider is making the sticky spiral in its orb web, a fiber emerges from the flagelliform spigot and is coated with an aqueous glue from the two aggregate spigots. If this was instead an adult male, the aggregate and flagelliform spigots would be represented by nubbins only. At the far left of the image parts of some of the many aciniform spigots can be viewed. The area shown in this micrograph can be seen at higher magnification in Plate 7.1D.

Plate 7.1D

Plate 7.2: *Gasteracantha geminata*

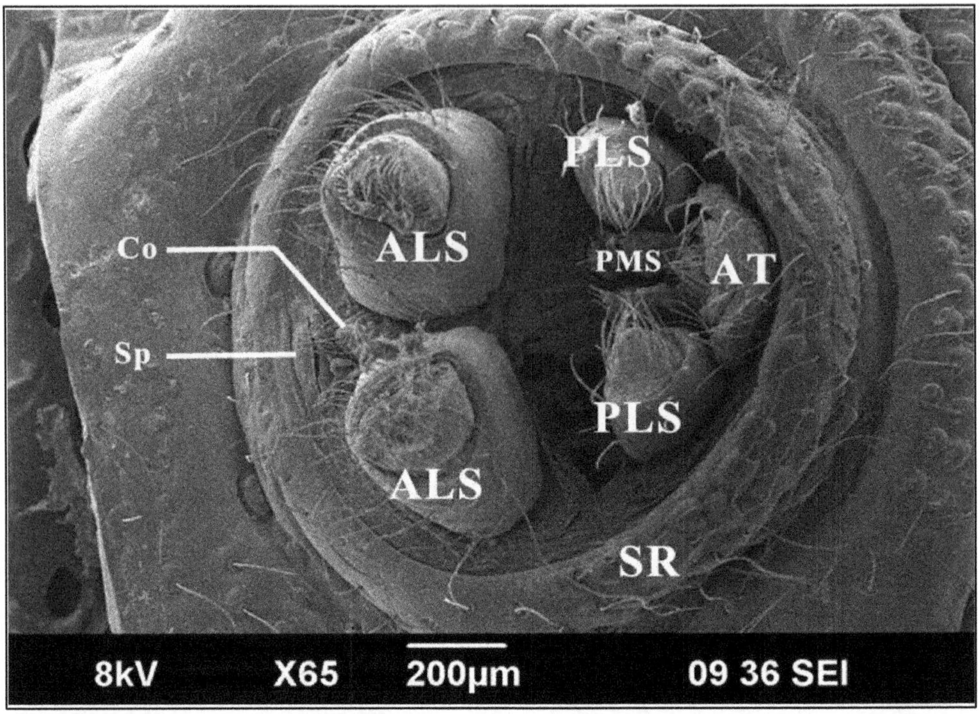

Plate 7.2A shows all six spinnerets, the anal tubercle (just posterior from the PMS), the colulus (between and a little anterior to the ALS), the spiracle (just anterior from the colulus), and the sclerotized ring that surrounds all of these structures.

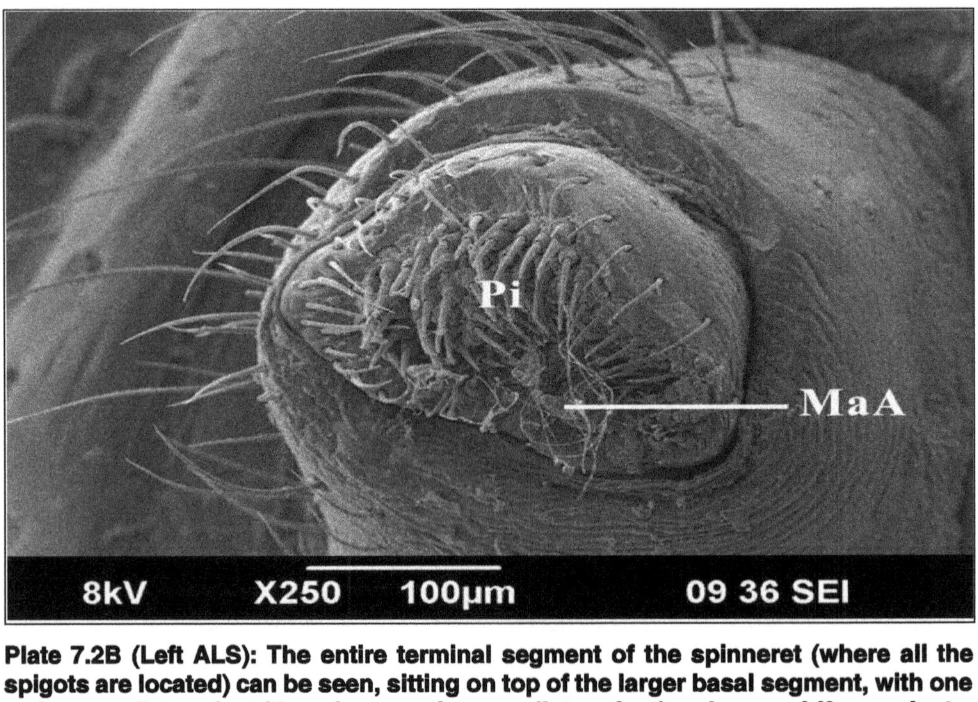

Plate 7.2B (Left ALS): The entire terminal segment of the spinneret (where all the spigots are located) can be seen, sitting on top of the larger basal segment, with one major ampullate spigot (the primary major ampullate spigot) and many piriform spigots.

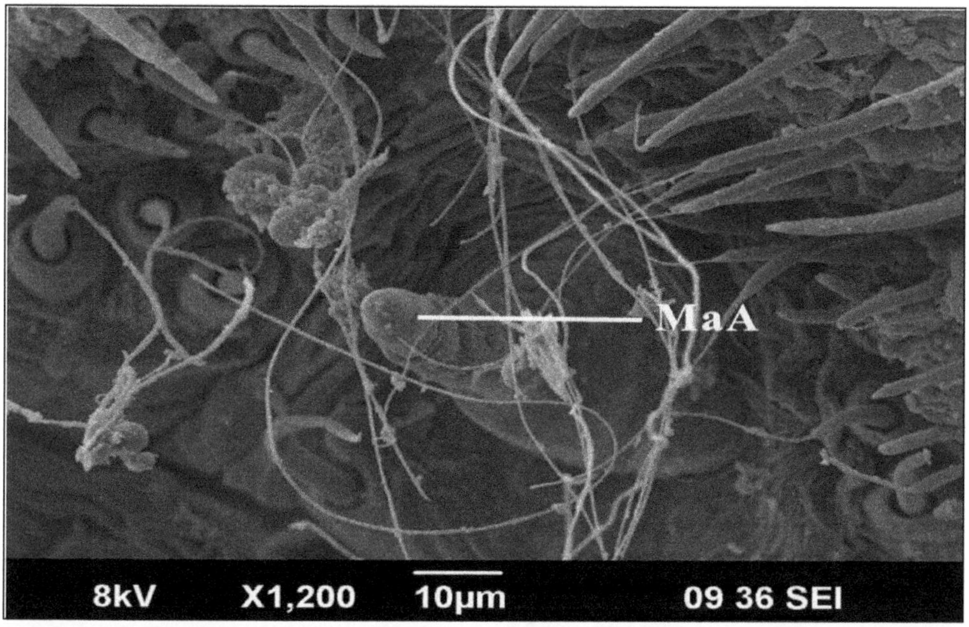

Plate 7.2C (Left ALS): The primary major ampullate spigot in the middle and some of the surrounding piriform spigots are observed. The thin fibres seen in the field are piriform silk fibres. The major ampullate nubbin is immediately to the left of the primary major ampullate spigot, but it is out of focus and some silk fibers cover part of it. The major ampullate tartipore is not visible from this angle.

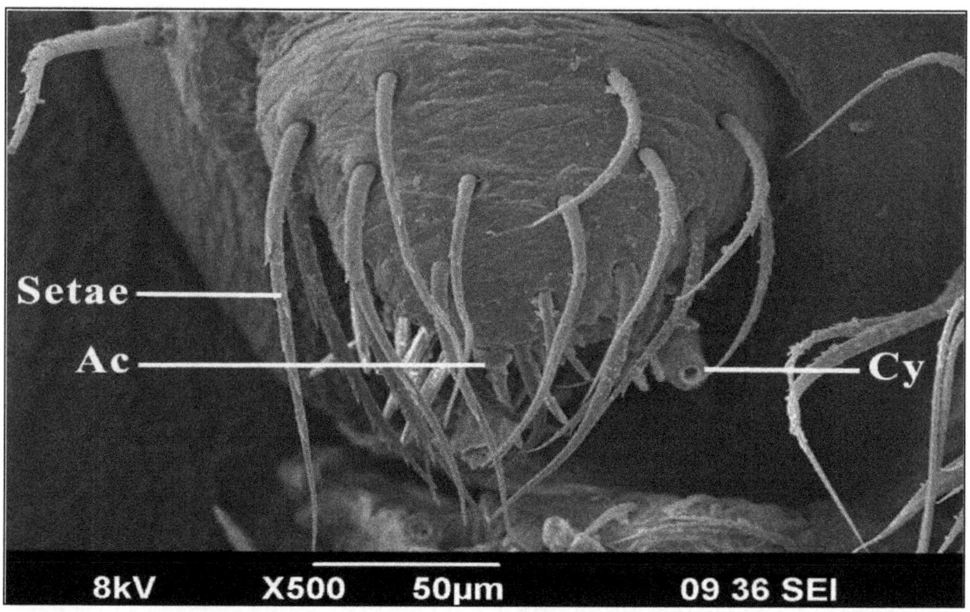

Plate 7.2D (Left PLS): The long serrated protrusions are setae. The aciniform spigots are on the opposite face of the spinneret, though a terminal part of one of these can be seen as projected from the other side. A cylindrical spigot too is partly visible.

Plate 7.3: ***Cyrtophora cicatrosa***

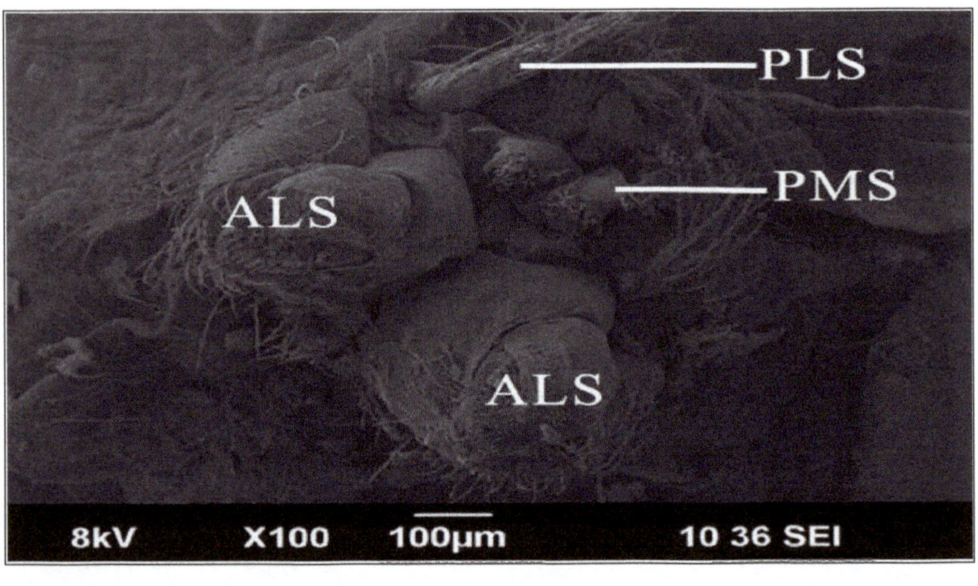

Plate 7.3A: All six spinnerets are shown, though the two PLS are especially collapsed.

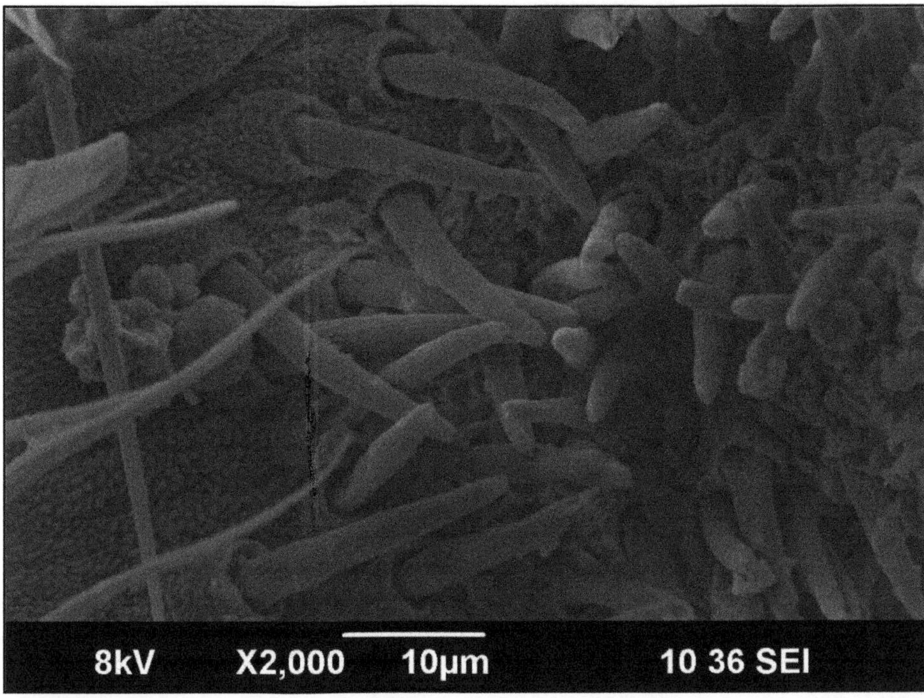

Plate 7.3B (Left ALS) shows many piriform spigots. The protuberances that spot the spigots and the surrounding exoskeleton (cuticle) are thicker little islands of cuticle. The protuberances, ridges, scales, etc. that occur on the cuticle are sometimes referred to as sculpturing of the cuticle. They are probably of taxonomic value and have sometimes been considered in taxonomic papers.

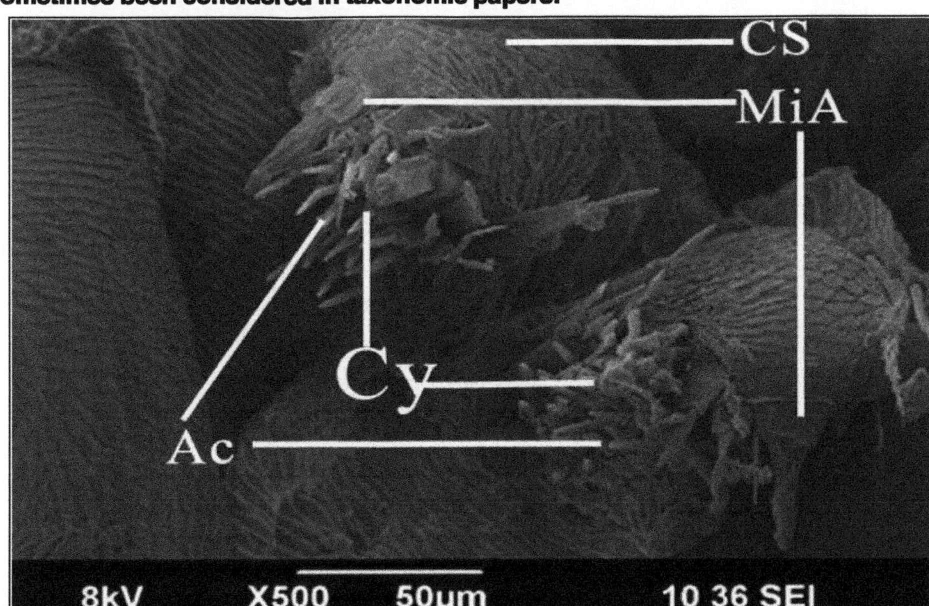

Plate 7.3C (Left and Right PMS): Each shows a cylindrical spigot, a minor ampullate spigot and numerous aciniform spigots.

Plate 7.3D (Right PMS) shows the one cylindrical spigot on the PMS (a little above the middle of the micrograph), surrounded by many aciniform spigots.

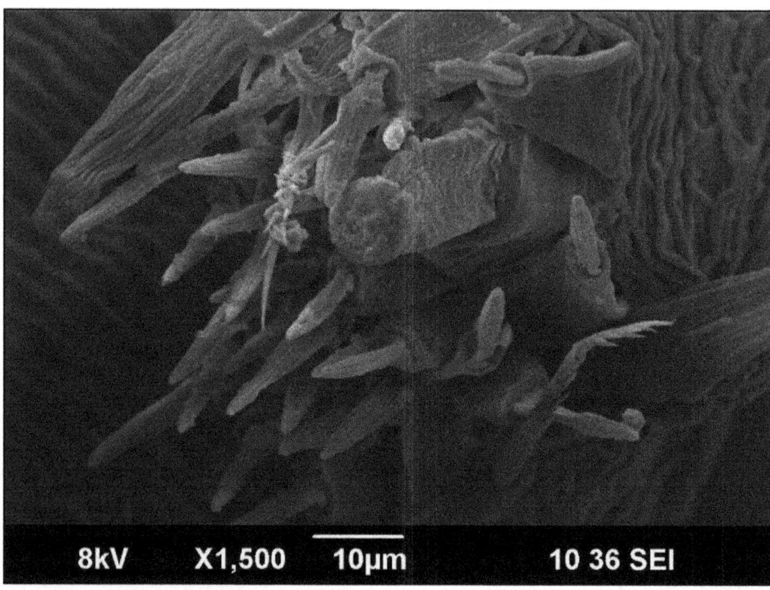

Plate 7.3E (Left PMS) is a similar micrograph, with the one cylindrical spigot shown slightly above and to the right of the middle region. Part of the primary minor ampullate spigot is sticking down from the top center, pointing toward the left. Many aciniform spigots are also present. The areas in these two micrographs can be seen at lower magnification in plate 3C.

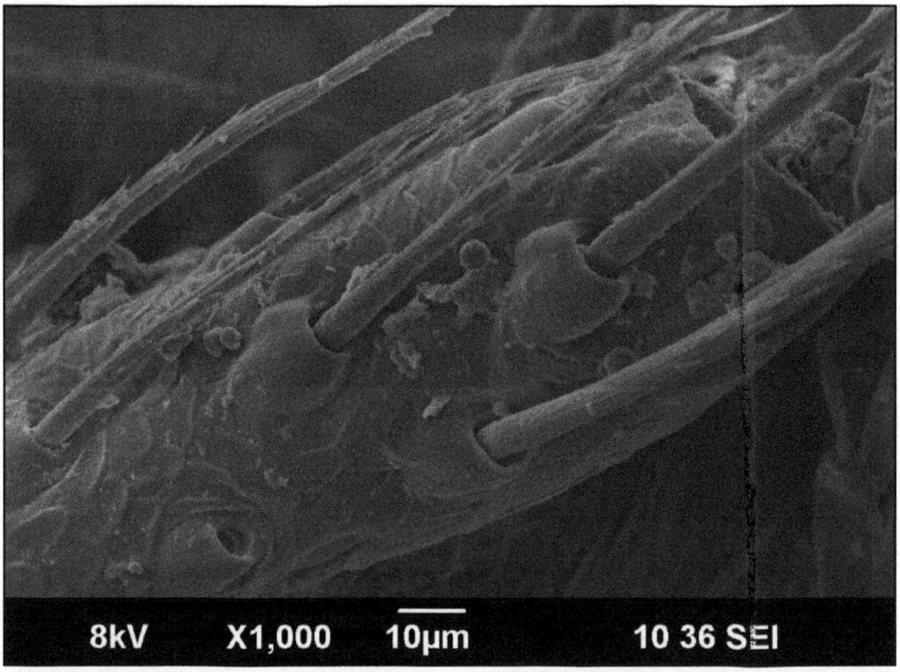

Plate 7.3F (Left PLS): Shows several setae, though one aciniform spigot can be seen at the bottom right. This area can be seen at lower magnification in Plate 7.3G, where additional aciniform spigots can be seen to the right of the one seen in Plate 7.3F.

Plate 7.3G

Plate 7.3H (Right PLS): The middle portion shows setae on the PLS. But a little left of top center in the image some spigots can be seen on the PLS, including one of the two cylindrical spigots (female), with a blunt shaft that is wider than the thin shafts of the surrounding aciniform spigots. On the far left is the primary minor ampullate spigot on the right PMS.

Plate 7.4: *Nephila pilipes*

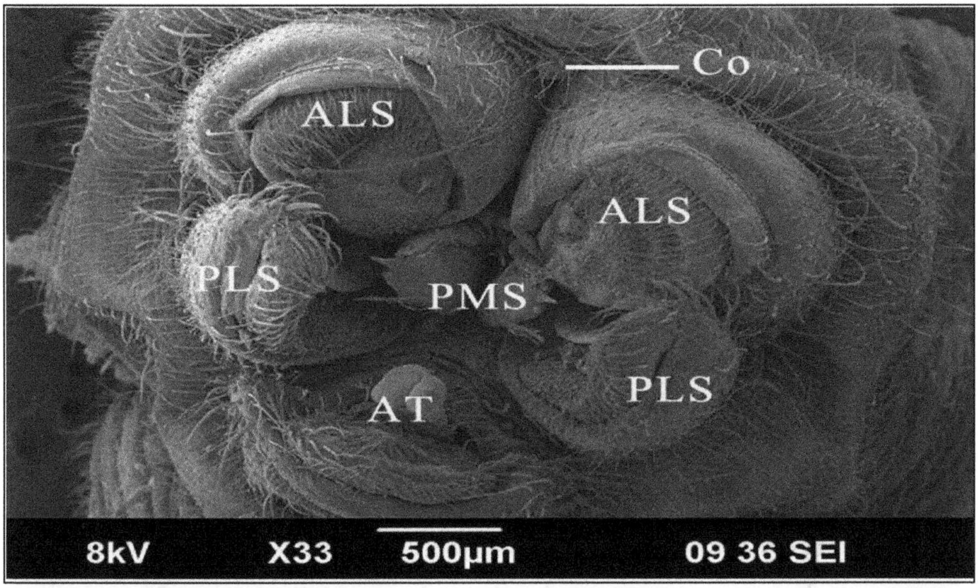

Plate 7.4A: Shows the six spinnerets, anal tubercle (at bottom), and little colulus (between the anterior part of the two ALS).

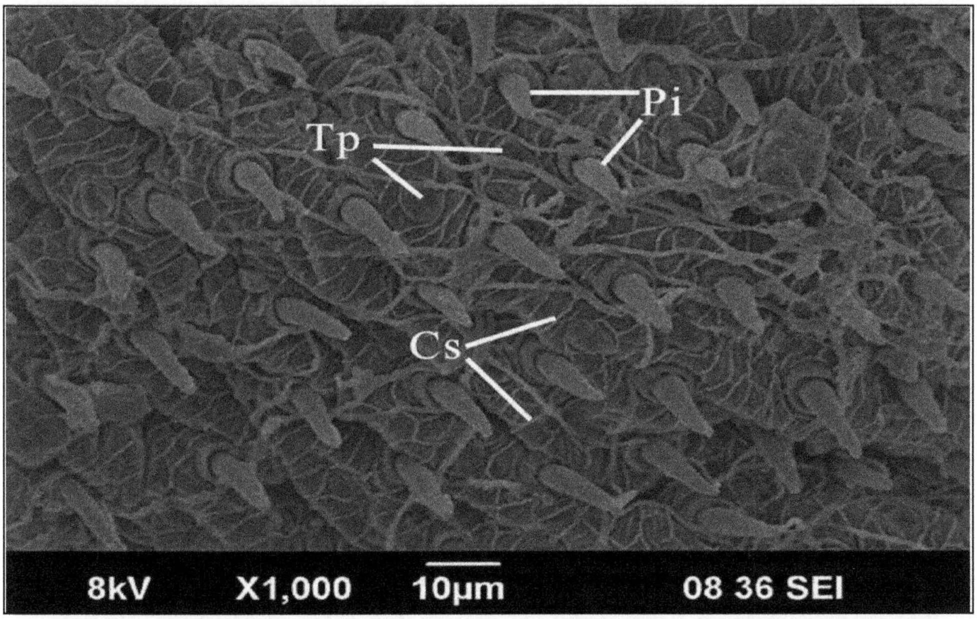

Plate 7.4B: Shows piriform spigots, and roughly circular or oval piriform tartipores can be nicely seen scattered amongst them. There are two network-like patterns running between the spigots. There are relatively thick strands of material that look as though sitting on the cuticle (exoskeleton). These strands are piriform silk. But then there are also thinner strands that look more like they are part of the cuticle. And in fact they are part of the cuticle. It is another example of cuticle sculpturing. In this case these thin ridges of slightly thicker cuticle run over the surface of the piriform spinning field.

Plate 7.4C: Shows most of the primary major ampullate spigot on the right ALS with a short segment of silk emerging from the tip. The silk is the cylinder with a uniform diameter. Where it starts to widen is where the spigot shaft begins. Just below the base of this spigot is the major ampullate nubbin (a bit out of focus). In the bottom right quadrant of the image is a structure kind of like a horizontal line but not perfectly horizontal (pointing up at the right). This may be the major ampullate tartipore but can't be said for certain because it is out of focus.

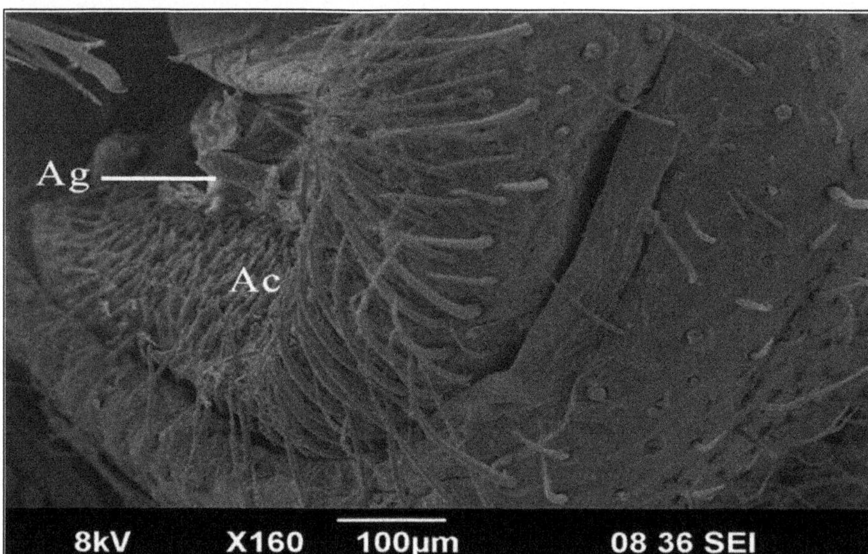

Plate 7.4D (Left PLS): Many aciniform spigots (out of focus) can be seen on the right. One of the aggregate spigots is also seen. The other aggregate spigot is pressing up against this one on the other side so that we can't see it, and the flagelliform spigot is between them (also not visible from this angle).

Plate 7.5: a Typical Orb Web

Abbreviations

ALS: Anterior Lateral Spinnerets
PMS: Posterior Median Spinnerets
PLS: Posterior Lateral Spinnerets
MaA: Major ampullate spigot
MiA: Minor ampullate spigot
Ac: Aciniform spigot
Pi: Piriform spigot
Cy: Cylindrical spigot
Ag: Aggregate spigot
Fl: Flagelliform spigot
Co: Colulus
Sp: Spiracle
AT: Anal Tubercle
Tp: Tartipore
N: Nubbin
SR: Sclerotized Ring
CS: Cuticular sculpturing

Discussion

This study explores the similarities and differences between four orb weavers *N. mukerjei, G. gasteracantha, C. cicatrosa* and *N. pilipes*. *Neoscona* is a typical araneoid. PLS of female *N. mukerjei* is most complex bearing many aciniform spigots, two cylindrical spigots, two aggregate spigots and a flagelliform spigot sticking out between the tips of the aggregate spigots. Its Posterior Median Spinneret (PMS) holds a cylindrical spigot and an ampullate spigot apart from the numerous aciniform gland spigots. The Anterior Lateral Spinneret (ALS) has a prominent primary major ampullate spigot (MaA) along with the major ampullate nubbin and many piriform spigots. This pattern conforms to that observed in other typical orb weavers (Scharff and Coddington, 1997).

As Griswold *et al.* (1998) established, ALS of *Gasteracantha* is with one major ampullate spigot (the primary major ampullate spigot) and many piriform spigots. Major ampullate nubbin can be seen in part, just to the left of the primary major ampullate spigot. *Gasteracantha* has a sclerotized ring that surrounds all six spinnerets, the anal tubercle, the colulus and the spiracle (Scharff and Coddington, 1997).

Coddington described *Cyrtophora* among other araneids, in detail. Though *C. cicatrosa* is in the family Araneidae, it is not a typical araneid because the aggregate and flagelliform glands and their spigots do not form in these spiders and, consequently, these spiders do not make sticky spirals in their webs (Coddington, 1989). The build tent webs which make them unique. A brush of aciniform spigots on the anterior face of the PMS is present and the aciniform spigot field as in cribellate orbweavers extends forward on anterior wall of the spinneret. Further aciniform spigots on the PMS are reduced in arrangement to a narrow ribbon on the posterior face of the PMS, but this couldn't be observed in the micrographs obtained, probably due to lack of proper spreading. The protuberances that spot the spigots and the surrounding cuticle are thicker little islands of cuticle referred to as sculpturing of the cuticle.

N. pilipes shows all characteristics of a typical araneoid in its spinneret morphology except that the PMS lodge too few aciniform spigots. MaA on ALS, MiA and cylindrical spigots on PMS and aggregate-flagelliform triad on PLS hold good for *N. pilipes* as for any araneoid. On ALS, apart from the piriform silk fibres, there are thinner strands that look more like they are part of the cuticle, which is another example of cuticle sculpturing. In this case these thin ridges of slightly thicker cuticle run over the surface of the piriform spinning field.

Thus all four araneoids have some features in common *viz.*, MaA and its nubbin on ALS, MiA on PMS, numerous piriform spigots on ALS and many aciniform spigots on PMS and PLS. They also show some conspicuous differences. *N. mukerjei* is the most typical of these orb-weavers. *G. geminata* was atypical for having a sclerotized ring around the spinnerets. *C. citricola* was atypical for lacking aggregate and flagelliform spigots. *N. pilipes* is different from the other three in that the number of aciniform spigots on the PMS is very small.

Conclusion

This study is the pioneering work on spinneret morphology of spiders in India and attempts to draw on the similarities and differences between four orb weavers. It has been observed that the spinneret morphology depends on the web building behaviour of the spider. At the same time, there can be atypical characteristics varying from one genus to the other among spiders of similar web building character. On comparing the current study with other published studies on same genera of araneoids, it is observed that spinneret morphology doesn't vary among different species of the same genus. Orb weavers are a thoroughly investigated group in terms of spinneret morphology. It is suggested that many more orb weavers be analysed at various instars to make a wider comparison.

Lack of proper spreading of spinnerets was a major lacuna confronted during this study. Experience and patience are the only virtues by which this problem can be overcome. Use of apt instruments including scissors for separating spinnerets is imperative in obtaining good results.

On the whole the study provided valuable insights into the spinneret morphology of these orb weaving spiders. Spinneret morphology varies with the type of web and hence can be relied upon as a behavioural indicator. More importantly it can serve as a taxonomic criterion as genera of the same family differ in spinneret structure. A wide gap prevails in this field of study in India and hence offers great scope for research.

Acknowledgements

We thank profusely Fr. (Dr.) Johnson X. Palackapillil, Principal, Sacred Heart College, Thevara, who kindly permitted us to use the library and lab facilities of the College.

We are immensely grateful to Kerala State Council for Science, Technology and Environment for facilitating this study by providing grant for student project.

This work would not have been possible but for the expert and kind guidance and inestimable help of Dr. Mark A. Townley, Department of Zoology, University of New Hampshire, Durham, USA, who is an authority in the field of Arachnology, particularly spinneret morphology. We thank him sincerely.

Heartfelt thanks are due to the Director of Molecular Biology Division of Amritha Institute of Medical Sciences (AIMS), Edapally for allowing us to use the SEM facility at their institute.

Dr. Sudhikumar A.V., Assistant Professor, Department of Zoology, Christ College, Iringalakkuda helped us with timely suggestions and inspiration. We thank him too.

References

Coddington, J.A., 1989. Spinneret silk spigot morphology: evidence for the monophyly of orbweaving spiders, Cyrtophorinae (Araneidae), and the group Theridiidae plus Nesticidae. *Journal of Arachnology*, 17: 71–95.

Coddington, J.A., 1990. Cladistics and spider classification: araneomorph phylogeny and the monophyly of orbweavers (Araneae: Araneomorphae, Orbiculariae). *Acta Zoologica Fennica*, 190: 75–87.

Eberhard, W.G., 1982. Behavioral characters for the higher classification of orbweaving spiders. *Evolution 36*: 1067– 1095.

Griswold, C.E., Coddington, J.A., Hormiga, G. and Scharff, N., 1998. Phylogeny of the orb–web building spiders (Araneae, Orbiculariae: Deinopoidea, Araneoidea). *Zoological Journal of the Linnean Society,* 123: 1–99.

Hormiga, G., Eberhard, W.G. and Coddington, J.A., 1995. Web construction behaviour in Australian *Phonognatha* and the phylogeny of nephiline and tetragnathid spiders (Araneae, Tetragnathidae). *Australian Journal of Zoology*, 43(4): 313–364.

Mittermeier, R.A., Patricio, R.G., Hoffman, M., Pilgrim, J., Brooks, T., Mittermeier, C.G., Lamoreux and Fonseca, G.A.B., 2005. Hotspots revisited: Earth's biologically richest and most endangered terrestrial ecoregions. *Conservation International*, pp. 1–432.

Myers, N., Mittermeier, R.A., Mittermeier, C.G., Fonseca, G.A.B. and Kent, J., 2000. Biodiversity Hotspots for Conservation Priorities. *Nature*, 403: 853–858.

Noss, R.F., 1990. Indicators for monitoring biodiversity: A hierarchical approach. *Conservation Biology*, 4: 355–364.

Peters, H.M., 1955. Uber den Spinnapparat von *Nephila madagascariensis* (Radnetzspinnen, Fam. Argiopidae). *Zeitschrift fur Naturforschung,* 10b: 395–404.

Peters, H.M., 1993. Functional organization of the spinning apparatus of *Cyrtophora citricola* with regard to the evolution of the web (Araneae, Araneidae). *Zoomorphology,* 113: 153–163.

Platnick, N.I., 2012. The World Spider Catalog, Version 13.5. *American Museum of Natural History* online at http://research.amnh.org/iz/spiders/catalog/COUNTS.html.

Sebastian, P.A. and Peter, K.V. (2009). *Spider of India*. Universities Press, Hyderabad, pp. 614+170 pl.

Spinneret evidence and the higher classification of the haplogyne spiders (Araneae, Araneomorphae). *American Museum Novitates,* 3016: 1–73.

Platnick, N.I., Coddington, J.A., Forster, R.F. and Griswold, C.E., 1991. Spinneret evidence and the higher classification of the haplogyne spiders (Araneae, Araneomorphae). *American Museum Novitates,* 3016: 1–73.

Riedl, R., 1979. *Order in Living Organisms*. John Wiley and Sons, New York.

Scharff, N. and Coddington, J.A., 1997. A phylogenetic analysis of the orb weaving spider family Araneidae (Arachnida. Araneae). *Zoological Society of the Linnean Society*, 120: 355–434.

Sekiguchi, K., 1955. Differences in the spinning organs between male and female adult spiders. *Science Reports of the Tokyo Kyoiku Daigaku, Section B,* 8: 23–32.

Shear, W.A., Palmer, J.M., Coddington, J.A. and Bonamo, P.M., 1989. A Devonian spinneret: Early evidence of spiders and silk use. *Science, Wash.*, 246: 479–481.

Townley, M.A., Horner, N.V., Cherim, N.A., Tugmon, C.R. and Tillinghast, E.K., 1991. Selected aspects of spinning apparatus development in *Araneus cavaticus* (Araneae, Araneidae). *Journal of Morphology*, 208: 175–191, C5–C7.

Townley, M.A., 1993. Silk use during proecdysis in the spider *Araneus cavaticus* (Araneae, Araneidae). Doctoral dissertation, University of New Hampshire.

Townley, M.A. and Tillinghast, E.K., 2003. On the use of ampullate gland silks by wolf spiders (araneae, lycosidae) for attaching the egg sac to the spinnerets and a proposal for defining nubbins and tartipores. *Journal of Arachnology*, 31: 209–245.

Yu, L. and Coddington, J.A., 1990. Ontogenetic changes in the spinning fields of *Nuctenea cornuta* and *Neoscona theisi* (Araneae, Araneidae). *Journal of Arachnology*, 18: 331–345.

2013, Animal Diversity, Natural History and Conservation, Vol. 3 *Pages* **95–118**
Editors: **V.K. Gupta and Anil K. Verma**
Published by: **DAYA PUBLISHING HOUSE, NEW DELHI**

Chapter 8

Status, Diversity and Distribution of Freshwater Mollusca of India

Amit Mukhopadhyay and Basudev Tripathy*
Zoological Survey of India,
M-Block, New Alipore, Kolkata – 53, W.B., India

ABSTRACT

The documentation on the freshwater mollusca in India dates back to surveys conducted since pre independence India and thereafter published in different volumes of Fauna of British India. In the past couple of decades, the Zoological Survey of India along with various research institutions and Universities has conducted surveys and research on freshwater mollusca and that led to discovery of many new species from various localities. The Zoological Survey of India is the repository for fauna of India and has vast animal collections in the National Zoological Collections, largest in the subcontinent. The present paper deals with complete revisions on the information available on the status, diversity and distribution of freshwater mollusc of India. We also dealt with updated checklist along with synonymy of mollusc species occurring in the various freshwater systems.

Keywords: *Freshwater mollusca, Diversity, Endemism, Western Ghats, Rare species, India.*

---

* *Corresponding author.* E-mail: tripathyb@yahoo.co.uk

Introduction

Molluscs are an important group for freshwater biodiversity, and where abundant play an important role in ecosystem functioning (Vaughn *et al.*, 2004). They form an important component of most biological monitoring programmes that rate water quality and status of aquatic systems based on invertebrate assemblages (Ponder 1994; Seddon 1998; Strong *et al.*, 2008). Bivalves in particular, as they accumulate toxic substances to a greater extent than other organisms, are used to monitor water quality (Salanki *et al.*, 2003). Molluscs show a great specialization of ecological niches in freshwater environments, making them more vulnerable to modifications in their environment (Bouchet, 1992; Lydeard *et al.*, 2004). Freshwater environments are some of the most fragile and highly threatened ecosystems in the world. Consequently, molluscs in the freshwater systems have suffered a severe decline in the diversity, distribution and abundance due to human induced alteration of habitats, pollution, siltation, deforestation, poor agricultural practices, the destruction of riparian zones and invasion by introduced species (Biggins *et al.*, 1995; Pimm *et al.*, 1995). Freshwater molluscs are essential to the maintenance of wetland ecosystems, primarily due to their control of water quality and nutrient balance through filter-feeding and algal-grazing and to a lesser degree, as a food source for predators including a number of fish species, and in some parts of the world they compose a significant food resource, especially for the rural poor. In some regions they are one of the most threatened groups of freshwater taxa (Kay, 1995). The impact of developments such as dams, and siltation caused by deforestation and agricultural clearance has not been adequately researched and there is little awareness of the complex life histories of some groups such as unionid mussels that rely on the maintenance of migratory fish runs to carry their parasitic larvae to the river headwaters. Many species are also restricted to microhabitats, such as the riffles (areas of fast current velocity, shallow depth, and broken water surface) between pools and runs (areas of rapid non-turbulent flow).

There are an estimated 5,000 freshwater molluscs for which valid descriptions exist, in addition to a possible additional 10,000 undescribed species. Different authors given different number of species availability but certain number is not possible to ascertain. Of these currently known species, only around 27.7 per cent (1,374) have had their conservation status assessed for the IUCN Red List globally in 2009.

An array of literature is available on the freshwater molluscs of India. Preston's work –Fauna of British India-Freshwater Gastropoda and Plecypoda is nearly hundred years old. Subba Rao started to consolidate all information results in a good book 'Handbook on Freshwater Molluscs of India' in 1989. In the intervening period a vast amount of knowledge has accumulated on freshwater molluscs. A number of new species, new excellent information on their biology, bionomics and distribution have been added. This chapter serves as a guide and it is hoped, based on this to publish a revised fauna of freshwater molluscs of India in near future.

The freshwater molluscs are distinguished into two natural groups namely primary freshwater and secondary freshwater. The primary freshwater molluscs are confined exclusively to the freshwater habitats. There are many species which are distributed both in estuarine and freshwaters. Such genera as *Neritina, Clithon, Septaria, Scaphula* and *Novaculina* belong to this category. These genuses primarily freshwater with tolerance to saline waters. The family Stenothyridae and subfamily Iravadiinae include primarily estuarine or saline water species, occasionally occurring in freshwaters. Among bivalve the genera *Villoritta, Geloina, Batissa* and *Tanysiphon*are primarily estuarine, rarely occurring in freshwaters. One species of the family Assiminedae is also an estuarine species occasionally found in freshwater also.

Freshwater molluscs are common in ponds, lakes, quiet water pools, and flowing waters like perennial rivers, irrigation canals etc. The malacofauna of the streams may be somewhat different from that of the stagnant waters. Most of the endemic freshwater molluscs are found in the Western Ghats and Nilgiris and are remarkable for their characteristics.

Gastropods are generally found attached to submerged vegetation, rocks, sticks, bricks etc. but bivalves live partly buried in the sand or mud. Similar to the gastropods, the bivalves exhibit variations in shell depending upon the ecology of the species and may produce ecophenotypes (Subba Rao, 1989).

Freshwater molluscs of the tropics are usually faced with an annual dry season. Gastropods and bivalves show varying capacity to survive desiccation. Many snails can overcome this by aestivation either under dead vegetation, rocks, bricks or debris or actively burrowing at least the aperture of the shell in the mud. Some of the shell is capable of anaerobic respiration. With the advent of favourable conditions the molluscs resume to their normal activities. The discontinuous and transient nature of freshwater bodies leads to isolation of snails or bivalves into small local populations. Much remains to be worked on the ecology of Indian freshwater molluscs.

Status of Freshwater Molluscs in India

There are an estimated 5,000 freshwater molluscs for which valid descriptions exist in the World, in addition to a possible additional 10,000 undescribed species through the world (Balian *et al.*, 2008). In India freshwater molluscs are 202 (Subba Rao,1989) in number which is almost 4 per cent of world population. Some are endemic and available from specialized niche. Freshwater environments are some of the most fragile and highly threatened in the world. These ecosystem have been constantly manipulated by humans to satisfy their needs with little or no thought to the long terms effect on them (Kaufman, 1992). Bivalves can consume huge toxic substances that are why it can be used as the key species to judge the water quality of a certain place. But recently freshwater molluscan diversity are facing severe extinction due to human induced alteration of habitats, pollution, siltation, deforestation, poor agricultural practice, destruction of riparian zones and invasion by introduced species.

Restricted Distribution of Freshwater Molluscs from India

In India, some of the species of freshwater molluscs shows discontinuous distribution. Some the species of gastropods *viz. Angulyagra oxytropis, A. microchaetophora, Bellmya bengaliensis, f. balteata, Lymnae aovalior, L. horae, Paludomus oricatus, P. stephanus, P. reticulata, P. pustulosa, P. ornatus, P. conica v. kopiliensis, Taia crassicallosa, Bithynia textum, Pila theobaldi, Corbicula assamensis, Parreysia theobaldi,*

P. burmanus, P. nuttalliana, P. involuta, P. smaragdites, P. gowhattensis, P. corbis, Trapizoideus exolescens, Sphaerium austeni, S. kuiperi, Camptoceras lineatum, Solenaia soleniformis, Scabis crispata, Physania velaris, Lamellidens jenkinsianus sub sp. *obesa, L. phenchooganjensis, Ferrissia viola* are the bivalves which are exclusively distributed from North eastern region of India. Their restricted distribution makes them vulnerable, threat prone and very rare.

Species like *Bithynia troscheli, Hippeutis fontanus, Gyraulus pankogensis, Planorbis rotundatus, P. tangitarensis, Radix auricularia, R. brevicauda, R. lagotis f.* (*form*) *costulata, R. lagotis f. defilippi, R. lagotis f. solidissima, R. lagotis f. striata, Galba truncatula, Bithynia tentaculata kashmirensis, Lymnaea stagnalis, Valvate piscinalis, Sphaerium mitchelli, S. kashmirensis, Pisidium casertanum, Corbicula cashmiriensis* are the bivalves which shows restricted distribution from high altitude lakes, rivers of Jammu and Kashmir.

The water bodies of Andaman and Nicobar Islands having freshwater and estuarine water which is the home of some isolated species which are no longer reported from mainland of India. We may call them "exclusively Islands species". Species such as *Potamopyrgus nevilli, Tarebia semigranosa, Stenomelania punctata, S. plicaria, S. aspirans, Melanoides nicobarica, M. nevilli, M. crebra, Thiara amarula, Stenothyra hangerfordiana, Neritodryas subsulcuta, Neritina variegate, N. turrita, Segmintina taia, Septaria porcellana, Theodoxus reticularis* and *T.bicolor*are among gastropods and *Battissa similis, B. violecia, B. inflata* are the bivalves which are restricted in their distribution within Andaman and Nicobar group of Islands.

Gangetic basin of West Bengal is also shows isolated distribution of estuarine and freshwater molluscs. *Novaculina gangetica, Tanysiphon rivalis* are the bivalves and *Stenothyra woodmasoniana, S. solata, S. ornate, S. nana, S. deltae, Iravadia princeps, Assiminea francesi, Neritina smithi, Pila globosa, Pila incrassatula, Mainwaringia paludomoidea* are the gastropods which are restricted their distribution in West Bengal only.

Some of the species that is *Lymnaea brevissima,Paludomus obesus, P. inflatus, Physa acuta, Cremnoconchus conicus, C.carinatus, C.syhadrensis* and *Bithynia everzardi* among gastropods and *Parreysia khadakvaslaensis* and *Parreysia cylindrical*among bivalves which are only reported from Maharashtra.

Many species which are known from their type locality and represented by only a few species; further data regarding their population, biology, ecology, vulnerability and rarity is very much unknown. *Corbicula krishnaea* from Krishna river, *C. annandalei* from Kerala backwater, *Tricula graveli* from the bed of Narmada river near Hosangabad of MadhyaPradesh(Ramkrishna *et al.*, 2007), *Bellamya micron, Bellamya bengalensisf colairensis* from Kolleru lake and *Gyraulus saltensis* from Punjab are some of the examples.

Some of the major wetland *viz.* Kerala Backwater, a unique area which is also a home of some restricted species such as *Paludomus neritoides, P.stomatodom, P. sulcatus, P. annandalei, Iravadia funereal, I. annandalei, Villorita cornucopia* and *V. prasadhi* which are not reported from other parts of India. Species such as *Ferrissia tenuis* from Nilgiri Hills, *Stagnicola tugabhadraensis* from Kurnool, *Scaphula nagarjunai*from Andhra Pradesh, *Faunus ater* from Goa and Andaman and Nicobar Islands, *Iravadia ennurensis* from Tamil Nadu have restricted distribution.

As a whole Western Ghats, a unique biodiversity hotspot recorded 77 species of freshwater molluscs of which *Arcidopsis footei, Paludomus inflatus, Pila saxea, Neritina perottetiana, Neritna platyconche, N. reticulata, Paracrostoma tigrina, Paracrostoma martini, Cremnochonchus* spp. and *Psedomulleria dalyii* are some of the species which are endemic in this region. *Pseudomullaria dalyii* (Family: Etheridae) endemic cemented bivalve confined to couple of rivers in the central Western Ghats that is also a rare Gondwanaland relict (Madhyastha, 2001). This family shows unique discontinuous distribution, with recognized genera *viz., Acostea* (South America), *Pseudomulleria* (India) and *Etheria* (Africa) (Smith, 1898).Some of these species especially *Cremnochonus,* a spring habitat species occasionally occur at high altitude of Western Ghats.

Endemism

A detailed study on endemism of freshwater molluscs available reveals that 47 species are endemic from India, of which 33 are gastropods and 14 are bivalves. 3 species are endemic from Andhra Pradesh, 1 from Punjab, 7 species from Assam, 3 from Manipur, 1 from Meghalaya, 2 from Mizoram, 13 from West Bengal, 2 from Tamil Nadu, 6 from Maharashtra, 3 from Andaman and Nicobar Islands, 5 from Kerala Backwater and 3 from Jammu and Kashmir are endemic species.

Endemism from Western Ghats

A total of 77 species of freshwater molluscs studied of which 52 gastropods (12 families and 23 genera) and 25 bivalve species (five families and eight genera) of which 28 species are endemic from this region (Arvind *et al.,* 2010), but insufficient data that is list of the endemic species as studied by them not provided and still there is confusion of such high endemicity from this region. But the number is not as high as predicted by the earlier author, there are some taxonomic confusion regarding identification of species.

Conservation Status (IUCN Red List Category)

According to a report of IUCN, seven species (12 per cent) are assessed as threatened which are *Cremnochonchus syhadrensis, C. carrinatus, Arcidopsis footei, Pseudomulleria dalyi* are assessed as Endangered and *Cremnochonchus conicus, Parreysia khadakvasiensis* and *Scaphula nagarjunai* are assed as Vulnerable. The majority (88 per cent) are assed as Least Concern (Arvind *et al.,* 2010).

Distribution

Table 8.1: Distribution of Freshwater Molluscs State-wise

Sl.No.	*States*	*Families*	*Genera*	*Species*	*Per cent*
1.	Andaman and Nicobar Islands	10	20	51	25.24
2.	West Bengal	14	21	65	32.17
3.	Bihar	13	19	55	27.22
4.	Orissa	11	16	40	19.80

Contd...

Table 8.1–*Contd...*

Sl.No.	*States*	*Families*	*Genera*	*Species*	*Per cent*
5.	Andhra Pradesh	13	24	43	21.28
6.	Kerala	13	23	60	29.70
7.	Maharashtra	10	19	60	29.70
8.	Jammu and Kashmir	8	16	35	17.32
9.	Delhi	9	12	17	8.41
10.	Madhya Pradesh	7	7	25	12.37
11.	Jahrkhand	9	8	23	11.38
12.	Sikkim	5	5	9	4.45
13.	Arunachal Pradesh	4	5	5 (partly Worked out)	2.4
14.	Meghalaya	10	14	43	21.28
15.	Manipur	11	16	52	26
16.	Nagaland	9	16	21	10.5
17.	Rajasthan	6	8	13	6.4
18.	Tripura	8	14	30	14.85

From the above Table 8.1 it has been clearly seen high species diversity from West Bengal followed by Kerala and Maharashtra.

Economically and Commercially Important Freshwater Molluscs from India

Freshwater molluscs species*viz.*, *Batissa* spp., *Parreysia* spp., *Lamelliens* spp., *Corbicula* spp., *Villorita cornucopia* and *Villoritta cyprinoides* are extensively used as food and to sell by low income groups for whom freshwater resources are often vital importance in sustaining livelihood and food security. For examples Andaman Jarwas tribes are often used *Batissa s*pp. as a staple food in their everyday meal. Shell fishes are consumed throughout the North East *i.e.* Arunachal Pradesh, Meghalaya, Mizoram, Manipur, Tripura and Assam. Shells are more popular in the Garo hills than fishes. Fishing villages around Vembanad Lake in Kerala, about 6500 people are involved in the black calm fishery (*Villorita cornucopia* and *V. cyprinoides*). For most people in these villages, the black clam is their main source of income (Kripa *et al.*, 2004).

A Checklist of Freshwater Molluscs of India

Class GASTROPODA

Subclass PROSOBRANCHIA

Order ARCHAEOGASTROPODA

Superfamily NERITOIDEA

Family NERITIDAE

Subfamily NERITINAE

Genus ***Neritina*** s. st.

1. *Neritina* (*Neritina*) *pulligera* (Llinnaeus, 1767)

Subgenus ***Dostia*** Gray, 1847

2. *Neritina* (*Dostia*) *platyconcha* (Annandale and Prashad, 1919)
3. *Neritina* (*Dostia*) *violacea* (Gmelin)

Subgenus ***Neripteron*** Lesson, 1830

4. *Neritina* (*Neripteron*) *auriculata* Lamarck, 1826

Subgenus ***Vittina*** H.B Baker, 1923

5. *Neritina* (*Vittina*) *perottetiana* (Recluz, 1841)
6. *Neritina* (*Vittina*) *smithi* Wood, 1828
7. *Neritina* (*Vittina*) *turrita* (Gmelin, 1791)
8. *Neritina* (*Vittina*) *variegata* Lesson, 1831

Genus ***Neritodryas*** von Martens, 1869

9. *Neritodryas subsulcata* (Sowerby, 1836)

Genus ***Thedoxus*** Montfort, 1810

Subgenus ***Clithon*** Montfort, 1810

10. *Theodoxus* (*Clithon*) *bicolor* (Recluz, 1842)
11. *Theodoxus* (*Clithon*) *corona* (Linnaeus, 1758)
12. *Theodoxus* (*Clithon*) *reticularis* (Sowerby, 1836)

Family SEPTARIIDAE

Genus ***Septaria*** Ferrussac, 1807

13. *Septaria lineata* (Lamarck, 1876)
14. *Septeria porcellana* (Linnaeus, 1758)

Order MESOGASTROPDA

Superfamily VIVIPARIDAE

Family VIVIPARIDAE

Subfamily VIVIPARINAE

Genus ***Taia*** Annandale, 1918

15. *Taia craaicallosa* Annandale and Rao, 1925

    Subfamily BELLAMYINAE

Genus ***Bellamya*** Jousseaume, 1886

16. *Bellamya bengalensis* (Lamarck1822)
    - form *t ypica*
    - form *annandalei* (Kobelt, 1908)
    - form *balteata* (Benson)
    - form *colairensis* (Annandale, 1921)
    - form *doliaris* (Gould, 1843)
    - form *eburnean* (Annandale)

form *gigantea* (Reeve)

phase *halophila* (Kobelt)

form *mandiensis* (Kobelt, 1909)

17. *Bellamya crassispiralis* (Annandale, 1921)
18. *Bellamya crassa* (Benson, 1836)
19. *Bellamya dissimilis* (Mueller, 1774)
20. *Bellamya micron* (Annandale, 1921)

Genus ***Angulyagra*** Rao, 1931

21. *Angulyagra microchaetophora* (Annandale, 1921)
22. *Angulyagra oxytropis* (Benson, 1836)

Genus ***Cipangopaludina*** Hannibal, 1912

23. *Cipangopaludina lecythis* (Benson, 1836)

Family AMPULLARIIDAE

Genus ***Pila*** (Bolten) Roeding, 1798

24. *Pila globosa* (Swainson, 1822)

var. *incrassatula* (Nevill)

var. *minor* (Nevill)

25. *Pila scutata* (Mousson, 1848)

var. *compacta* Reeve

26. *Pila theobaldi* (Hanley, 1875)
27. *Pila virens* (Lamarck, 1822)

Subgenus ***Turbinicola*** Annandale and Prashad, 1921

28. *Pila* (*Turbinicola*) *saxea* (Reeve, 1856)

Superfamily VALVATOIDEA

Family VALVATIDAE

Genus ***Valvata*** Mueller, 1774

29. *Valvata piscinalis* (Mueller, 1784)

Super family LITTORINOIDEA

Family LITTORINIDAE

Subfamily LITTORININAE

Genus ***Mainwaringia*** Nevill, 1884

30. *Mainwaringia paludomoidea* (Nevill, 1884)

Subfamily BEMBICIINAE

Genus ***Cremnoconchus*** Blanford,1869

31. *Cremnoconchus syhadrensis* (Blanford, 1863)

Subgenus ***Lissoconchus*** Thiele, 1931

32. *Cremnoconchus* (*Lissoconchus*) *carinatus* (Layard, 1854)
33. *Cremnoconchus* (*Lissoconchus*) *conicus* (Blanford, 1870)

Superfamily RISSOIDEA

Family HYDROBIIDAE

Subfamily POTAMOPYRGINAE

Genus ***Potamopyrgus*** Stimpson, 1865

Subgenus ***Indopyrgus*** Thiele, 1928

34. *Potamopyrgus* (*Indopyrgus*) *nevilli* Thiele, 1928

Family BITHYNIIDAE

Genus ***Bithynia*** Leach, 1818

35. *Bithynia tentaculata kashmirensis* (Nevill, 1884)
36. *Bithynia troscheli* (Paasch, 1842)

Subgenus ***Digoniostoma*** Annandale, 1920

37. *Bithynia* (*Digoniostoma*) *cerameopoma* (Benson, 1830)
38. *Bithynia* (*Digoniostoma*) *evezardi* (Blanford, 1880)
39. *Bithynia* (*Digoniostoma*) *pulchella* (Benson, 1836)
40. *Bithynia* (*Digoniostoma*) *textum* Annandale, 1921

Genus ***Gabbia*** Tryon, 1865

41. *Gabbia orcula* Frauenfeld var. *producta* (Nevill, 1884)
42. *Gabbia stenothyroides* (Dohrn, 1857)
43. *Gabbia travancorica* (Benson, 1860)

Genus ***Mysorella*** Annandale, 1919

44. *Mysorella costigera* (Kuester, 1852)

Family POMATIOPSIDAE

Subfamily TRICULINAE

Tribe ***Triculini***

Genus ***Tricula*** Benson, 1843

45. *Tricula gravely* Prashad, 1921
46. *Tricula montana* Benson, 1843

Family STENOTHYRIDAE

Genus ***Stenothyra*** Benson, 1856

47. *Stenothyra blanfordiana* Nevill, 1880
48. *Stenothyra deltae* (Benson, 1836)
49. *Stenothyra foveolata* Benson, 1856
50. *Stenothyra hungerfordiana* Nevill, 1880

51. *Stenothyra minima* (Sowerby, 1837)
52. *Stenothyra nana* Prashad, 1921
53. *Stenothyra or nata* Annandale and Prashad, 1921
54. *Stenothyra soluta* Annandale and Prashad, 1919
55. *Stenothyra woodmasoniana* Nevill, 1880

Genus ***Gangetia***Ancey, 1890

56. *Gangeti amiliacea* (Nevill, 1880)

Family IRAVADIIDAE

Genus ***Iravadia*** Blanford, 1867

57. *Iravadia annandalei* Preston, 1916
58. *Iravadia ennurensis* Preston, 1916
59. *Iravadia funerea* Preston, 1916
60. *Iravadia princeps* Preston, 1915

Family ASSIMINEIDAE

Subfamily ASSIMINEINAE

Genus ***Assiminea*** Fleming, 1828

61. *Assiminea francessi* (Wood, 1826)

Superfamily CERITHIOIDEA

Family THIARIDAE

Subfamily THIARINAE

Genus ***Thiara*** Roeding, 1798

Subgenus ***Thiara*** s.st.

62. *Thiara* (*Thiara*) *amarula* (Linnaeus, 1758)
63. *Thiara* (*Thiara*) *rudis* (Lea, 1850)
64. *Thiara* (*Thiara*) *scabra* (Mueller, 1774)

Subgenus ***Sermyla*** H. and A. Adams, 1854

65. *Thiara* (*Sermyla*) *riqueti* (Grateloup)

Genus ***Faunus*** Montfort, 1810

66. *Faunus ater* (Linnaeus, 1758)

Genus ***Melanoides*** Olivier, 1804

67. *Melanoides crebra* Lea, 1850
68. *Melanoides nevillei* (Brot, 1874)
69. *Melanoides nicobarica* (Moerch, 1859)
70. *Melanoides peddamunigalensis* Ray and Roy Chowdhury
71. *Melanoides tuberculata* (Muller, 1774)

Genus ***Stenomelania*** P. Fischer, 1885

72. *Stenomelania aspirans* (Hinds, 1874)
73. *Stenomelania plicaria* (Born, 1780)
74. *Stenomelania punctata* (Lamarck, 1822)
75. *Stenomelania torulosa* (Bruguiere, 1789)

Genus ***Tarebia*** H.A. Adams, 1854

76. *Tarebia granifera* (Lamarck, 1822)
77. *Tarebia lineata* (Gray, 1828)
78. *Tarebia semigranosa* von dem Busch, 1842

Family PLEUROCERIDAE

Subfamily MELANATRIINAE

Genus ***Brotia*** H.Adams, 1866

Subgenus ***Brotia*** s.st.

79. *Brotia* (*Brotia*) *fuscata* (Born)

Subgenus ***Antimelania*** Fischer and Crosse, 1892

80. *Brotia* (*Antimelania*) *costula* (Rafinesque, 1833)

Genus ***Sulcospira*** Troschel, 1858

81. *Sulcospira huegeli* (Philippi, 1841)

Subfamily PALUDOMINAE

Genus ***Paludomus*** Swainson, 1840

Subgenus ***Paludomus*** s.st.

82. *Paludomus* (*Paludomus*) *annandalei* Preston, 1909
83. *Paludomus* (*Paludomus*) *blanfordiana* Nevill, 1877
84. *Paludomus* (*Paludomus*) *conica* (Gray, 1834)
    var. *kopilensis* Nevill
85. *Paludomus* (*Paludomus*) *inflatus* Brot, 1880
86. *Paludomus* (*Paludomus*) *obesus* (Philippi, 1842)
87. *Paludomus* (*Paludomus*) *ornatus* Benson, 1856
88. *Paludomus* (*Paludomus*) *pustulosa* Annandale, 1925
89. *Paludomus* (*Paludomus*) *regulata* Benson, 1856
90. *Paludomus* (*Paludomus*) *reticulate* Blanford, 1870
91. *Paludomus* (*Paludomus*) *rotunda* Blanford, 1870
92. *Paludomus* (*Paludomus*) *stephanus* (Benson, 1836)
93. *Paludomus* (*Paludomus*) *transchauricus* (Gmelin)

Subgenus ***Philopotamis*** Layard, 1855

94. *Paludomus* (*Philopotamis*) *sulcatus* Reeve, 1847

Subgenus ***Stomatodon*** Benson, 1862

95. *Paludomus* (*Stomatodon*) *stomatodon* Benson, 1862

Subgenus ***Tanalia*** Gray, 1847

96. *Paludomus* (*Tanalia*) *loricatus* Reeve, 1817
97. *Paludomus* (*Tanalia*) *neritoides* Reeve, 1847

Subclass PULMONATA

Order BASOMMATOPHORA

Superfamily LYMNOIDEA

Family LYMNAEIDAE

Genus ***Lymnaea*** Lamarck, 1799

Subgenus ***Lymnaea*** s.st.

98. *Lymnaea stagnalis* (Linnaeus, 1758)

Subgenus ***Pseudosuccinea*** F.C. Baker, 1908

99. *Lymnaea* (*Pseudosuccinea*) *acuminata* Lamarck, 1822

form typica Lamarck, 1822

form *brevissima* Annandale and Rao, 1925

form *chlamys* Benson, 1836

Form *gracilior* Martens, 1881

form *hians* Sowerby, 1873

form *malleata* Annandale and Rao, 1925

form *patul*a (Troschel, 1837)

form *rufescens* Gray, 1822

100. *Lymnaea* (*Pseudosuccinea*) *biacuminata* Annandale and Rao, 1925
101. *Lymnaea* (*Pseudosuccinea*) *horae* Annandale and Rao, 1925
102. *Lymnaea* (*Pseudosuccinea*) *luteola* Lamarck, 1822

from *typica* Lamarck, 1822

from *australis* Annandale and Rao, 1925

from *impura* Troschel, 1837

from *ovalis* Gray, 1822

from *succinea* Deshayes, 1834

103. *Lymnaea* (*Pseudosuccinea*) *ovalior* Annandale and Prashad, 1821

Genus ***Galba*** Schrank, 1803

104. *Galba andersoniana* (Nevill, 1887)
105. *Galba hookeri* (Reeve, 1850)
106. *Galba truncatula* (Mueller, 1774)

Genus ***Radix*** Montfort, 1810

107. *Radix auricularia* (Linnaeus, 1758)

108. *Radix brevicauda* (Sowerby, 1758)
109. *Radix lagotis*
    form *costulata* (von Martens, 1874)
    form *defilippi* (Issel, 1867)
    form *solidissima* (Kobelt, 1872)
    Form *striata* (Andreae, 1911)
110. *Radix persica* (Issel, 1865)

Genus ***Stagnicola*** Leach in Jeffreys, 1830

111. *Stagnicola tungabhadraensis* Ray, 1967

Superfamily PHYSOIDEA

Family PHYSIDAE

Genus ***Physa*** Draparnaud, 1801

112. *Physa acuta* Draparnaud, 1805

Superfamily PLANORBIOIDEA

Family PLANORBIDAE

Subfamily PLANORBINAE

Tribe **Planorbeae**

Genus ***Planorbis*** Mueller, 1774

113. *Planorbis planorbistangitarensis* Germain, 1918
114. *Planorbis rotundatus* Poiret, 1801

Genus ***Gyraulus*** Agassiz in Charpentier, 1837

115. *Gyraulus barrackporensis* (Clessin, 1886)
116. *Gyraulus convexiusculus* (Hutton, 1849)
117. *Gyraulus euphraticus* (Mousson, 1874)
118. *Gyraulus labiatus* (Benson, 1850)
119. *Gyraulus ladacensis* Nevill, 1878
120. *Gyraulus pankongensis* (von Martens, 1882)
121. *Gyraulus rotula* (Benson, 1850)
122. *Gyraulus saltensis* Germain, 1922

Tribe **Camptocerateae**

Genus ***Camptoceras*** Benson, 1843

123. *Camptoceras lineatum* Blanford, 1871
124. *Camptoceras subspinosum* Annandale and Prashad, 1920
125. *Camptoceras terebra* Benson, 1843

Tribe **Segmentinineae**

Genus ***Segmentina*** Fleming, 1822

Subgenus ***Polypylis*** Plisbry, 1906

126. *Segmentina* (*Polypylis*) *calatha* (Benson, 1850)
127. *Segmentina* (*Polypylis*) *taia* Annandale and Rao, 1925
128. *Segmentina* (*Polypylis*) *trochoidea* (Benson, 1836)

Genus ***Hippeutis*** Agassiz in Charpentier, 1837

129. *Hippeutis fontanus* (Lightfoot, 1786)

Subgenus ***Helicorbis*** Benson, 1855

130. *Hippeutis* (*Helicorbis*) *umbilicali sumbilicalis* (Benson, 1836)

Family BULLINIDAE

Subfamily BULLININAE

Genus ***Indoplanorbis*** Annandale and Prashad, 1920

131. *Indoplanorbis exustus* (Deshayes, 1834)

Family ANCYLIDAE

Subfamily FERRISSINAE

Genus ***Ferrissia*** Walker, 1903

132. *Ferrissia baconi* (Bourguignat, 1853)
133. *Ferrissia ceylanica* (Benson, 1864)
134. *Ferrissia tenuis* (Bourguignat, 1862)
135. *Ferrissia verruca* (Benson, 1855)
136. *Ferrissia viola* Annandale and Prashad, 1923

Class BIVALVIA

Subclass PTERIOMORPHIA

Order ARCOIDA

Superfamily ARCOIDEA

Family ARCIDAE

Subfamily ANADARINAE

Genus ***Scaphula*** Benson, 1834

137. *Scaphula celox* Benson, 1836
138. *Scaphula deltae* Blanford, 1867
139. *Scaphula nagarjunai* Janaki Ram and Radhakrishna, 1984

Subclass PALEOHETERODONTA

Order UNIONOIDA

Superfamily UNIONOIDEA

Family UNIONIDAE

Subfamily UNIONINAE

Genus ***Arcidopsis*** Simpson, 1900

140. *Arcidopsis footei* (Theobald, 1876)

Genus ***Physunio*** Simpson, 1900

Subgenus ***Velunio*** Hans, 1914

141. *Physunio* (*Velunio*) *velaris* (Swerby, 1868)

Genus ***Scabies*** Hans, 1911

142. *Scabies crispata* (Gould, 1843)

Genus ***Solenaia*** Conard, 1868

143. *Solenaia soleniformis* (Benson, 1836)

Subfamily AMBLEMINAE (Quadrulinae)

Tribe **Amblemini**

Genus ***Lamellidens*** Simpson, 1900

144. *Lamellidens consobrinus* (Lea, 1859)
145. *Lamellidens corrianus* (Lea, 1834)
146. *Lamellidens generosus* (Gould, 1847)
147. *Lamellidens jenkinsianus* (Benson, 1862)
    sub sp. *daccaensis* (Preston, 1912)
    sub sp. *obesa* (Hanley and Theobald, 1877)
148. *Lamellidens marginalis* (Lamarck, 1819)
149. *Lamellidens phenchooganjensis* Preston, 1912

Genus ***Parreysia*** Conrad, 1853

Subgenus ***Parreysia*** s.st.

150. *Parreysia* (*Parreysia*) *annandalei* Preston, 1912
151. *Parreysia* (*Parreysia*) *burmanus* (Blanford, 1869)
152. *Parreysia* (*Parreysia*) *corbis* (Benson, 1856)
153. *Parreysia* (*Parreysia*) *corrugata* (Mueller, 1774)
    sub sp. *laevirostris* (Benson, 1862)
    sub sp. *nagpoorensis* (Lea, 1859)
154. *Parreysia* (*Parreysia*) *favidens* (Benson, 1862)
    var. *assamensis* Preston, 1912
    var. *chrysis* (Benson, 1862)

var. *deltae* (Benson, 1862)

var. *marcens* (Benson, 1862)

var. *plagiosoma* (Benson, 1862)

var. *viridula* (Benson, 1862)

155. *Parreysia* (*Parreysia*) *gowhattensis* (Theobald, 1873)
156. *Parreysia* (*Parreysia*) *rajahensis* (Lea, 1841)
157. *Parreysia* (*Parreysia*) *sikkimensis* (Lea, 1859)
158. *Parreysia* (*Parreysia*) *smaragdites* (Benson, 1862)
159. *Parreysia* (*Parreysia*) *triembolus* (Benson, 1855)

Subgenus ***Radiatula*** Simpson, 1900

160. *Parreysia* (*Radiatula*) *andersoniana* (Nevill, 1877)
161. *Parreysia* (*Radiatula*) *bonneaudi* (Eydoux, 1838)
162. *Parreysia* (*Radiatula*) *caerulea* (Lea, 1831)

subsp. *gaudichaudi* (Eydoux, 1838)

163. *Parreysia* (*Radiatula*) *cylindrica* Annandale and Prashad, 1919
164. *Parreysia* (*Radiatula*) *involuta* (Benson, 1856)
165. *Parreysia* (*Radiatula*) *khadakvaslaensis* (Ray, 1966)
166. *Parreysia* (*Radiatula*) *lima* (Simpson, 1900)
167. *Parreysia* (*Radiatula*) *nuttalliana* (Lea, 1856)
168. *Parreysia* (*Radiatula*) *occata* (Lea, 1860)
169. *Parreysia* (*Radiatula*) *olivaria* (Lea, 1831)
170. *Parreysia* (*Radiatula*) *pachysoma* (Benson, 1862)
171. *Parreysia* (*Radiatula*) *shurtleffiana* (Lea, 1856)
172. *Parreysia* (*Radiatula*) *theobaldi* (Preston, 1912)

Genus ***Trapezoideus*** Simpson, 1900

173. *Trapezoideus exolescens exolescens* (Gould, 1843)

Superfamily MUTELOIDEA

Family ETHERIIDAE

Genus ***Pseudomulleria*** Anthony, 1907

174. *Pseudomulleria dalyi* (E.A. Smith)

Subclass HETERODONTA

Order VENEROIDEA

Superfamily SOLENOIDEA

Family CULTELLIDAE

Genus ***Neosolen*** Ghosh, 1920

175. *Neosolen aquaedulcioris* Ghosh, 1920

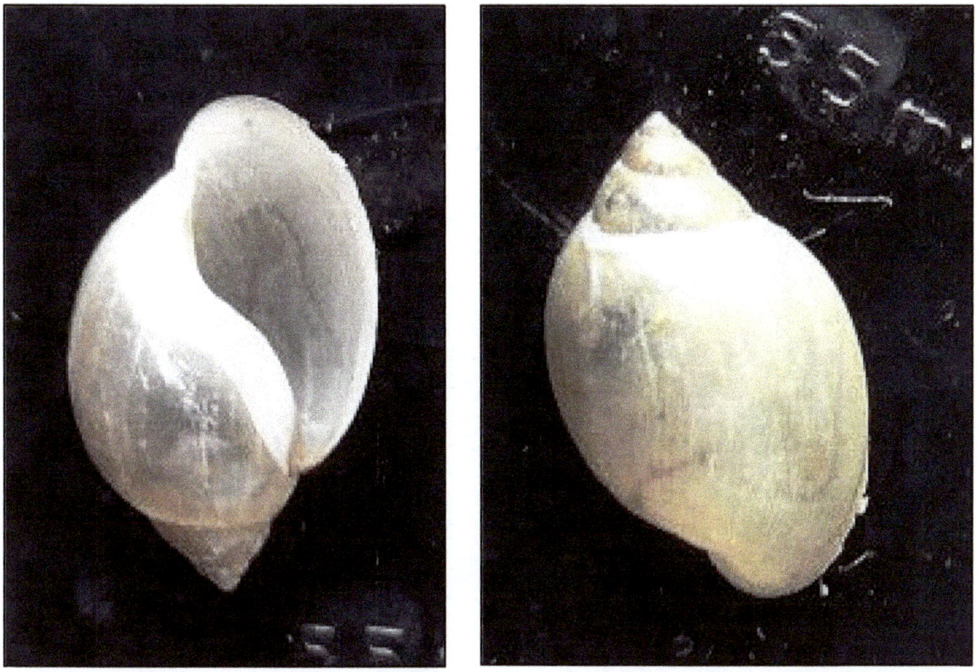

Plate 8.1: *Lymnaea* (*Pseudosuccinea*) *luteola* f. *typica* Lamarck

Plate 8.2: *Lymnaea* (*Pseudosuccinea*) *accuminata* Lamarck

Plate 8.3: ***Pila virens*** **Lamarck**

Plate 8.4: ***Lymnaea*** **(*Pseudosuccinea*)** ***accuminata*** **f.** ***typica*** **Lamarck**

Plate 8.5: *Lymnaea* (*Pseudosuccinea*) *accuminata* f. *rufescens* Gray

Plate 8.6: *Parreysia* (*Radiatula*) *pachysoma* (Benson)

Genus ***Tanysiphon*** Benson, 1858

176. *Tanysiphon rivalis* Benson, 1858

Superfamily TYELLINOIDEA

Family PSAMMOBIIDAE

Subfamily NOVACULININAE

Genus ***Novaculina*** Benson, 1830

177. *Novaculina gangetica* Benson, 1830

Plate 8.7: ***Bellamya bengalensis***

Plate 8.8: ***Indoplanorbis exustus***

Plate 8.9: *Tarebia lineata*

Superfamily CORBICULOIDEA

Family CORBICULIDAE

Genus ***Corbicula*** Mergerle, 1811

178. *Corbicula annandalei* Prashad, 1928
179. *Corbicula assamensis* Prashad, 1928
180. *Corbicula bensoni* Deshayes, 1854
181. *Corbicula cashmiriensis* Deshayes, 1854
182. *Corbicula krishnaea* Ray, 1967
183. *Corbicula peninsularis* Prashad, 1928
184. *Corbicula striatella* Deshayes, 1854

Genus ***Batissa*** Gray, 1853

185. *Batissa inflata* Prime, 1860
186. *Batissa similis* Prime, 1859
187. *Batissa violacea* Prime, 1859

Genus ***Polymesoda*** Radinesque, 1820

Subgenus ***Geloina*** Gray, 1842

188. *Polymesoda* (*Geloina*) *bengalensis* (Lamarck, 1818)
189. *Polymesoda* (*Geloina*) *erosa* (Solander, 1786)

Genus ***Villorita*** Gray in Griffith and Pidgeon, 1834

190. *Villorita cornucopia* Prashad, 1921

Family PISIDIIDAE (Sphaeriidae)

Genus ***Pisidium*** L. Pfeiffer, 1821

Subgenus ***Pisidium*** s.st.

191. *Pisidium* (*Pisidium*) *casertanum* (Poli, 1791)

Subgenus ***Afropisidium*** Kuiper, 1962

192. *Pisidium* (*Afropisidium*) *clarkeanum* G. and H. Nevill, 1871
193. *Pisidium* (*Afropisidium*) *ellisi* Dance, 1967
194. *Pisidium* (*Afropisidium*) *nevillianum* Theobald, 1876

Subgenus ***Europisidium*** Stadnichenko, 1984

195. *Pisidium* (*Europisidium*) *mitchelli* Prashad, 1925

Subgenus ***Odhnerpisidium*** Kuiper, 1982

196. *Pisidium* (*Odhnerpisidium*) *atkinsonianum* Theobald, 1876
197. *Pisidium* (*Odhnerpisidium*) *kuiperi* Dance, 1967

Genus ***Sphaerium*** Scopoli, 1777

198. *Sphaerium austeni* Prashad, 1921
199. *Sphaerium indicum* Deshayes, 1854
200. *Sphaerium kashmirensis* Prashad, 1937

Photographs of some of the above species are given in Plates 8.1–8.9.

Note: Additional two new species have been described recently that is *Paracrostoma tigrina* and *Paracrostoma martini* by Kohler and Glaubrecht (2007). Due to taxonomic confusion these two are not included in the Check list.

Conservation and Research for Freshwater Molluscs of India

Taxonomic research is central to ecological studies and conservation, but it is one of the most neglected disciplines, especially in biodiversity rich areas. Taxonomic knowledge of freshwater molluscs is severely lacking and a greater degree of taxonomic research on freshwater mollusca is required. Research into the distribution, population status, specific threats and taxonomy of freshwater molluscs of India is urgently required. Many type localities need to be resurveyed to confirm if described range-restricted freshwater molluscs are still present or have already become extinct and to confirm the taxonomic status of previously described species. The lack of trained malacologists, funding and the socio-political situation in the region has greatly hampered research on freshwater molluscs. Also except for a few commonly occurring species, information on ecology, population structure and dynamics, distribution, and habitat preference is not known. A greater degree of taxonomic research and training is also required to ensure that widely accepted taxonomic concepts are adopted.

Acknowledgements

We are thankful to the Director, Zoological Survey of India, Kolkata for permission and encouragement for publishing this article. We are grateful to Dr. A. Anand for providing photographs of freshwater molluscs.

References

Aravind, N.A., Madhyastha, N.A., Rajendra, G.M. and Dey, Anirudha, 2010. *The Status and Distribution of Freshwater Molluscs of the Western Ghats*, IUCN.

Biggins, R.G., Neves, R.J. and Dohner, C.K., 1995. *Draft NationalStrategy for the Conservation of Native Freshwater Mussels*. U.S.Fish and Wildlife Service, Washington, DC, p. 26.

Bouchet, P., 1992. Extinction and preservation of species inthe tropical world: What future for molluscs? *American Conchocologist,* 20: 20–24.

Kohler, F. and Glaubrecht, M., 2007. Out of Asia and into India, on the molecular phylogeny and biogeography of the endemic freshwater gastropod *Paracrostoma* Crossman, 1900.*Biological Journal of the Linnean Society*, 91: 627–651.

Kripa, V., Velayudhan, T.S., Shoji, J., Alloycious, P.S.,Radhakrishnan, P. and Sharma, J., 2004. Clam fisheriesin Vembanad Lake, Kerala, with observations on the Socio economic conditions of the clam fishers. *Marine Fisheries Information Service.* T and E Series, 178: 14, 15 and 16.

Lydeard, C., Cowie, R.H., Bogan, A.E., Bouchet, P., Cummings, K.S., Frest, T.J., Herbert, D.G., Hershler, R., Gargominy, O., Perez, K., Ponder, W.F., Roth, B., Seddon, M., Strong, E.E. and Thompson, F.G., 2004. The global decline of non-marine mollusks. *Bio Science,* 54: 321–330.

Madhyastha, N.A., 2001. *Pseudomulleria dalyi* (*Acostea dalyi*): A rare cemented bivalve of Western Ghats. *Zoos' Print Journal,* 16: 573.

Pimm, S.L., Russel, G.L., Gittleman, J.L. and Brooks, T.M., 1995. The future of biodiversity. *Science,* 269: 347–350.

Ponder, W.F., 1994. Australian freshwater mollusca: Conservation priorities and indicator species. *Memoirs of the Queensland Museum,* 36: 1919–196.

Ramakrishna and Day, A., 2007. *Handbook on Indian Freshwater Molluscs.* Zoological Survey of India, Kolkata, India.

Ramkrishna and Alfred, J.R.B., 2007. *Faunal Resources in India.* Zool. Surv. India, Kolkata, pp. 1–427.

Seddon, M.B., 1998. Red Listing for molluscs: a tool forconservation? pp. 27–44. In: *Molluscan Conservation: A Strategy forthe 21st Century,* (Eds.) I.J. Killeen, M.B. Seddon and A. Holmes. *Journal of Conchology* Special Publication 2. Conchological Society of Great Britain and Ireland.

Strong, E.E., Gargominy, O., Ponder, W.P. and Bouchet, P., 2008. Global diversity of gastropods (Gastropoda, Mollusca) in freshwater. *Hydrobiologia,* 595: 149–166.

Subba Rao, N.V., 1989. *Handbook of Freshwater Molluscs of India.* Zoological Survey of India, Kolkata, p. 289.

Vaughn, C.C., Gido, K.B. and Spooner, D.E., 2004. Ecosystemprocesses performed by unionid mussels in streammesocosms: Species roles and effects of abundance. *Hydrobiologia,* 527: 35–47.

2013, Animal Diversity, Natural History and Conservation, Vol. 3 *Pages* **119–133**
Editors: **V.K. Gupta and Anil K. Verma**
Published by: **DAYA PUBLISHING HOUSE, NEW DELHI**

Chapter 9

Distribution of the Invasive Giant African Snail, *Achatina fulica* (Bowdich) in Bangalore Urban District

M. Jayashankar*
Department of Zoology,
Bangalore University, Bangalore – 560 056, Karnataka

ABSTRACT

According to the results analyzed during the present survey *A.fulica* was detected in 42 of the 57 villages surveyed in Bangalore Urban district during 2008-10. Human aided agency is reported to be the main means of dispersal for *A. fulica*. But, its impact as pest seems more recent and still unaware of in many villages. There is a rich density of the snails in the nursery studied. An average of 83 snails per meter square during season and 14.06 snails per meter square during off-season in four quadrates of one meter square was recorded. The trees and stone pillars in the Dhanavanthari vana nursery serve as sites of resting and feeding for single or an aggregate of snails.

Keywords: *Giant African Snail, Achatina fulica, Bangalore Urban District, Dispersal.*

Introduction

The Giant African Snail, *Achatina fulica* (Bowdich) has been recorded in every continent except Antarctica and is a major crop pest across the globe since 1800's

* E-mail: jay_zoology@rediffmail.com

spread primarily through human activities (Mead, 1961, 1979; Raut and Barker, 2002; Silvana, 2007; Fontinilla *et al.,* 2007). The International Union for Conservation of Nature and Natural resources (IUCN) has listed *A. fulica* as one of the world's 100 most invasive species (Lowe *et al.,* 2000). It is a classic example of an introduced species (Zanol *et al.,* 2010; Steffek and Kusik, 2010). In India, it was believed to have been introduced in Chouringhie gardens of Calcutta in 1847 by the British Conchologist William Henry Benson and from there on spread to many states of the country in course of time (Mead, 1961). According to Malacologist Naggs (1997), although Benson had got the snail to India from Mauritius, he had handed them to a friend before leaving India and it was the friend who released them in his garden. This perhaps has lead to the large scale invasion of the snail in different parts of the country. Recent genetic studies indicate that all *A. fulica* now occurring throughout South Asia, Southeast Asia and the Pacific region are derived from one haplotype (Fontinilla *et al.,* 2007), the source being a single pair of specimens released in Chouringhie in 1847.

The GALS has been thriving in some areas of India for a period of 100-150 years, with no clear evidence of abatement in its pest status (Raut and Ghose, 1984). In spite of this long thriving history, reports on the agri-horticultural pest status of the snail are minimal. So far, the control measures of all types have failed to eradicate this pest, hence an integrated approach to contain the snail is recommended as an ultimate remedy (Srivastava, 1992). The snail has been reported for the first time causing damage to ornamental plants and vegetables in Bangalore during kharif (Veeresh *et al.,* 1979). Current spread and distribution of the snail in agri-horticultural regions of Bangalore is unknown. Hence, the survey on occurrence of *A. fulica* was undertaken in agri-horticultural areas of Bangalore Urban District.

The Giant African Snail is the biggest–sized and economically the most important land snail pest known in the world.

Systematic Position of *Achatina fulica* (Bowdich, 1822)

Kingdom	:	Animalia
Phylum	:	Mollusca
Class	:	Gastropoda
Subclass	:	Pulmonata
Order	:	Stylommatophora
Suborder	:	Sigmurethra
Superfamily	:	Achatinoidea
Family	:	Achatinidae
Subfamily	:	Achatininae
Genus:	:	*Achatina*
Subgenus	:	*Lissachatina*
Species	:	*fulica*

Material and Methods

Study Region

The study was conducted in the Bangalore region (12°58'0"N, 77°34'0"E) of Karnataka State. Bangalore region comprises of two districts namely Bangalore Urban and Bangalore Rural. Bangalore Urban district has four taluks *viz.,* Bangalore North, Bangalore South, Bangalore East and Anekal. Bangalore region comes under the eastern dry agro climatic zone of Karnataka with 35°C maximum and 14°C minimum temperature; 679.1 to 888.9 mm of rainfall and situated at an elevation of 800-1500 m. Soil type is red loamy, clay lateritic. The climate of Bangalore is that of a tropic savannah type, with summer season (March to May), rainy (June to September and November and December) and winter (December to January). The annual rainfall of Bangalore is 859mm spread over 6 months. It receives rains during both South-west and North-east monsoons.

The present study encompasses survey conducted during 2008-2010 in villages within the Bangalore urban region. Bangalore Urban has peripheral agriculture and horticulture practicing areas and has 558 inhabited villages with a total population (2001 census) of 65, 37,124 geographical area of 2, 17,410 hectares with 60,814 hectares total area sown, total food grains in 3, 28,487 hectares, oil seeds in 1,364 hectares, fruits 3,845 hectares and vegetables in 2,077 hectares (Source: Office of the District Statistical Office, Bangalore Rural and Urban Districts, Records of the year 2008-09).

Data Collection

Results of survey conducted in 57 villages (Figure 9.1) in four taluks of Bangalore Urban district of Karnataka state is presented in the present paper. The formal interview with agri-horticulturists involved the use of structured interview questionnaires. Continuous questionnaire–Opinionnaire survey and field observations were conducted during the study period in the study region, involving discussions with local people in the areas visited. The interviewees were provided with informational photographs, shells and contact numbers to communicate on the status of the snails. Questionnaire was prepared consulting earlier surveys (Asamoah, 1999, USDA–APHIS, 2005) in *A.fulica* infested Countries.

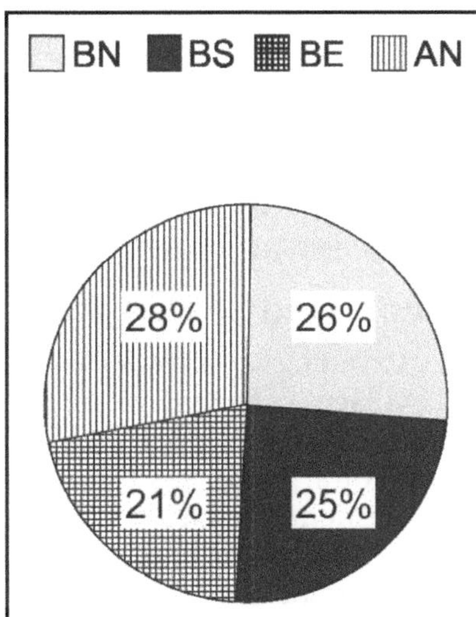

Figure 9.1: Percentage of Villages Surveyed in Bangalore Urban District

In the field, exact location of each sample collection site was recorded with a handheld Garmin e-trex, 12 Channel global positioning system (GPS) device to determine coordinates for each study point and of district and taluk map (Land records of 2001) of Karnataka State Remote Sensing Applications Centre (KSRAC), Government of Karnataka. All points were within the administrative boundary of the locations. Arc GIS Software (8.1 version) was used for the data analysis and projection.

Biodiversity Indices

Relative abundance and species diversity of random samples of pilot surveys and also multiple replicates of $1m^2$ quadrants were laid in the selected study area in nurseries of Bangalore north and South taluks. The samples were collected in early hours of the day during monsoon from the sites. Simpson's diversity index and Shannon-Weiner index have been calculated for the same.

Site 1

Hebbal Forest Department Nursery (Bangalore North): A total number of 1,994 plastic bags with saplings (six month old) belonging to eight plant species was *viz., Azadirachta indica* (84 bags), *Grevilia robust* (368), *Dolichandra* sp. (520), *Enterolobium saman* (232), *Abrus precatorius* (328), *Bauhinia* sp. (152), *Phyllanthus embilica* (10) and *Ficus glomerata* (100) were counted.

Site 2

Department of Zoology, Bangalore University (Bangalore South): Malacofauna were collected during the early hours of the day continuously in the lawn and corridors of the Department.

Site 3

Gandhi Bhavan, Forest Department Nursery, Bangalore University Campus (Bangalore South): *Azadirachta indica* containing sapling bags (84 bags) were searched to locate the different species of molluscs.

Trees and Stone Pillars

Visual surveillance of vegetation in the Dhanavanthari vana Nursery was undertaken to locate the snails. Information on the preference of flora, height of climbing and aggregation of snails were collected and analyzed

Results and Discussion

According to the results analyzed during the present survey *A. fulica* was detected in 42 villages in Bangalore Urban district. Human aided agency is reported to be the main means of dispersal for *A. fulica*. Collection and dispersal linked to agri-horticultural goods was often given as a mechanism for dispersal by respondents to the Questionnaire-Opinionnaire during interviews. But, its impact as pest seems more recent and still unaware of in many villages. The Global Positioning System (GPS) coordinates of villages of the detection survey are represented in Tables 9.1–9.4 and Maps 9.1–9.5 (Geo-tagging), indicating the occurrence status (present or absent) of Giant African land snails (GALS) and other malacofauna (OMF) in the villages surveyed.

Table 9.1: Geotagging of Villages and Malaco Faunal Occurrence in Bangalore North Taluk

Sl.No	*Village*	*North*			*East*			*AF*	*OMF*
1.	Nellukunte	13°	13'	15.7"	77°	32'	28.5"	P	P
2.	Seekote	13°	7'	42.8"	77°	30'	42.6"	P	P
3.	Byalakere	13°	7'	2.8"	77°	31'	22.1"	P	A
4.	Gadenahalli	13°	11'	16.2"	77°	37'	7.6"	P	P
5.	Kudurnahalli	13°	11'	9.1"	77°	37'	46.8"	P	P
6.	Tharabnalli	13°	11'	15.7"	77°	37'	52.9"	P	P
7.	Doddajaala	13°	10'	40.4"	77°	38'	53.5"	P	P
8.	Mediagrahara	13°	6'	19.6"	77°	31'	43.7"	P	P
9.	Meenakunte	13°	10'	49.9"	77°	38'	18.2"	A	A
10.	Gudadahalli	13°	8'	23.9"	77°	27'	47.8"	A	A
11.	Hesaraghata	13°	8'	14"	77°	28'	18.1"	P	P
12.	Maddagirihalli	13°	7'	35"	77°	30'	32"	P	P
13.	Aivarakandapura	13°	8'	22.1"	77°	30'	2.3"	P	P
14.	Singanayakanahalli	13°	8'	50.7"	77°	33'	51.9"	P	P
15.	Honnenahalli	13°	19'	21.6"	77°	50'	1.1"	P	P

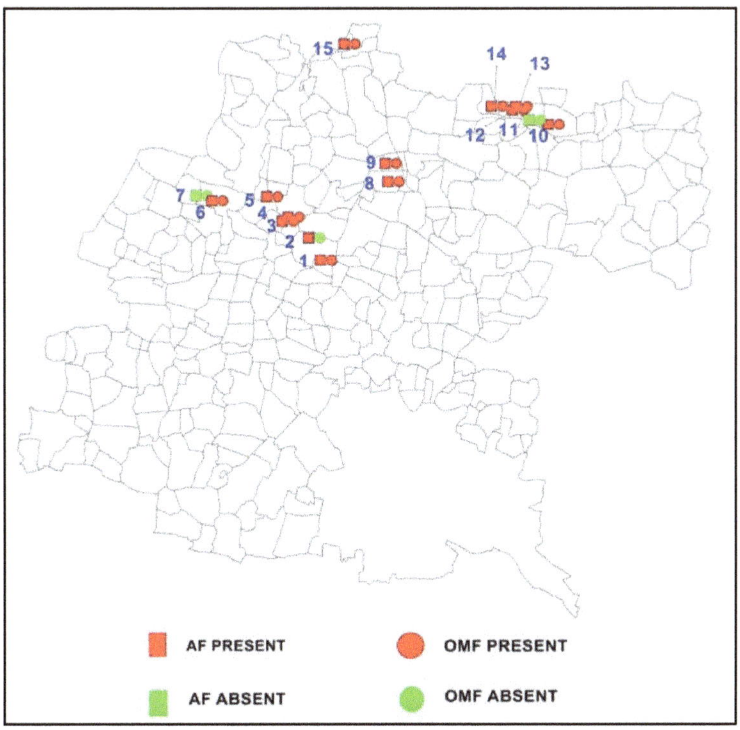

Map 9.1: Geotagging of Villages and Malaco Faunal Occurrence in Bangalore North Taluk

Table 9.2: Geotagging of Villages and Malaco Faunal Occurrence in Bangalore South Taluk

Sl.No	*Village*	*North*			*East*			*AF*	*OMF*
1.	Ramasandra	12°	56'	39.9"	77°	28'	5"	P	P
2.	Dhananayakanahalli	12°	56'	11.8"	77°	27'	53.6"	A	P
3.	Komghatta	12°	56'	11.8"	77°	27'	53.6"	P	A
4.	Bairohalli	12°	54'	52.7"	77°	27'	8.5"	P	P
5.	Hosapalya	12°	54'	32.5"	77°	27'	4.8"	P	P
6.	Malagondanahalli	12°	54'	53.1"	77°	24'	58.6"	A	A
7.	Mudhenapalya	12°	55'	50.4"	77°	25'	6.3"	P	P
8.	Maragondanahalli	12°	56'	2.7"	77°	25'	38.6"	P	P
9.	Gulganjalli	12°	56'	7.7"	77°	25'	24.1"	P	P
10.	Krishnasagar	12°	56'	37.7"	77°	25'	28.4"	P	P
11.	Sulekere	12°	56'	13.6"	77°	26'	55.4"	P	P
12.	Yelacheguppe	12°	57'	12.7"	77°	25'	24.1"	P	P
13.	Bycheguppe	12°	57'	53.9"	77°	23'	17.6"	P	P
14.	Devamachinalli	12°	58'	4.6"	77°	22'	44"	P	P

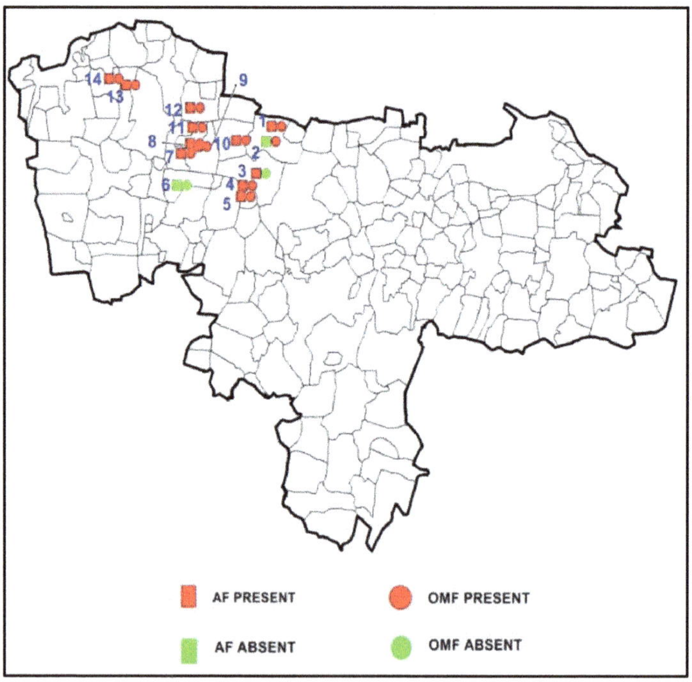

Map 9.2: Geotagging of Villages and Malaco Faunal Occurrence in Bangalore South Taluk

Table 9.3: Geotagging of Villages and Malaco Faunal Occurrence in Bangalore East Taluk

Sl.No	*Village*	*North*			*East*			*AF*	*OMF*
1.	Beleshivalli	13°	2'	49.2"	77°	40'	52.2"	P	P
2.	Rampura	13°	3'	1.7"	77°	41'	14.8"	P	P
3.	Channasandra	13°	2'	26.6"	77°	40'	51.6"	A	P
4.	Hormavu	13°	2'	22.6"	77°	40'	36.5"	A	P
5.	Kalkere	13°	1'	53.5"	77°	40'	26.3"	A	P
6.	Aduru	13°	3'	11.3"	77°	42'	5.4"	P	P
7.	Bidrahalli	13°	3'	44.8"	77°	42'	58.9"	A	A
8.	Hirandahalli	13°	2'	57"	77°	43'	16.3"	P	P
9.	Avalahalli	13°	2'	21.7"	77°	43'	47.4"	P	P
10.	Manduru	13°	5'	24.8"	77°	44'	33.8"	A	A
11.	Kattugollahalli	13°	5'	36.8"	77°	44'	19.8"	A	A
12.	Hancharahalli	13°	5'	36.2"	77°	44'	37.3"	P	A

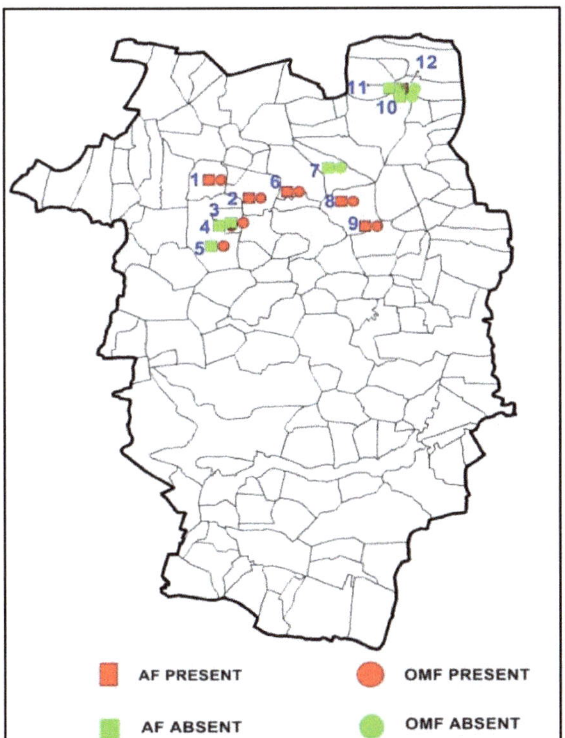

Map 9.3: Geotagging of Villages and Malaco Faunal Occurrence in Bangalore South Taluk

Table 9.4: Geotagging of Villages and Malaco Faunal Pccurrence in Anekal Taluk

Sl.No	*Village*	*North*			*East*			*AF*	*OMF*
1.	Hargadde	12°	15'	13"	77°	39'	30.4"	A	P
2.	Kumbaranahalli	12°	45'	13"	77°	39'	30.4"	A	P
3.	Suragajakkanahalli	12°	44'	13.4"	77°	4'	12.5"	A	P
4.	Honnakalasapura	12°	43'	41.6"	77°	40'	30.6"	P	P
5.	Gowrenahalli	12°	43'	17.3"	77°	40'	59.1"	A	P
6.	Kaval hosahalli	12°	43'	52.2"	77°	42'	27.1"	P	P
7.	Channena agrahara	12°	44'	31.9"	77°	42'	36.7"	P	P
8.	Marasooru	12°	45'	55.8"	77°	42'	16.54"	P	P
9.	Iggaluru	12°	46'	37.8"	77°	42'	10"	P	P
10.	Chandapura	12°	48'	10.4"	77°	42'	24.5"	P	P
11.	Heelalige	12°	48'	29.4"	77°	42'	40"	A	A
12.	Narayanghatta	12°	15'	11.3"	77°	43'	32.1"	A	A
13.	Chokasandra	12°	52'	29.4"	77°	43'	33.2"	A	A
14.	Dommasandra	12°	52'	53"	77°	44'	44.4"	P	P
15.	Ramsagara	12°	48'	36.5"	77°	44'	49.9"	P	P
16.	Mayasandra	12°	45'	5.89"	77°	44'	57"	P	P

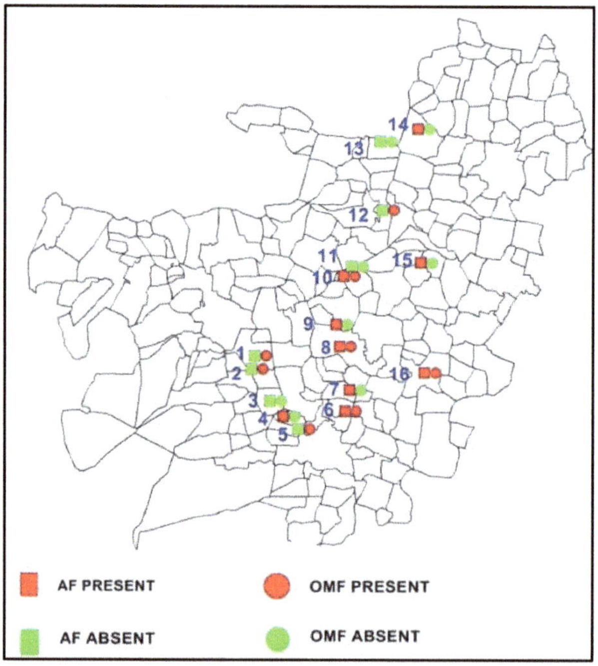

Map 9.4: Geotagging of Villages and Malaco Faunal Occurrence in Bangalore Urban District

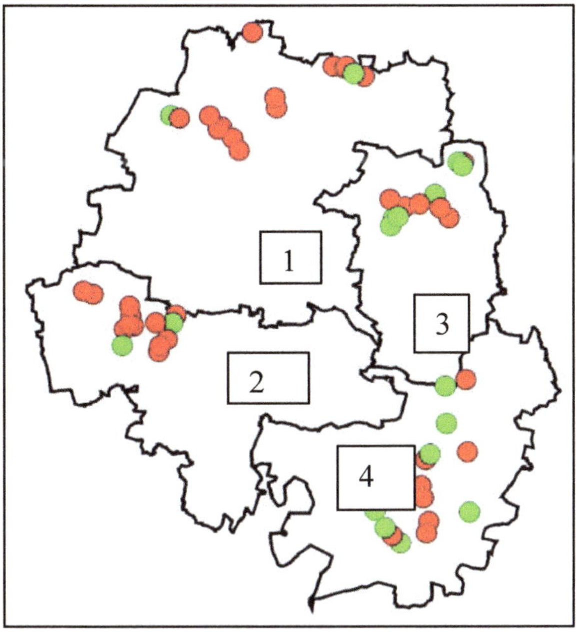

Map 9.5: Geotagging of Villages and *A. fulica* Occurrence in Bangalore Urban District

1: Bangalore North; 2: Bangalore South; 3: Bangalore East; 4: Anekal

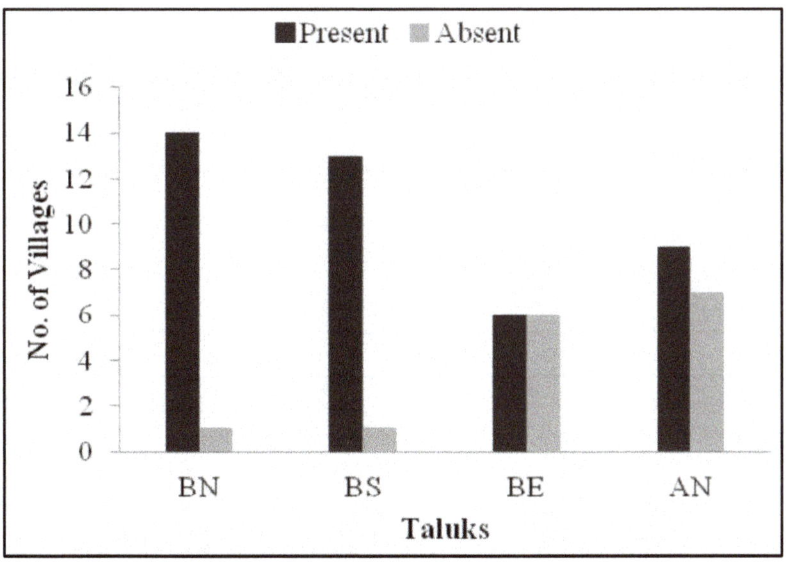

Figure 9.2: Multiple Bar Plots with Occurrence Percentage of *A. fulica* in Bangalore Urban District

BN: Bangalore North; BS: Bangalore South; BE: Bangalore East; AN: Anekal

Biodiversity Indices

There is a rich density of the snails in the nursery studied. An average of 83 snails per meter square during season and 14.06 snails per meter square during off-season in four quadrates of one meter square was recorded. The present observations (Figure 9.2) are comparable to data obtained by Raut and Ghose (1984) in Bangalore with 9 snails/m^2. *Azadirachta indica* recorded a highest of 144 individuals while *Bauhinia* sp recorded the least of 7 individuals in Hebbal nursery. The relative abundance, Simpson's diversity index and Shannon-Weiner index have been calculated as follows (Tables 9.5–9.7).

Table 9.5: Relative Abundance of different Malaco Fauna Occurring in Hebbal Forest Department Nursery

Malacofauna	*Counts (n)*	*Relative (per cent) Abundance*
Achatina fulica	45	17.5
Macrochlamys indica	82	32.0
Filicaulis alte	19	7.4
Mariella dussumeri	7	2.7
Deroceros laevi	12	4.6
Glessula bravis	91	35.5
Total (n)	256	100

Simpson diversity (D) = 0.27 and Shannon-Weiner (H') = 0.63 was recorded in the Hebbal Forest Department nursery.

Table 9.6: Relative Abundance of different Malaco Fauna Occurring in Gandhi Bhavan Forest Department Nursery

Malacofauna	*Counts (n)*	*Relative (per cent) Abundance*
Glessula bravis	109	74.7
Achatina fulica	9	6.2
Mariela dussumeri	21	14.4
Filicaulis alte	7	4.8
Total (n)	146	100

Simpson diversity (D) = 0.58 and Shannon-Weiner (H') index = 0.35 was recorded in the Gandhi Bhavan Forest Department nursery in Bangalore University Campus.

Trees and Stone Pillars

The trees and stone pillars in the Dhanavanthari vana nursery serve as sites of resting and feeding for single or an aggregate of snails. Flora belonging to 10 families was observed in the nursery randomly to locate the snails. Details of the approximate tree height (ft), number of snails at different heights were also recorded (Table 9.8). The tree species included *Santalum album, Pongamia pinnata, Azadirachta indica, Phyllanthus embilica, Adenanthera pavonia, Tamarindus indica, Tectona grandiflora,*

Mangifera indica, Spathodea campanulata, Musa paradisiaca, Aelanthes excels, Hibiscus chinensis, Nyctanthes arbor-tristis, Ficus religiosa and *F.glomerata*. A maximum of five snails of 4-6 cm long were located at 12 feet from ground on *Aelanthes excels.*

Table 9.7: Relative Abundance of different Malaco Fauna Occurring in the Premises of Department of Zoology

Malacofauna	*Counts (n)*	*Relative (per cent) Abundance*
Filicaulis alte	127	70.2
Macrochlamys indica	34	18.8
Cryptozona bistralis	6	3.3
Glessula bravis	1	0.6
Achatina fulica	9	5.0
Mariella dussumeri	4	2.2
Total (n)	181	100

Simpson diversity (D) = 0.53 and Shannon-Weiner (H') index = 0.40 was recorded in the Zoology Department in Bangalore University Campus.

Based on frequency index, of the 54 stone pillars searched for presence of snail signs, it was found that 47 (87 per cent) pillars showed snail faecal matter only and 11 (20 per cent) with snails also resting on them, while 7 (13 per cent) showed no signs of snails.

Table 9.8: Details of Snails Observed on the Flora at the Nursery

Tree Sl.No.	*Scientific Name*	*Family*	*Approximate Tree Height (ft)*	*No. Snails Observed*	*Faecal Matter*	*Height at which snails were Located*
1.	*Adenanthera pavonia*	Papillionaceae	30	2	0	14
				2	0	16
2.	*Nyctanthes arbor-tristis*	Oleaceae	30	2	0	6
				2	0	25
				2	0	20
3.	*Santalum album*	Santalaceae	25	0	1	12
4.	*Adenanthera pavonia*	Papillionaceae	25	1	1	20
5.	*Adenanthera pavonia*	Papillionaceae	30	1	0	25
6.	*Aelanthes exceles*	Symorbaceae	35	5	0	12
7.	Palm	Palmaceae	18	1	0	1
8.	*Pongamia bipinnata*	Papillionaceae	10	0	1	10
9.	Palm	Palmaceae	10	0	1	8
10.	Palm	Palmaceae	18		1	3
11.	Palm	Palmaceae	18	4	0	0
				2	0	8
				0	1	8

Contd...

Table 9.8–*Contd...*

Tree Sl.No.	*Scientific Name*	*Family*	*Approximate Tree Height (ft)*	*No. Snails Observed*	*Faecal Matter*	*Height at which snails were Located*
12.	*Santalum album*	Santalaceae	30	1	0	20
13.	*Santalum album*	Santalaceae	25	0	1	20
14.	*Santalum album*	Santalaceae	25	0	1	6
15.	*Santalum album*	Santalaceae	30	0	1	11
16.	*Tamarindus indica*	Papillionaceae	40	0	1	8
17.	*Tamarindus indica*	Papillionaceae	40	0	1	16
18.	*Pongamia bipinnata*	Papillionaceae	25	0	1	12
19.	*Pongamia bipinnata*	Papillionaceae	25	0	1	12
20.	*Pongamia bipinnata*	Papillionaceae	30	0	1	18
21.	*Pongamia bipinnata*	Papillionaceae	30	0	1	6
22.	*Azadirachta indica*	Meliaceae	40	1	0	7
			40	0	1	15
			25	2	0	18
				1	0	12
23.	*Santalum album*	Santalaceae	12	1	1	10
24.	*Santalum album*	Santalaceae	20	0	1	12
25.	*Santalum album*	Santalaceae	40	0	1	12
26.	*Santalum album*	Santalaceae	20	0	1	10
27.	*Azadirachta indica*	Meliaceae	30	0	1	8
28.	*Azadirachta indica*	Meliaceae	20	1	0	8
29.	*Azadirachta indica*	Meliaceae	30	0	1	6
30.	*Tectona grandifloris*	Teliaceae	30	0	1	12
31.	*Tectona grandifloris*	Teliaceae	40	1	0	10
				2	0	6
32.	*Santalum album*	Santalaceae	10	1	0	10
33.	*Santalum album*	Santalaceae	20	1	0	15
34.	*Santalum album*	Santalaceae	10	0	1	4
35.	*Santalum album*	Santalaceae	7	0	1	3
36.	*Santalum album*	Santalaceae	30	0	1	10
37.	*Santalum album*	Santalaceae	40	2	1	7
38.	*Santalum album*	Santalaceae	20	1	0	5
39.	*Santalum album*	Santalaceae	20	0	1	6
40.	*Santalum album*	Santalaceae	30	0	1	12
41.	*Santalum album*	Santalaceae	30	0	1	0
42.	*Santalum album*	Santalaceae	25	0	1	12

Contd...

Table 9.8–*Contd...*

Tree Sl.No.	*Scientific Name*	*Family*	*Approximate Tree Height (ft)*	*No. Snails Observed*	*Faecal Matter*	*Height at which snails were Located*
43.	*Santalum album*	Santalaceae	30	1	0	4
44.	*Santalum album*	Santalaceae	20	0	1	8
45.	*Santalum album*	Santalaceae	4	1	0	1
46.	*Santalum album*	Santalaceae	6	0	1	4
47.	*Santalum album*	Santalaceae	18	0	1	1
48.	*Azadirachta indica*	Meliaceae	30	0	1	3
49.	*Santalum album*	Santalaceae	30	1	0	14
50.	*Mangifera indica*	Anacardiaceae		0	0	0
51.	*Mangifera indica*	Anacardiaceae		0	0	0
52.	*Musa paradisica*	Musaceae	8	0	1	6
53.	*Rosa chinensis*	Malvaceae	0	0	0	0
54.	*Santalum album*	Santalaceae	0	0	0	0
55.	*Santalum album*	Santalaceae	40	0	1	3
56.	*Ficus religiosa*	Moraceae	30	1	0	8
57.	*Ficus religiosa*	Moraceae	30	0	1	6
58.	*Ficus religiosa*	Moraceae	30	0	1	6
				1	0	6
				1	0	12
59.	*Santalum album*	Santalaceae	15	0	1	6
60.	*Santalum album*	Santalaceae		0	0	0
61.	*Tectona grandifloris*	Teliaceae		0	1	4
62.	*Phyllanthus sp.*	Rutaceae	15	0	0	0
				0	0	0
				0	0	0
63.	*Ficus glomerata*	Moraceae	40	0	1	7

The present study was geared towards determining the distribution pattern and abundance level of the giant African snails occurring in and around Bangalore. *A. fulica* is one of the important intermediate hosts of *Angiostrongylus cantonensis* (Strongylida: Metastrongylidae) and is commonly eaten by humans in some countries (Alicata, 1966; 1991; Prociv *et al.*, 2000). *A. cantonensis*, a rat lungworm, can cause eosinophilic meningitis and angiostrongyliasis in humans following ingestion of contaminated foods or intermediate/paratenic hosts with infective larvae. Thus, in addition to its status as an agricultural and environmental pest, the role of the snail in the epidemiology of the transmission of helminthoses of medical and veterinary interest should be considered. In conclusion monitoring spread of *A. fulica* is necessary considering the fact that invasive or introduced species can prey upon, infect, outcompete or hybridize with native species or even alter habitat at the expense of native species (Simberloff *et al.*, 2005; Meyer *et al.*, 2008).

Acknowledgements

The author is grateful to Dr.M.S.Reddy (Bangalore University) and Dr.Abraham Verghese (IIHR, Bangalore) for their valuable guidance and suggestions. The author is also thankful to Dr. Balakrishna, Dr.Francis and Mr. Rajashekar (KSRSAC) for extending laboratory space and facility for plotting the maps required.

References

Alicata, J., 1966. The presence of *Angiostrongylus cantonensis* in the islands of the Indian Ocean and probable role of the giant African snail, *Achatina fulica*, in the dispersal of the parasite to the Pacific islands. *Canadian J. Zool.*, 44: 1041–1049.

Alicata, J., 1991. The discovery of *Angiostrongylus cantonensis* as a cause of human eosinophilic meningitis. *Parasitol. Today*, 7: 151–153.

Asamoah, S.A., 1999. Ecology and Status of the giant African snails in the Bia Biosphere Reserve in Ghana. Final Report (September, 1999): Ecological Studies on the Giant African Snails, Submitted to: The division of Ecological sciences, UNESCO (MAB) Young Scientist Research Award Scheme, *Paris Cedex 15 France*. 1–31.

Fontanilla, I.K.C., Hudelot, C., Naggs, F. and Wade, C.M., 2007. *Achatina fulica*: its molecular phylogeny and genetic variation in global populations. *Abstracts. World Congress of Malacology*, Antwerp, Belgium. 63.

Lowe, S., Browne, S.M., Boudjrlas, and De Poorter, M., 2000. 100 of the world's worst invasive alien species: A selection from the global invasive species database. The Invasive Species Specialists Group of the Species Survival Commission of the World Conservation Union. 12 pp. First published in Aliens 12, December 2000. Reprinted November 2004. Hollands Printing, Auckland, New Zealand.

Mead, A.R., 1961. The Giant African Snail. *The University of Chicago Press*, 257.

Mead, A.R., 1979. Economic malacology with particular reference to *Achatina fulica*. In: *Pulmonates, 2B*, (Eds.) V. Fretter and J. Peake. Academic Press, London, 150.

Meyer, W.M., Hayes, K.A. and Meyer, A.L., 2008. Giant African snail *Achatina fulica* as a snail predator. *Amer. Malac. Bull.*, 24: 117–119.

Naggs, F., 1997. William Benson and the early study of land snails in British India and Ceylon. *Archives of Natural History*, 24: 37–88.

Prociv, P., Spratt, D.M. and Carlisle, M.S., 2000. Neuro-angiostrongyliasis: unresolved issues. *Int. J. Parasitol.*, 30: 1295–1303.

Raut, S.K. and Barker, G.M., 2002. *Achatina fulica* Bowdich and other Achatinidae as Pests in Tropical Agricultural. In: *Mollusc as Crop Pests*, (Eds.) G.M. Barker. CABI Publishing, Wallingford: 55–114.

Raut, S.K. and Ghose, K.C., 1984. Pestiferous Land Snails of India, *Z.S.I. Tech. Monog.* 11: 151. Zoological Survey of India, Calcutta.

Silvana, C.T., Fabio, A.F., Norma, C.S., Cowie, R.H. and Monica, A F., 2007. Rapid spread of an invasive snail in South America: The giant African snail, *Achatina fulica*, in Brazil. *Biol. Invasions,* 9: 693–702.

Simberloff, D., Parker, I.M. and Windle, P.N., 2005. Introduced species policy, management, and future research needs. *Frontiers in Ecology and Environment,* 3(1): 12–20.

Srivastava, P.D., 1992. *Problem of Land Snail Pests in Agriculture: A Study of the Giant African Snail.* Concept Publishing Company, New Delhi, p. 234.

Steffek, J. and Kusik, P., 2010. Giant African land snail (*Achatina fulica*): new entry on a list of allochtonous mollusc species in Slovakia. *Aliens: The Invasive Species Bulletin,* 29: 55–56.

USDA-APHIS, 2005. New Pest Response Guidelines. Giant African Snails: Snail Pests in the Family Achatinidae. USDA – APHIS – PPQ – Emergency and Domestic Programs – Emergency Planning, Riverdale, Maryland. *http: // www.aphis.usda.gov/import_export/plants/manuals/index.shtml.*

Veeresh, G.K., Rajagopal, D. and Puttarudraiah, M., 1979. First record of African giant snail, *Achatina fulica* (Bowdich) (Mollusca: Gastropoda) as a serious pest of ornamental crops in Bangalore. *Curr. Res.,* 8: 202–204.

Zanol, J., Fernandez, M.A., Oliveira, A.P.M., de Russo, C.A. de M. and Thiengo, S.C., 2010. The exotic invasive snail *Achatina fulica* (Stylommatophora, Mollusca) in the State of Rio de Janeiro (Brazil): Current status. *Biota Neotropica,* 10(3): 447–451.

2013, Animal Diversity, Natural History and Conservation, Vol. 3 *Pages* **135–176**
Editors: **V.K. Gupta and Anil K. Verma**
Published by: **DAYA PUBLISHING HOUSE, NEW DELHI**

Chapter 10

Distribution and Conservation Status of Syngnathids Fish, Family Syngnathidae: A Review

Bishnupriya Saikia and Umesh C. Goswami*
Department of Zoology, Fish Biology and Fishery Sciences, DST-FIST (Govt of India) and UGC-SAP Sponsored Department, Gauhati University, Guwahati – 781 014, Assam, India

ABSTRACT

Syngnathidae is a very interesting family due to its peculiar morphological and structural characters found within each member. This paper attempts to elucidate the geographical distribution and conservation status of all the species of family Syngnathidae. In the present review we have enumerated the findings of earlier works as well as a report or the reference of a freshwater species *Microphis deocata,* a vulnerable species from the north eastern Himalayan rivers. Among all 296 species, 16 species are Least Concern (LC), 41 Data Deficient (DD), 8 Vulnerable (VU), 1 Endangered (EN), 3 Near Threatened (NT) and 1 is Critically Endangered (CR). There is no information on the rest of the species concerned with the conservation status.

Keywords: *Syngnathids fish, Syngnathidae, Geographical distribution of fish, Conservation status.*

* *Corresponding author.* E-mail: ucgoswami@rediffmail.com

Introduction

The family Syngnathidae ('Syn' means fused; 'gnathus' means jaws) includes fishes commonly called pipefishes, seahorses, pipehorses and seadragons which are characterized by the presence of dermal plates covering the body, tufted gills, pore-like gill opening and the absence of pelvic fins and true-jaw teeth (Dawson, 1985). The family consists of about 215 species in 52 genera.

Fossil records suggest that the oldest syngnathoid fossils date from the early Eocene [48–50 million years before present (B.P.)], and the majority of these fossil taxa have been collected from the rich fossil beds of the Monte Bolca formation in northern Italy (Blot, 1980). Six fossil syngnathoids have been collected from Eocene deposits in this region, representing five distinct taxonomic lineages. Species of the now extinct Ramphosidae, a family which has been suggested to be a sister group to the extant Pegasidae (Pietsch, 1995), have been collected from Eocene sites in Monte Bolca (Blot, 1980) and from a second locality in Jutland, Denmark (Nielsen, 1960). Three additional Monte Bolca fossils are believed to represent morphologically-distinct syngnathoid lineages, and *Solenorhynchus elegans* and *Calamostoma breviculum* Agassiz 1833 are thought to be intermediate forms between the Solenostomidae and the Syngnathidae (Orr, 1995). The oldest syngnathid fossil is believed to be *Prosolenostomus lessenii* (Blot, 1980), a species that has been placed in the stem lineage of the Syngnathidae on the basis of its absence of dorsal and pelvic fins (Orr, 1995). The age of *Prosolenostomus* indicates that the family Syngnathidae is at least 50 million years old.

Pipefishes are the most abundant and diverse group of this family. They derive the name from their peculiarly long and slender angular body. They are mostly stick-like with head in line with the body. Pipefishes are cosmopolitan in distribution between 71°N and 56°S latitudes (Dawson, 1985). Body is semi-flexible due to the presence of bony plates and rings. Jaws are fused; snout is long and tubular with a small mouth at the tip. Fins are rayed and variably present. Ventral fins are absent and the anal fin is small or degenerated. Some species have a moderately sized caudal fin, and if caudal fin is absent the tail is prehensile.

The seahorse is the another popular member of the Syngnathidae family owing to its peculiar form. The head and neck of a seahorse resemble those of a horse and the body that of a caterpillar. Besides, it has a tail like that of a monkey, a brood pouch like that of a kangaroo and independently moving eyes like those of a chameleon. It has a prehensile tail by which it anchors itself on to a holdfast. It is exclusively marine and occurs in shallow coastal habitats amidst seagrasses, seaweeds, corals and mangroves (Foster and Vincent, 2004). It is master of camouflage, which facilitates prey capture and avoidance from predators. Most seahorse species are monogamous during a breeding season (Vincent, 1995; Vincent and Sadler, 1995) but polygamy has also been reported in *Hippocampus abdominalis* (Woods, 2000).

Materials and Methods

Available literature of pipefishes was studied for geographical distribution. For conservation status, IUCN Red List (IUCN, 2010) has been considered. Further, the information about the fishes were confirmed from the database of FishBase.

Results

The family Syngnathidae consists of 4 subfamilies–Hippocampinae, Solegnathinae, Syngnathinae and Doryhamphinae, each with various genera. There are 296 species of which only one species, *Syngnathus watermeyeri* is Critically Endangered (CR). In north-eastern Himalayan region, only one species of freshwater pipefish, *Microphis deocata* is found and its conservation status is Near Threatened (NT).

Distribution and Conservation Status of all the Species of Family Syngnathidae

1. *Acentroneura dendritica* (pipehorse)

Distribution: Western Atlantic: New Brunswick, Canada and Bermuda to Brazil (Robins and Ray, 1986)

IUCN Red List Status: Not evaluated

2. *Acentronura gracilissima* (Bastard Seahorse)

Distribution: Western Pacific: Restricted to areas around Japan and Indochina but there are records from the central and western Indian ocean (Dawson, 1985)

IUCN Red List Status: Not Evaluated

3. *Acentronura tentaculata* (Short pouch pigmy pipehorse)

Distribution: Indo-West Pacific: northern Red Sea and Mozambique to Torres and Queensland, Australia (Dawson, 1986)

IUCN Red List Status: Not Evaluated

4. *Anarchopterus criniger* (Fringed pipefish)

Distribution: Western Atlantic: North Carolina, USA and northern Gulf of Mexico to Brazil. Absent from west of Mississippi Delta and in Caribbean (Robins *et al.*, 1991)

IUCN Red List Status: Not Evaluated

5. *Anarchopterus tectus* (Insular pipefish)

Distribution: Western Atlantic: Florida, USA and Bahamas to Argentina. Reported in northeastern Brazil (Robins *et al.*, 1991)

IUCN Red List Status: Not Evaluated

6. *Apterygocampus epinnulatus* (Minute pipefish)

Distribution: Western Central Pacific: Indonesia, the Philippines and Papua New Guinea (Dawson, 1985)

IUCN Red List Status: Not Evaluated

7. *Bhanotia fasciolata* (Berbed pipefish)

Distribution: Eastern Indian Ocean and Western Pacific: Andaman Sea to Vanuatu (Dawson, 1985)

IUCN Red List Status: Not Evaluated

8. *Bhanotia nuda* (Naked pipefish)

Distribution: Western Central Pacific: Palau and southeastern Papua New Guinea (Dawson, 1985)

IUCN Red List Status: Least Concern (LC)

9. *Bhanotia Pauciradiata* (Few-rayed pipefish)

Distribution: Western Central Pacific: Indonesia (Allen and Kuiter, 1995)

IUCN Red List Status: Not Evaluated

10. *Bryx analicarens* (Pink pipefish)

Distribution: Western Indian Ocean: Hermits Island (Eritrea), ZanZibar (Tanzania), Seychelles, Makram coast, Kuwait to Abu Dhabi (United Arab Emirates) and Gulf of Oman (Dawson, 1985)

IUCN Red List Status: Not Evaluated

11. *Bryx dunckeri* (Pugnose pipefish)

Distribution: Western Atlantic: Canada to North Carolina, USA and Bermuda to the West Indies; extending to northern South America (Robins and Ray, 1986). Lesser Antilles (Smith, 1997)

IUCN Red List Status: Not Evaluated

12. *Bryx randalli* (Ocellated pipefish)

Distribution: Western Atlantic: Beize, Isla de Provi dencia, Lesser Antilles (Smith, 1997)

IUCN Red List Status: Not Evaluated

13. *Bryx veleronis* (Offshore pipefish)

Distribution: Eastern Pacific: Revillagigedo Islands to the Galapagos Islands, including the offshore islands of Costa Rica and Panama (Dawson, 1985)

IUCN Red List Status: Data Deficient (DD)

14. *Bulbonaricus brauni* (Pugheaded pipefish)

Distribution: Eastern Indian Ocean: off Sumatra, Indonesia and from Western Australia. Western Central Pacific: Palau (Dawson, 1985)

IUCN Red List Status: Not Evaluated

15. *Bulbonaricus brucei*

Distribution: Western Indian Ocean: Known only from Pangani, Miziwi Islands, Tanzania (Dawson, 1985)

IUCN Red List Status: Not Evaluated

16. *Bulbonaricus davaoensis* (Davao Pughead pipefish)

Distribution: Indo-West Pacific: Kenya (near Mombasa) and from the Philippines south eastward to Fiji (Dawson, 1985)

IUCN Red List Status: Not Evaluated

17. *Campichthys galei* (Gales pipefish)

Distribution: Eastern Indian Ocean: endemic to Australia (Dawson, 1985)

IUCN Red List Status: Not Evaluated

18. *Campichthys nanus*

Distribution: Western Indian Ocean: Known only from the vicinity of Pinda, Mozambique (Dawson, 1985)

IUCN Red List Status: Not Evaluated

19. *Campichthys tricarinatus* (Three-keel pipefish)

Distribution: Western Central Pacific: Northern Australia including Baleine Bank and Montebello Island in western Australia (Dawson, 1985)

IUCN Red List Status: Not Evaluated

20. *Campichthys tryoni* (Tryon's pipefish)

Distribution: Western Central Pacific: endemic to Australia(Dawson, 1985)

IUCN Red List Status: Not Evaluated

21. *Choeroichthys brachysoma* (Short-bodied pipefish)

Distribution: Indo-Pacific: Red Sea and East Africa to the Society Islands, north to the Philippines and Guam, south to northern Australia (Dawson, 1985)

IUCN Red List Status: Not Evaluated

22. *Choeroichthys cinctus* (Barred shortbody pipefish)

Distribution: Western Pacific: Indonesia to Samoa (Dawson, 1985)

IUCN Red List Status: Not Evaluated

23. *Choeroichthys latispinosus* (Muiron pipefish)

Distribution: Eastern Indian Ocean: known only from South Murion Island, Western Australia (Dawson, 1985)

IUCN Red List Status: Not Evaluated

24. *Choeroichthys sculptus* (Sculptured pipefish)

Distribution: Indo-Pacific: East Africa to the Line, Society (Dawson, 1986), and Gambier islands, north to southern Japan, south to Tonga.

IUCN Red List Status: Not Evaluated

25. *Choeroichthys smithi* (Shortfin pipefish)

Distribution: Western Indian Ocean: Durban, South Africa (Heemstra, 1995), Relagoa Bay and Pinda in Mozambique. Records from Mozambique Islands and Zanzibar, Tanzania are uncertain.

IUCN Red List Status: Not Evaluated

26. ***Choeroichthys suillus*** **(Pig-snouted pipefish)**

Distribution: Western Central Pacific: endemic to northern Australia, from Western Australia to Queensland and McCluer Island, Northern Territory (Dawson, 1985).

IUCN Red List Status: Not Evaluated

27. ***Corythoichthys amplexus*** **(Brown-banded pipefish)**

Distribution: Indo-West Pacific: East Africa to Samoa, north to Ryukyu Islands, south to thr Great Barrier Reef. Record from the Mascarenes (Kuiter and Debelius, 1994) is apparently erroneous (Fricke, 1999).

IUCN Red List Status: Least Concern (LC)

28. ***Corythoichthys benedetto***

Distribution: Western-Pacific: from Bali and southern Sulawesi, Indonesia westward to northern Papua, New Guinea (Allen and Erdmann, 2008).

IUCN Red List Status: Not Evaluted

29. ***Corythoichthys flavofasciatus*** **(Network pipefish)**

Distribution: Indo-Pacific: Red Sea and East Africa (Fricke, 1999) to the Tuamoto Islands, north to Ryukyu Islands, south to northern Australia (Fricke, 1999) and the Austral Islands.

IUCN Red List Status: Not Evaluated

30. ***Corythoichthys haematopterus*** **(Massmate pipefish)**

Distribution: Indo- Pacific: East Africa to Vanuatu, north to southern Japan (Dawson, 1985)

IUCN Red List Status: Not Evaluated

31. ***Corythoichthys insularis***

Distribution: Western Indian Ocean: Amirante and Comoros Islands. Also reported from the Maldives (Randall and Anderson, 1993)

IUCN Red List Status: Not Evaluated

32. ***Corythoichthys intestinalis*** **(Scribbled pipefish)**

Distribution: Western Central Pacific: Borneo to Samoa, north to Marshall and Mariana Islands, south to Rowley Shoals and New Caledonia; throughout Micronesia (Dawson, 1985)

IUCN Red List Status: Not Evaluated

33. ***Corythoichthys nigripectus*** **(Black-breasted pipefish)**

Distribution: Indo-Pacific: northern Red Sea and from Ulithi Atoll (Caroline Islands) eastward to the Society Islands (Myers, 1991)

IUCN Red List Status: Not Evaluated

34. *Corythoichthys ocellatus* (Ocellated pipefish)

Distribution: Western Central Pacific: Celebes and Philippines to Palau, the Solomon Islands, and Australia (Myers, 1991)

IUCN Red List Status: Not evaluated

35. *Corythoichthys paxtoni* (Paxtoni's pipefish)

Distribution: Western Pacific: Australia. More recently reported from the Chesterfield Islands (Kulbicki *et al.,* 1994)

IUCN Red List Status: Not Evaluated

36. *Corythoichthys polynotatus* (Many-spotted pipefish)

Distribution: Western Central Pacific: known only from Palau and the Philippines. Recorde from Indonesia (Allen and Adrim, 2003)

IUCN Red List Status: Not Evaluated

37. *Corythoichthys schultzi* (Schultzi's pipefish)

Distribution: Indo-Pacific: Red Sea and East Africa to Tonga, south to northern Australia and New Caledonia (Myers, 1991)

IUCN Red List Status: Not Evaluated

38. *Cosmocampus albirostris* (Whitenose pipefish)

Distribution: Western Atlantic: northeastern Florida and Northern Gulf of Maxico in USA, and Bahamas to Curacao. Reported from south-eastern Brazil (Viana and Bockmann, 1995)

IUCN Red List Status: Not Evaluated

39. *Cosmocampus arctus arctus* (Snubnose pipefish)

Distribution: Eastern Pacific: Tomales Bay, California to Mazatlan, Mexico; throughout the Gulf of California (Dawson, 1985)

IUCN Red List Status: Not Evaluated

40. *Cosmocampus arctus coccineus*

Distribution: Eastern Pacific: Galapagos Islands and from Bahia deBanderas in Mexico, Costa Rica, Panama and southward to Bahia Independencia, Peru (Dawson, 1985)

IUCN Red List Status: Not Evaluated

41. *Cosmocampus balli* (Balli's pipefish)

Distribution: Eastern Central Pacific: Hawaii (Dawson, 1985)

IUCN Red List Status: Not Evaluated

42. *Cosmocampus banner* (Roughridge pipefish)

Distribution: Indo-West Pacific: Gulf of Aqaba and East Africa to Fiji, north to the Ryukyu and Marshall islands. Recently recorded from Tonga (Randall *et al.,* 2003)

IUCN Red List Status: Not Evaluated

43. *Cosmocampus brachycephalus* (Crested pipefish)

Distribution: Western Atlantic: southern Florida, USA and Bahamas to northern South America (Robins and Ray, 1986)

IUCN Red List Status: Not Evaluated

44. *Cosmocampus darrosanus* (D'Arros pipefish)

Distribution: Indo-West Pacific: Mozambique to Queensland, Australia; Guam in Micronesia (Myers, 1991)

IUCN Red List Status: Not Evaluated

45. *Cosmocampus elucens* (Shortfin pipefish)

Distribution: Western Atlantic: North Carolina (USA), Bermuda and the northern Gulf of Mexico to the Leser Antilles (Bohlke and Chaplin, 1993) and Suriname

IUCN Red List Status: Least Concern (LC)

46. *Cosmocampus heraldi*

Distribution: Southeast Pacific: known only from Isla San Felix and Islas Juan Fernandez, Chile (Dawson, 1985)

IUCN Red List Status: Not Evaluated

47. *Cosmocampus hildebrandi* (Dwarf pipefish)

Distribution: Western Atlantic: southern and western Florida in USA (Robins *et al.,* 1991)

IUCN Red List Status: Not Evaluated

48. *Cosmocampus howenssis* (Lord Howe pipefish)

Distribution: Indo-Pacific: mostly south of 24°S, and from Jervis Bay, New South Wales, Australia to Easter Island (Dawson, 1985)

IUCN Red List Status: Not Evaluated

49. *Cosmocampus investigatoris*

Distribution: Indo-West Pacific: Kuwait to Ko Samui Islnd, Gulf of Thailand (Dawson, 1985)

IUCN Red List Status: Not Evaluated

50. *Cosmocampus maxweberi* (Maxweber's pipefish)

Distribution: Indo-Pacific: Gulf of Aqaba; then from off Sumatra, Indonesia to the Marshall Islands and Samoa (Myers, 1991)

IUCN Red List Status: Not Evaluated

51. *Cosmocampus profundus* (Deepwater pipefish)

Distribution: Western Atlantic: known only from eastern Florida in USA and western Caribbean (Robins and Ray, 1986)

IUCN Red List Status: Not Evaluated

52. *Cosmocampus retropinnis*

Distribution: Eastern Atlantic: off south Morocco to Gambia (24°35t30t‹...N, 13°31t N) (Dawson, 1990)

IUCN Red List Status: Not Evaluated

53. *Doryichthys boaja* (Long-snouted pipefish)

Distribution: Asia: Indonesia, Malaysia, Thailand, Cambodia and Viet Nam. Records from China and Taiwan need confirmation and verification by additional collections (Saikia and Goswami, 2012).

IUCN Red List Status: Not Evaluated

54. *Doryichthys contiguous*

Distribution: Asia: Mekong basin in Laos, Thailand and Cambodia (Saikia and Goswami, 2012)

IUCN Red List Status: Not Evaluated

55. *Doryichthys deokhatoides*

Distribution: Asia: Mekong basin in Laos and Thailand, Chao Phraya basin, Malay Peninsula, Sumatra and Borneo (Saikia and Goswami, 2012)

IUCN Red List Status: Not Evaluated

56. *Doryichthys heterosoma*

Distribution: Asia: known only from the Sambas River (West-Kalimantan) and Natuna Islands in Indonesia (Saikia and Goswami, 2012)

IUCN Red List Status: Not Evaluated

57. *Doryichthys martensii*

Distribution: Asia: Indonesia, Malaysia, Brunei, Darussalam and Thailand (Saikia and Goswami, 2012)

IUCN Red List Status: Not Evaluated

58. *Doryrhamphus aurolineatus*

Distribution: Western Indian Ocean: Arabian Sea, Oman, Masirah Island (Randall and Earle, 1994)

IUCN Red List Status: Not Evaluated

59. *Doryrhamphus baldwini* (Redstripe pipefish)

Distribution: Eastern Central Pacific: endemic to Hawaii (Dawson, 1985)

IUCN Red List Status: Not Evaluated

60. *Doryrhamphus bicarinatus* (Narrowstripe pipefish)

Distribution: Western Indian Ocean: Sodwana Bay, South Africa and Inhaca Island, Mozambique. More recently recorded from the Maldives (Anderson *et al.*, 1998)

IUCN Red List Status: Not Evaluated

61. *Doryrhamphus chapmani* (Glowtail pipefish)

Distribution: Southwest Pacific: known only from the vicinity of Noumea, New Caledonia (Dawson, 1985)

IUCN Red List Status: Not Evaluated

62. *Doryrhamphus dactyliophorus* (Ringed pipefish)

Distribution: Indo-Pacific: Red Sea and East Africa to Samoa, north to Japan (Masuda *et al.*, 1984), south to Australia

IUCN Red List Status: Data Deficient (DD)

63. *Doryrhamphus excisus abbreviates*

Distribution: Western Indian ocean: endemic to the Red Sea from the Guif of Suez south to Massawa and ajiddah on the east coast (Dawson, 1985)

IUCN Red List Status: Not Evaluated

64. *Doryrhamphus excisus excisus* (Bluestripe pipefish)

Distribution: Indo-Pacific and Eastern Pacific: Persian Gulf and East Africa to the west coast of the Americas (Dawson, 1985)

IUCN Red List Status: Not Evaluated

65. *Doryrhamphus excises paulus*

Distribution: Eastern Central Pacific: Revillagigedo Islands (Dawson, 1985)

IUCN Red List Status: Not Evaluated

66. *Doryrhamphus janssi* (Janss' pipefish)

Distribution: Western Central Pacific: Gulf of Thailand to the Solomon Islands, north to the Philippines, south to Queensland, Belau and Trunk in Micronesia (Myers, 1991)

IUCN Red List Status: Least Concern (LC)

67. *Doryrhamphus japonicus* (Honshu pipefish)

Distribution: Northwest Pacific: Japan and Taiwan (Masuda *et al.*, 1984)

IUCN Red List Status: Not Evaluated

68. *Doryrhamphus multiannulatus* (Many-banded pipefish)

Distribution: Western Indian Ocean: Red Sea and Sodwana Bay, South Africa to the Chagos islands and Maldives (Dawson, 1985)

IUCN Red List Status: Not Evaluated

69. *Doryrhamphus negrosensis malus* (Masthead Island pipefish)

Distribution: Western Pacific: endemic to Australia from Mackay Reef to Masthead Islands on the Great Barrier Reef, Queensland (Dawson, 1985)

IUCN Red List Status: Not Evaluated

70. *Doryrhamphus negrosensis negrosensis* (Negros pipefish)

Distribution: Western Pacific: Borneo to Vanuatu, north to the Yaeyama Islands, south to Rowley Shoals, Belau and Yap in Micronesia (Myers, 1991)

IUCN Red List Status: Not Evaluated

71. *Doryrhamphus pessuliferus* (Yellowbanded pipefish)

Distribution: Western Central Pacific: known off Sulade Islands in the Sulu Archipelago (Ref.5316). Eastern Indian Ocean: Australia (Hoese *et al.*, 2006)

IUCN Red List Status: Not Evaluated

72. *Dunckerocampus boylei* (Broad-banded pipefish)

Distribution: Indian Ocean: Red Sea, Mauritius and Tulamben, north coast of Bali, Indonesia. Expected to be wide ranging in the Indian Ocean and adjacent waters (Kuiter, 1998)

IUCN Red List Status: Not Evaluated

73. *Dunckerocampus naja*

Distribution: Central Pacific: Fiji (Allen and Kuiter, 2004)

IUCN Red List Status: Not Evaluated

74. *Enneacampus ansorgii*

Distribution: Africa: Gambia river drainage to the Cuanza river, Angola ([1]Dawson, 1986)

IUCN Red List Status: Least Concern (LC)

75. *Enneacampus kaupi*

Distribution: Africa: Liberia ([1]Dawson, 1986) to the Congo River estuary, Democratic Republic of the Congo (Anonymous, 1997). Records of a specimen from marine waters off Punta Durnford are questionable.

IUCN Red List Status: Least Concern (LC)

76. *Entelurus aequoreus* (Snake pipefish)

Distribution: Eastern Atlantic: Iceland and Norway to Azores and also enters Baltic Sea ([2]Dawson, 1986)

IUCN Red List Status: Not Evaluated

77. *Festucalex cinctus* (Girdled pipefish)

Distribution: West Pacific: tropical and temperate endemic of Australia (Dawson, 1985)

IUCN Red List Status: Not Evaluated

78. *Festucalex erythraeus*

Distribution: Indo-Pacific: Mozambique to Hawaii (Dawson, 1985)

IUCN Red List Status: Not Evaluated

79. *Festucalex gibbsi* (Gibbs' pipefish)

Distribution: Western Pacific: Irian Jaya and the Great Barrier Reef. Recently reported from the Chesterfield Islands (Kulbicki *et al.*, 1994)

IUCN Red List Status: Not Evaluated

80. *Festucalex kulbickii* (Kulbicki's pipeish)

Distribution: Western Pacific: New Caledonia (Fricke, 2004)

IUCN Red List Status: Not Evaluated

81. *Festucalex prolixus*

Distribution: Western Central Pacific: Sulu and Celebes sea and from off northwestern Irian Jaya (Dawson, 1985)

IUCN Red List Status: Not Evaluated

82. *Festucalex scalaris* (Ladder pipefish)

Distribution: Eastern Indian Ocean: endemic to Australia (Dawson, 1985)

IUCN Red List Status: Not Evaluated

83. *Festucalex wassi* (Wass' pipefish)

Distribution: Western Central Pacific: Fiji and American Samoa (Dawson, 1985)

IUCN Red List Status: Not Evaluated

84. *Filicampus tigris* (Tiger pipefish)

Distribution: Eastern Indian Ocean and Western Pacific: Australia (Western Australia, South Australia, Queensland and New South Wales) (Dawson, 1985)

IUCN Red List Status: Not Evaluated

85. *Halicampus boothae* (Booth's pipefish)

Distribution: Western Indian Ocean: Comoro Islands. Western Pacific: Honshu in Japan, Lord Howe and Norfolk Islands (Francis, 1993), Chesterfield Islands (Kulbicki *et al.*, 1994) and Fiji. Recently recorded from Tonga (Randall *et al.*, 2003)

IUCN Red List Status: Not Evaluated

86. *Halicampus brocki* (Brock's pipefish)

Distribution: Western Pacific: Philippines and northwestern Australia to Queensland, north to the Yaeyama Islands; Marshall and Mariana Islands in Micronesia (Myers, 1991)

IUCN Red List Status: Not Evaluated

87. *Halicampus dunckeri* (Duncker's pipefish)

Distribution: Indo-West Pacific: Red Sea to the Solomon Islands, north to southern Japan, south to the southern Great Barrier Reef. Belau [=Palau] and Pohnpei in Micronesia (Myers, 1991)

IUCN Red List Status: Not Evaluated

88. *Halicampus edmondsoni* (Edmondson's pipefish)

Distribution: Eastern Central Pacific: endemic to the Hawaiian Islands (Dawson, 1985)

IUCN Red List Status: Not Evaluated

89. *Halicampus grayi* (Gray's pipefish)

Distribution: Indo-West Pacific: Gulf of Aden, off Sri Lanka, and from the Gulf of Thailand and Australia to Kyushu Islands, Japan (Dawson, 1985)

IUCN Red List Status: Not Evaluated

90. *Halicampus macrorhynchus* (Ornate pipefish)

Distribution: Indo-West Pacific: northern Red Sea (Gulf of Suez and Aqaba), Indonesia off Sumbawa Islands, Queensland in Australia, Port Moresby in Papua New Guinea, New Britain Islands and Guadalcanal Islands in the Solomon Islands (Dawson, 1985)

IUCN Red List Status: Not Evaluated

91. *Halicampus mrquesensis*

Distribution: Eastern Central Pacific: Marquesan Islands (Dawson, 1985)

IUCN Red List Status: Not Evaluated

92. *Halicampus mataafae* (Samoan pipefish)

Distribution: Indo-Pacific: Red Sea and East Africa to Samoa (Dawson, 1986), south to Queensland, Australia ([3]Dawson, 1986)

IUCN Red List Status: Not Evaluated

93. *Halicampus nitidus* (Glittering pipefish)

Distribution: Western Pacific: Viet Nam to Fiji, north to Ryukyu Islands, south to Rowley Shoals and New Caledonia; Belau in Micronesia (Myers, 1991)

IUCN Red List Status: Not Evaluated

94. *Halicampus punctatus* (Starry pipefish)

Distribution: Northwest Pacific: endemic to Japan (Dawson, 1985)

IUCN Red List Status: Not Evaluated

95. *Halicampus spinirostris* (Spinysnout pipefish)

Distribution: Indo-Pacific: Sri Lanka, Western Australia and Queensland (Australia), and Tutuila Island, American Samoa (Dawson, 1985)

IUCN Red List Status: Not Evaluated

96. *Halicampus zavorensis* (Zavora pipefish)

Distribution: Western Indian Ocean: known only from one specimen from Zavora, Mozambique and two from Sur, Oman (Dawson, 1985)

IUCN Red List Status: Not Evaluated

97. *Haliichthys taeniophorus* (Ribboned pipefish)

Distribution: Indo-Pacific: Indonesian waters around Irian Jaya (northern coast and Tanamerah Bay) and Australian waters from Shark Bay, western Australia to the Torres Straits in northern Queensland (Dawson, 1985)

IUCN Red List Status: Not Evaluated

***98. Heraldia nocturna* (Upside-down pipefish)**

Distribution: Eastern Indian Ocean: southern Australia, from Geographe Bay, Western Australia to Terrigal, New South Wales (Dawson, 1985)

IUCN Red List Status: Not Evaluated

99. *Hippichthys albomaculosus*

Distribution: Oceania: Vanua Levu Island, Fiji (Saikia and Goswami, 2012)

IUCN Red List Status: Not Evaluated

100. *Hippichthys cyanospilos* (Blue-spotted pipefish)

Distribution: Indo-Pacific: northern Red Sea and East Africa (Dawson, 1985)

IUCN Red List Status: Not Evaluated

101. *Hippichthys hepatogonas* (Belly pipefish)

Distribution: Africa, Asia and Oceania: South Africa and Kenya to the Solomon Islands and in some Philippine lakes (Dawson, 1985)

IUCN Red List Status: Least Concern (LC)

102. *Hippichthys parvicarinatus* (Short-keel pipefish)

Distribution: Oceania: endemic to Australia (Dawson, 1985)

IUCN Red List Status: Not Evaluated

103. *Hippichthys penicillus* (Beady pipefish)

Distribution: Indo-West Pacific: western Persian Gulf (Kuwait, Saudi Arabia), north central Indian Ocean and eastward to Japan and Queensland, Australia (Dawson, 1985)

IUCN Red List Status: Not Evaluated

104. *Hippichthys spicifer* (Bellybarred pipefish)

Distribution: Indo-Pacific: Red Sea and East Africa to Sri Lanka and Samoa ([1]Dawson, 1986). Recorded upstream in freshwater of the Mekong, but this record may represent another species of this genus or another genus among the 15 genera of pipefishes expected in Mekong estuary and plume (Rainboth, 1996)

IUCN Red List Status: Least Concern (LC)

105. *Hippocampus abdominalis* (Big-belly seahorse)

Distribution: Southwest Pacific: Australia and New Zealand. Occurrence in Thailand and Philippines (Jeyaseelan, 1998) needs verification.

IUCN Red List Status: Data Deficient (DD)

106. *Hippocampus alatus* (Winged seahorse)

Distribution: Western Central Pacific: northern Australia (including Dampier Archipelago, Western Australia) and southeastern Papua New Guinea (Kuiter, 2001)

IUCN Red List Status: Not Evaluated

107. *Hippocampus algiricus* (West African seahorse)

Distribution: Eastern Central Atlantic: Senegal to Angola (Lourie *et al.*, 1999)

IUCN Red List Status: Data Deficient (DD)

108. *Hippocampus angustus* (Narrow- bellied seahorse)

Distribution: Eastern Indian Ocean: northwestern Australia only, from Torres Strait to North West Cape. Records from Queensland need verification (Lourie *et al.*, 1999)

IUCN Red List Status: Data Deficient (DD)

109. *Hippocampus barbouri* (Barbour,s seahorse)

Distribution: Western Central Pacific: southern Sulu Sea (Lourie *et al.*, 1999)

IUCN Red List Status: Vulnerable (VU) (A4cd)

110. *Hippocampus bargibanti* (Pygmy seahorse)

Distribution: Western Central Pacific: Indonesia, Papua New Guinea, Queensland and New Caledonia (Lourie *et al.*, 1999)

IUCN Red List Status: Data Deficient (DD)

111. *Hippocampus biocellatus* (False-eyed seahorse)

Distribution: Eastern Indian Ocean: known only from the Shark Bay region of Western Australia (Kuiter, 2011)

IUCN Red List Status: Not Evaluated

112. *Hippocampus barboniensis* (Reunion seahorse)

Distribution: Western Indian Ocean: Mauritius, Reunion and east coast of southern Africa (Lourie *et al.*, 1999)

IUCN Red List Status: Data Deficient (DD)

113. *Hippocampus breviceps* (Short-head seahorse)

Distribution: Indo-Pacific: endemic to southwest Australia (Lourie, 1999)

IUCN Red List Status: Data Deficient (DD)

114. *Hippocampus camelopardalis* (Gireffe seahorse)

Distribution: Western Indian Ocean: Zanzibar (Tanzania), Inhaca, Delagoa Bay (Mozambique) and False Bay (South Africa) (Lourie, 1999)

IUCN Red List Status: Data Deficient (DD)

115. *Hippocampus capensis* (Knysna seahorse)

Distribution: Southeast Atlantic: endemic to South Arica (Lourie, 1999)

IUCN Red List Status: Endangered (EN) (B1+2c+3d)

116. *Hippocampus colemani*

Distribution: Southwestern Pacific: Lord Howe Island (Eschmeyer, 2004)

IUCN Red List Status: Not Evaluated

117. *Hippocampus comes* (Tiger tail seahorse)

Distribution: Western Central Pacific: Malaysia, Singapore, Viet Nam and Philippines (lourie *et al.*, 1999)

IUCN Red List Status: Vulnerable (VU) (A2cd)

118. *Hippocampus coronatus*

Distribution: Northwest Pacific: Endemic to Japan (Lourie *et al.*, 1999)

IUCN Red List Status: Data Deficient (DD)

119. *Hippocampus curvicuspis* (New Caledonian thorny seahorse)

Distribution: Western Pacific: New Caledonia (Fricke, 2004)

IUCN Red List Status: Not Evaluated

120. *Hippocampus debelius* (Softcoral seahorse)

Distribution: Red Sea: Egypt, Hurghada, Erg Camel. Apparently endemic in the Red Sea (Gommon and Kuiter, 2009)

IUCN Red List Status: Not Evaluated

121. *Hippocampus denise* (Denise's pygmy seahorse)

Distribution: Western Pacific: Apparently more widespread than Indonesia, Vanuata and Palau. Observed from Malaysia, Solomon Islands and Pohn Pei, Micronesia (Lourie and Randall, 2003)

IUCN Red List Status: Data Deficient (DD)

***122. Hippocampus erectus* (Lined seahorse)**

Distribution: Western Atlantic: Nova Scotia, Canada and Venezuela. A southern form that may be prove to be a separate species is known from Rio de Jeneiro, Brazil and questionably from Suriname (Lourie *et al.*, 1999)

IUCN Red List Status: Vulnerable (VU) (A4cd)

123. *Hippocampus fisheri* (Fisher's seahorse)

Distribution: Western Pacific: Hawaii and questionably from Lord Howe Island and New Caledonia (Lourie *et al.*, 1999)

IUCN Red List Status: Data Deficient (DD)

124. *Hippocampus fuscus* (Sea Pony)

Distribution: Indian Ocean: Red Sea, Saudi Arabia, Djibouti and Sri Lanka. Records from South Africa, Madagascar, Mauritius and Reunion are questionable (Lourie *et al.*, 1999)

IUCN Red List Status: Data Deficient (DD)

125. *Hippocampus grandiceps* (Big-head seahorse)

Distribution: Western Central Pacific: Gulf of Carpentaria, Australia (Kuiter, 2001)

IUCN Red List Status: Not Evaluated

126. *Hippocampus guttulatus* (Long-snouted seahorse)

Distribution: Eastern Atlantic: British Isles to Morocco, Canary Islands (Dawson, 1990), Madeira and the Azores including the Mediterranean. Black Sea specimens may represent a different species.

IUCN Red List Status: Data Deficient (DD)

127. *Hippocampus hendriki* (Eastern spiny seahorse)

Distribution: Western Central Pacific: Queensland, Australia (Kuiter, 2001)

IUCN Red List Status: Data Deficient (DD)

128. *Hippocampus hippocampus* (Short snouted seahorse)

Distribution: Eastern Atlantic: Wadden Sea southward to the Gulf of Guinea, Canary Islands along the African coast to Guinea. Also in the Mediterranean (Lourie *et al.*, 1999)

IUCN Red List Status: Data Deficient (DD)

129. *Hippocampus histrix* (Thorny seahorse)

Distribution: Indo-Pacific: Tanzania and South Africa to Hawaii and Tahiti, north to Japan, south to New Caledonia. Reported from the Arafura Sea (Russel and Houston, 1989)

IUCN Red List Status: Data Deficient (DD)

130. *Hippocampus ingens* (Pacific seahorse)

Distribution: Eastern Pacific: San Diego in California, USA to Peru including the Galapagos Islands (Lourie *et al.*, 1999)

IUCN Red List Status: Vulnerable (VU) (A4cd)

131. *Hippocampus jayakari* (Jayakar's seahorse)

Distribution: Western Indian Ocean: Red Sea and Arabian Sea (Lourie *et al.*, 1999)

IUCN Red List Status: Data Deficient (DD)

132. *Hippocampus jugumus* (Collared seahorse)

Distribution: Southwest Pacific: known from a single specimens from Lord Howe Island (Kuiter, 2001)

IUCN Red List Status: Not Evaluated

133. *Hippocampus kelloggi* (Great seahorse)

Distribution: Indo-West Pacific: East Africa and the Red Sea to Japan and Lord Howe Island, Australia (Lourie *et al.*, 1999)

IUCN Red List Status: Data Deficient (DD)

134. *Hippocampus kuda* (Spotted seahorse)

Distribution: Indo-Pacific: Pakistan and India to southern Japan, Hawaii and the Society Islands (Lourie *et al.*, 1999)

IUCN Red List Status: Vulnerable (VU) (A4cd)

135. *Hippocampus lichtensteinii* (Lichtenstein's seahorse)

Distribution: Western Indian Ocean: Red Sea. No exact localities are stated (Lourie *et al.*, 1999)

IUCN Red List Status: Data Deficient (DD)

136. *Hippocampus minotaur* (Bullneck seahorse)

Distribution: Southwest Pacific: known only from southeastern Australia (Lourie *et al.*, 1999)

IUCN Red List Status: Data Deficient (DD)

137. *Hippocampus mohnikei* (Japanese seahorse)

Distribution: Western Pacific: Japan and Viet Nam; occurrence along Chinese coast needs confirmation (Lourie *et al.*, 1999)

IUCN Red List Status: Data Deficient (DD)

138. *Hippocampus montebelloensis* (Monte Bello seahorse)

Distribution: Eastern Indian Ocean: known only from Monte Bello Islands and Exmouth Gulf, western Australia (Kuiter, 2001)

IUCN Red List Status: Not Evaluated

139. *Hippocampus multispinus* (Northern spiny seahorse)

Distribution: Western Central Pacific: northern Australia (Danpier Archipelago, Western Australia to Northern Territory) and southern Papua New Guinea (Kuiter, 2001)

IUCN Red List Status: Not Evaluated

140. *Hippocampus paradoxus* (Paradoxical seahorse)

Distribution: Eastern Indian Ocean: Western Australia (Foster and Gommon, 2010)

IUCN Red List Status: Not Evaluated

141. *Hippocampus patagonicus*

Distribution: Southwest Atlantic: Argentina (Eschmeyer, 2006)

IUCN Red List Status: Not Evaluated

142. *Hippocampus pontohi*

Distribution: South East Asia: Indonesia (Lourie and Kuiter, 2008)

IUCN Red List Status: Data Deficient (DD)

143. *Hippocampus procerus* (High-crown seahorse)

Distribution: Western Central Pacific: Queensland, Australia (Kuiter, 2011)

IUCN Red List Status: Not Evaluated

144. *Hippocampus pusillus* (Pygmy thorny seahorse)

Distribution: Western Pacific: New Caledonia (Fricke, 2004)

IUCN Red List Status: Not Evaluated

145. *Hippocampus queenslandicus* (Queensland seahorse)

Distribution: Western Central Pacific: Queensland, Australia (Kuiter, 2001)

IUCN Red List Status: Not Evaluated

146. *Hippocampus reidi* (Longsnout sea horse)

Distribution: Western Atlantic: North Carolina in USA, Bermuda (Lieske and Myers, 1994) and Bahamas to Santa Catarina, Brazil (Floeter *et al.,* 2003)

IUCN Red List Status: Data Deficient (DD)

147. *Hippocampus satomiae*

Distribution: South East Asia: Indonesia. Known from scattered localities including Derawan (type locality) and Lembeh Strait (northern Sulawesi) as well as northern Borneo, Malaysia (Lourie and Kuiter, 2008)

IUCN Red List Status: Data Deficient (DD)

148. *Hippocampus semispinosus* (Half-spined seahorse)

Distribution: Western Central Pacific: Indonesia (Kuiter, 2001)

IUCN Red List Status: Not Evaluated

149. *Hippocampus severnsi*

Distribution: Pacific Ocean: Indonesia, Japan, Papua New Guinea, Solomon Islands and Fiji (Lourie and Kuiter, 2008)

IUCN Red List Status: Data Deficient (DD)

150. *Hippocampus sindonis* (Dhiho's seahorse)

Distribution: Northwest Pacific: southern Japan and the southern Korean Peninsula (Lourie *et al.,* 1999)

IUCN Red List Status: Least Concern (LC)

151. *Hippocampus spinosissimus* (Hedgehog seahorse)

Distribution: Indo-Pacific: Sri Lanka to Taiwan and Australia (Lourie *et al.*, 1999)

IUCN Red List Status: Vulnerable (VU) (A4cd)

152. *Hippocampus subelongatus* (West Australian seahorse)

Distribution: Eastern Indian Ocean: southwest Australia (Lourie *et al.*, 1999)

IUCN Red List Status: Data Deficient (DD)

153. *Hippocampus trimaculatus* (Longnose seahorse)

Distribution: Indo-Pacific: southern India to Japan, Australia and Tahiti (Lourie *et al.*, 1999)

IUCN Red List Status: Vulnerable (VU) (A4cd)

154. *Hippocampus tyro*

Distribution: Western Indian Ocean: Seychelles (Randall and Lourie, 2009)

IUCN Red List Status: Not Evaluated

155. *Hippocampus waleananus* (Walea pygmy seahorse)

Distribution: Asia: Indonesia. Known only from the Togean Islands in Tomini Bay, central Sulawesi (Gammon and Kuiter, 2009)

IUCN Red List Status: Not Evaluated

156. *Hippocampus whitei* (White's seahorse)

Distribution: Southwest Pacific: Solomon Islands and Australia. Records from southern Mozambique and Natal, South Africa are misidentifications of *H. Camelopardalis* (Lourie *et al.*, 1999)

IUCN Red List Status: Data Deficient (DD)

157. *Hippocampus zebra* (Zebra seahorse)

Distribution: Western Central Pacific: northeastern Australia (Lourie *et al.*, 1999)

IUCN Red List Status: Data Deficient (DD)

158. *Hippocampus zosterae* (Dwarf seahorse)

Distribution: Western Atlantic: Bermuda, southern Florida (USA), Bahamas and the entire Gulf of Mexico (Lourie *et al.*, 1999)

IUCN Red List Status: Data Deficient (DD)

159. *Histogamphelus briggsii* (Briggs' pipefish)

Distribution: Eastern Indian Ocean: southern Australia, from South Australia to New South Wales and Tasmania (Dawson, 1985)

IUCN Red List Status: Not Evaluated

160. *Histogamphelus cristatus* (Macleay's crested pipefish)

Distribution: Eastern Indian Ocean: southern Australia (Dawson, 1985)

IUCN Red List Status: Not Evaluated

161. *Hypselognathus horridus* (Prickly pipefish)

Distribution: Eastern Indian Ocean: known only from the Great Australian Bight (Dawson, 1985)

IUCN Red List Status: Not Evaluated

162. *Hypselognathus rostratus* (Knife-snouted pipefish)

Distribution: Eastern Indian Ocean: southern Australia (Dawson, 1985)

IUCN Red List Status: Not Evaluated

163. *Ichthyocampus bikiniensis*

Distribution: Western Central Pacific: Marshall Islands (Breder and Rosen, 1966)

IUCN Red List Status: Not Evaluated

164. *Ichthyocampus carce*

Distribution: Indo-West Pacific: west coast of India to Indonesia (Dawson, 1985)

IUCN Red List Status: Not Evaluated

165. *Idiotropiscis australe* (Southern little pipehorse)

Distribution: Eastern Indian Ocean: South and Western Australia (Dawson, 1985)

IUCN Red List Status: Data Deficient (DD)

166. *Idiotropiscis larsonae* (Helen's pygmy pipehorse)

Distribution: Eastern Indian Ocean: known only off Alpha Island, Monte Bello Islands, Western Australia (Dawson, 1985)

IUCN Red List Status: Not Evaluated

167. *Idiotropiscis lumnitzeri* (Sydney,s pygmy pipehorse)

Distribution: Southwestern Pacific: known from the Sydney region (Clovelly, Gonulla and Botany Bay) and Jervis Bay (Kuiter, 2004)

IUCN Red List Status: Not Evaluated

168. *Kaupus costatus* (Deepbody pipefish)

Distribution: Eastern Indian Ocean: southern Australia (Dawson, 1985)

IUCN Red List Status: Not Evaluated

169. *Kimblaeus bassensis* (Trawl pipefish)

Distribution: Eastern Indian Ocean: Bass Strait and Tasman sea, Australia (Dawson, 1985)

IUCN Red List Status: Not Evaluated

170. *Kyonemichthys rumengani*

Distribution: Western Pacific: Indonesia (Gommon, 2007)

IUCN Red List Status: Not Evaluated

171. *Leptoichthys fistularius* (Brush-tailed pipefish)

Distribution: Eastern Indian Ocean: endemic to southern Australia (Dawson, 1985)

IUCN Red List Status: Not Evaluated

172. *Leptonotus blainvilleanus* (Deep-bodied pipefish)

Distribution: Southeast Pacific and Southwest Atlantic: Hornitos, Chile to Golfo San Jose, Argentina (Dawson, 1985)

IUCN Red List Status: Not Evaluated

173. *Leptonotus elevates*

Distribution: Southwest Pacific: mainland New Zealand and the Auckland Islands (Dawson, 1985)

IUCN Red List Status: Not Evaluated

174. *Leptonotus norae* (Longsnout pipefish)

Distribution: Southwest Pacific: endemic to New Zealand (Dawson, 1985)

IUCN Red List Status: Not Evaluated

175. *Lissocampus bannwarthi*

Distribution: Western Indian Ocean: known only from the Gulf of Suez and the Gulf of Aqaba (Dawson, 1985)

IUCN Red List Status: Not Evaluated

176. *Lissocampus caudalis* (Smooth pipefish)

Distribution: Eastern Indian Ocean: endemic to Australia (Dawson, 1985)

IUCN Red List Status: Not Evaluated

177. *Lissocampus fatiloquus* (Prophet's pipefish)

Distribution: Eastern Indian Ocean: endemic to Australia (Dawson, 1985)

IUCN Red List Status: Not Evaluated

178. *Lissocampus filum* (Shortsnout pipefish)

Distribution: Southwest Pacific: known only from New Zealand (including the Chatham Island) (Dawson, 1985)

IUCN Red List Status: Not Evaluated

179. *Lissocampus runa* (Javelin pipefish)

Distribution: Eastern Indian Ocean: endemic to southern Australia (Dawson, 1985)

IUCN Red List Status: Not Evaluated

180. *Maroubra perserrata* (Sawtooth pipefish)

Distribution: Eastern Indian Ocean: endemic to southern Australia (Dawson, 1985)

IUCN Red List Status: Not Evaluated

181. *Maroubra yasudai* (Orange pipefish)

Distribution: Northwest Pacific: Honshu Island, Japan (Dawson, 1985)

IUCN Red List Status: Not Evaluated

182. *Micrognathus andersonii* (Shortnose pipefish)

Distribution: Indo-Pacific: East Africa and the Red Sea to Samoa, north to southern Japan, south to Tonga, Belau, to the eastern Caroline and Mariana in Micronesia (Myers, 1991)

IUCN Red List Status: Not Evaluated

183. *Micrognathus brevicorpus* (Short-bodied pipefish)

Distribution: Western Pacific: New Caledonia, apparently has a wider distribution in the area (Fricke, 2004)

IUCN Red List Status: Not Evaluated

184. *Micrognathus brevirostris brevirostris* (Thorntail pipefish)

Distribution: Western Indian Ocean: endemic to the Red Sea, recorded from the Gulfs of Suez and Aqaba in the north to the Strait of Perim in the South (Dawson, 1985)

IUCN Red List Status: Not Evaluated

185. *Micrognathus brevirostris pygmaeus* (Thorntailed pipefish)

Distribution: Indo-Pacific: Moluccas to Society Islands, north to Marshall Islands in Micronesia. Reported from Christmas Island (Allen and Steene, 1988)

IUCN Red List Status: Not Evaluated

186. *Micrognathus crinitus* (Banded pipefish)

Distribution: Western Atlantic: Bermuda and the Bahamas to Santa Catarina, Brazil (Floeter *et al.*, 2003) including lower east Florida coast but absent from the Gulf of Mexico expect for a single record from Cayo Arcas, Mexico. Dawson (1982) noted that these are two distinct color patterns of this species with no definite intergrades. Some authors are of the opinions that these color patterns represent two different species and should this be true, Dawson stated that Kaup's *vittatus* has priority over *ensenadae* with AMNH 7792 documenting the Bermuda occurrence of *Micrognathus vittatus*.

IUCN Red List Status: Not Evaluated

187. *Micrognathus erugatus*

Distribution: Southwest Atlantic: Brazil (previously considered endemic to the country) (Smith, 1997)

IUCN Red List Status: Not Evaluated

188. *Micrognathus micronotopterus* (Tidepool pipefish)

Distribution: Western Pacific: Singapore, western Indonesia, Australia and the Philippines (Dawson,1985)

IUCN Red List Status: Not Evaluated

189. *Micrognathus natans* (Offshore pipefish)

Distribution: Western Central Pacific: Philippines, Queensland (Australia), off New Caledonia and the Fiji Islands (Dawson, 1985)

IUCN Red List Status: Not Evaluated

190. *Microphis argulus* (Flat-nosed pipefish)

Distribution: Western Indian Ocean: Mauritius and the Comoros. Pacific Ocean: Indonesia to the Marquesan Islands, north to New Guinea, south to the Society Islands (Dawson, 1985)

IUCN Red List Status: Not Evaluated

191. *Microphis brachyurus aculeatus*

Distribution: Africa: Dakar, Senegal to the vicinity of Ambrizette, Angola (Dawson, 1985)

IUCN Red List Status: Not Evaluated

192. *Microphis brachyurus brachyurus* (Short-tailed pipefish)

Distribution: Asia and Oceania: Sri Lanka, southern India, Nicobar Islands and eastward to Japan and the Society Islands. Records from Puerto Rico and USA need confirmation. Africa: east coast from Kenya south to Natal area. Also Madagascar and Mauritius (Skelton, 1993). Africa: Cote d'Ivoire (Teugels *et al.,* 1988). Western Atlantic: Atlantic coast of Florida, Bahamas, northern and southern Gulf of Mexico, Antilles, central and south American coast to Brazil (Smith, 1997)

IUCN Red List Status: Not Evaluated

193. *Microphis brachyurus lineatus* (Opossum pipefish)

Distribution: Western Atlantic: New Jersey and eastern Florida (USA), Bahamas, Antilles and from Veracruz, Mexico to Sao Paulo, Brazil (Dawson, 1984)

IUCN Red List Status: Not Evaluated

194. *Microphis brachyurus millepunctatus* (Shorttail pipefish)

Distribution: Western Indian Ocean: Kenya to Durban, South Africa and from Madagascar, Mauritius and Reunion (Dawson, 1985)

IUCN Red List Status: Not Evaluated

195. *Microphis brevidorsalis* (Stream pipefish)

Distribution: Asia and Oceania: Indonesia, western Caroline Islands and south eastward to Fiji (Saikia and Goswami, 2012)

IUCN Red List Status: Not Evaluated

196. *Microphis caudocarinatus* (Slender pipefish)

Distribution: Asia: known only from a juvenile specimen collected in the Tawarin River, northern Irian Jaya, Indonesia in 1903 (Saikia and Goswami, 2012)

IUCN Red List Status: Data Deficient (DD)

197. *Microphis cruentus* (Dumbea river pipefish)

Distribution: Oceania: endemic to New Caledonia (Saikia and Goswami, 2012)

IUCN Red List Status: Not Evaluated

198. *Microphis cuncalus* (Crocodile-tooth pipefish)

Distribution: Asia: India, Bangladesh and Sri Lanka (Talwar and Jhingran, 1991)

IUCN Red List Status: Least Concern (LC)

199. *Microphis deocata* (Deocata pipefish)

Distribution: Asia: India and Bangladesh (Talwar and Jhingran, 1991; Saikia and Goswami, 2012)

IUCN Red List Status: Near Threatened (NT)

200. *Microphis dunckeri*

Distribution: Asia: Restricted to Myanmar (Saikia and Goswami, 2012)

IUCN Red List Status: Least Concern (LC)

201. *Microphis fluviatilis* (Freshwater pipefish)

Distribution: Africa: Sabaki River, Kenya to Coffee Bay, South Africa; also Madagascar. Reported from Zimbabwe (Bell-Cross and Minshull, 1988)

IUCN Red List Status: Not Evaluated

202. *Microphis insularis* (Andaman pipefish)

Distribution: Asia: probably endemic to the southern Andaman Islands (Saikia and Goswami, 2012)

IUCN Red List Status: Vulnerable (VU) (B1ab(iii))

203. *Microphis jagorii*

Distribution: Asia: known only from the Philippines (Dawson, 1985). Reported from Okinawa Island, Japan (Yoshino and Yoshigou, 1998)

IUCN Red List Status: Not Evaluated

204. *Microphis leiaspis* (Barhead pipefish)

Distribution: Africa: occurs in Madagascar. Southeast Asia and Oceania: Japan south to Indonesia and east to New Caledonia and Fiji (Dawson, 1985)

IUCN Red List Status: Least Concern (LC)

205. *Microphis manadensis* (Menado pipefish)

Distribution: Asia and Oceania: Indonesia, Taiwan, Palau, New Guinea and Solomon Islands. Also reported from the Philippines (Kottelat, 1993). Reports from Australia requires verification by additional collections.

IUCN Red List Status: Not Evaluated

206. *Microphis mento* (Red pipefish)

Distribution: Asia: the Philippines and Indonesia (Sulawesi). Recorded from Papua New Guinea (Allen, 1991)

IUCN Red List Status: Not Evaluated

207. *Microphis ocellatus*

Distribution: Asia: Sri Lanka and Indonesia (Saikia and Goswami, 2012)

IUCN Red List Status: Not Evaluated

208. *Microphis pleurostictus*

Distribution: Asia: Batu (Bato) Lake and Yassot Creek, Luzon, Philippines (Saikia and Goswami, 2012)

IUCN Red List Status: Not Evaluated

209. *Microphis retzii* (Ragged-tail pipefish)

Distribution: Asia and Oceania: Indonesia and the Philippines southeastward to Ponape [=Pohnpei] and Samoa (Saikia and Goswami, 2012)

IUCN Red List Status: Not Evaluated

210. *Microphis spinachioides* (Spinach pipefish)

Distribution: Oceania: Sepik River and its midstream tributary (570km upstream), the May River, Papua New Guinea. Also known from Ramu river (Saikia and Goswami, 2012)

IUCN Red List Status: Data Deficient (DD)

211. *Minyichthys brachyrhinus*

Distribution: Eastern Indian Ocean: off Sumatra, Indonesia. Pacific Ocean: Philippines, Fiji, Kure Islands and the Hawaiian Islands (Dawson, 1985)

IUCN Red List Status: Not Evaluated

212. *Minyichthys inusitatus*

Distribution: Western Central Atlantic: Caribbean Sea (Dawson, 1983)

IUCN Red List Status: Not Evaluated

213. *Minyichthys myersi* (Myer's pipefish)

Distribution: Indo-West Pacific: Cargados Carajos, Philippines, Moluccas, Papua New Guinea and Guam (Myers, 1991)

IUCN Red List Status: Not Evaluated

214. *Minyichthys sentus*

Distribution: Eastern Atlantic: off Morocco, the Canary Islands and from of Algeria in the Mediterranean ([2]Dawson, 1986)

IUCN Red List Status: Not Evaluated

215. *Mitotichthys meraculus* (Western crested pipefish)

Distribution: Eastern Indian Ocean: Western Australia (Dawson, 1985)

IUCN Red List Status: Not Evaluated

216. *Mitotichthys mollisoni* (Mollison's pipefish)

Distribution: Eastern Indian Ocean: Tasmania, Australia (Paxton *et al.*, 1989)

IUCN Red List Status: Not Evaluated

217. *Mitotichthys semistriatus* (Half-banded pipefish)

Distribution: Eastern Indian Ocean: Victoria and Tasmania, Australia (Dawson, 1985)

IUCN Red List Status: Not Evaluated

218. *Mitotichthys tuckri* (Tucker's pipefish)

Distribution: Southwest Pacific: New South Wales and Tasmania, Australia (Dawson, 1985)

IUCN Red List Status: Not Evaluated

219. *Nannocampus elegans*

Distribution: Western Indian Ocean: Inhaa, Island, Mozambique to Great Fish Point, South Africa ([3]Dawson, 1986)

IUCN Red List Status: Not Evaluated

220. *Nannocampus pictus* (Reef pipefish)

Distribution: Indo-West Pacific: East Africa to the Great Barrier Reef, north to Sri Lanka (Paxton *et al.*, 1989)

IUCN Red List Status: Not Evaluated

221. *Nannocampus lindemanensis* (Painted pipefish)

Distribution: Eastern Indian Ocean and Western Pacific (Hoese *et al.*, 2006)

IUCN Red List Status: Not Evaluated

222. *Nannocampus subosseus* (Bony-headed pipefish)

Distribution: Eastern Indian Ocean: known only from Western Australia (Dawson, 1985)

IUCN Red List Status: Not Evaluated

223. *Nannocampus weberi*

Distribution: Western Central Pacific: known only from Bali and Sumba, Indonesia (Dawson, 1985)

IUCN Red List Status: Not Evaluated

224. *Nerophis lumbriciformis* (Worm pipefish)

Distribution: Eastern Atlantic: southern Norway, Kattegat and British Isles to Rio de Oro in Western Sahara (Dawson, 1990)

IUCN Red List Status: Not Evaluated

225. *Nerophis maculatus*

Distribution: Eastern Atlantic: Portugal and the Azores. Mediterranean Sea: apparently most common in the Adriatic Sea and the western Mediterranean ([2]Dawson, 1986)

IUCN Red List Status: Not Evaluated

226. *Nerophis ophidian* (Straightnose pipefish)

Distribution: Eastern Atlantic: Norway to Morocco (excluding region from Denmark to Netherlands), also throughout the Mediterranean and the Black Sea ([2]Dawson, 1986)

IUCN Red List Status: Not Evaluated

227. *Notiocampus ruber* (Red pipefish)

Distribution: Eastern Indian Ocean: southern Australia from western Australia to New South Wales and Tasmania (Dawson, 1985)

IUCN Red List Status: Not Evaluated

228. *Penetopteryx nanus* (Lost pipefish)

Distribution: Western Central Atlantic: Bahamas, Isla de Providencia in western Caribbean (Bohlke and Chaplin, 1993)

IUCN Red List Status: Not Evaluated

229. *Penetopteryx taeniocephalus*

Distribution: Indo-Pacific: Madagascar to Christmas Island (Line Island) (Dawson, 1985)

IUCN Red List Status: Not Evaluated

230. *Phoxocampus belcheri* (Rock pipefish)

Distribution: Indo-West Pacific: Red Sea and East Africa, eastward to Japan and Fiji Islands. Recently recorded from Tonga (Randall *et al.,* 2003)

IUCN Red List Status: Not Evaluated

231. *Phoxocampus diacanthus* (Spined pipefish)

Distribution: Indo-Pacific: Sri Lanka to Samoa, north to Hong Kong, south to New Caledonia (Myers, 1991)

IUCN Red List Status: Not Evaluated

232. *Phoxocampus tetrophthalmus* (Kampen's pipefish)

Distribution: Western Central Pacific: Indonesia, including western Irian Java, the Philippines and Guam (Kami, 1971)

IUCN Red List Status: Not Evaluated

233. *Phycodurus eques* (Leafy seadragon)

Distribution: Eastern Indian Ocean: endemic to southern Australia (Kuiter, 1993)

IUCN Red List Status: Near Threatened (NT)

234. *Phyllopteryx taeniolatus* (common seadragon)

Distribution: Eastern Indian Ocean: southern Australia, from southern Western Australia to New South Wales and Tasmania (May and Maxwell, 1986)

IUCN Red List Status: Near Threatened (NT)

235. *Pseudophallus elcapitanensis*

Distribution: Central America: Jimenez, Costa Rica to Rio Chico, Panama (Saikia and Goswami, 2012)

IUCN Red List Status: Not Evaluated

236. *Pseudophallus mindii*

Distribution: Central and South America: Atlantic and Caribbean coastal rivers (Greenfield and Thomerson, 1997)

IUCN Red List Status: Not Evaluated

237. *Pseudophallus starksii*

Distribution: Central and south America: Baja California Sur and Sinaloa, Mexico to Santa Rosa, Ecuador (Saikia and Goswami, 2012)

IUCN Red List Status: Not Evaluated

238. *Pugnaso curtirostris* (Pug-nosed pipefish)

Distribution: Eastern Indian Ocean: southern Australia (Dawson, 1985)

IUCN Red List Status: Not Evaluated

239. *Siokunichthys bentuviai*

Distribution: Western Indian Ocean: El Tur, Sinai Peninsula and Dahlak Archipelago (Ethiopia) in the Red Sea and in the vicinity of Musha Islands, near Djibouti, in the Gulf of Aden (Dawson, 1983)

IUCN Red List Status: Not Evaluated

240. *Siokunichthys breviceps* (Softcoral pipefish)

Distribution: Western Indian Ocean: Mozambique. Western Pacific: Indonesia, the Philippines and Queensland, Australia (Dawson, 1983)

IUCN Red List Status: Not Evaluated

241. *Siokunichthys herrei* (Herre,s pipefish)

Distribution: Indo-West Pacific: northern Red Sea to Indonesia, the Philippines, Solomon Islands and Fiji (Dawson, 1983)

IUCN Red List Status: Not Evaluated

242. *Siokunichthys nigrolineatus* (Mushroom-coral pipefish)

Distribution: Western Central Pacific: Indonesia and the Philippines (Dawson, 1983)

IUCN Red List Status: Least Concern (LC)

243. *Siokunichthys southwelli* (Southwell's pipefish)

Distribution: Western Indian Ocean: Sri Lanka. Western Central Pacific: Philippines (Dawson, 1983)

IUCN Red List Status: Not Evaluated

244. *Siokunichthys striatus* (New Caledonian soft-coral pipefish)

Distribution: Western Pacific: New Caledonia (Fricke, 2004)

IUCN Red List Status: Not Evaluated

245. *Solegnathus dunckeri* (Duncker's pipehorse)

Distribution: Western Pacific: Australia (Paxton *et al.*, 1989)

IUCN Red List Status: Data Deficient (DD)

246. *Solegnathus hardwickii* (Hardwicke's pipefish)

Distribution: Western Indian Ocean: Mauritius. Western Pacific: South China Sea north to southern Japan, south to northern Australia (Paxton *et al.*, 1989)

IUCN Red List Status: Data Deficient (DD)

247. *Solegnathus lettiensis* (Gunther's pipehorse)

Distribution: Eastern Indian Ocean: Indonesia and Western Australia (Dawson, 1985)

IUCN Red List Status: Data Deficient (DD)

248. *Solegnathus robustus* (Robust pipefish)

Distribution: Eastern Indian Ocean: endemic to southern Australia (Dawson, 1985)

IUCN Red List Status: Data Deficient (DD)

249. *Solegnathus spinosissimus* (Spiny pipehorse)

Distribution: Southwest Pacific: southern Australia and New Zealand (Paulin *et al.,* 1989)

IUCN Red List Status: Data Deficient (DD)

250. *Stigmatopora argus* (Spotted pipefish)

Distribution: Eastern Indian Ocean: Australia (Dawson, 1985)

IUCN Red List Status: Not Evaluated

251. *Stigmatopora macropterygia* (Long-snouted pipefish)

Distribution: Southwest Pacific: endemic to New Zealand and the Auckland Islands (Dawson, 1985)

IUCN Red List Status: Not Evaluated

252. *Stigmatopora narinosa* (Southern Gulf pipefish)

Distribution: Eastern Indian Ocean: South Australia (Browne and Smith, 2007)

IUCN Red List Status: Not Evaluated

253. *Stigmatopora nigra* (wide-bodied pipefish)

Distribution: Indo-West Pacific: southern Australia and New Zealand (Dawson, 1985)

IUCN Red List Status: Not Evaluated

254. *Stipecampus cristatus* (Ring-backed pipefish)

Distribution: Eastern Indian Ocean: southern Australia (Dawson, 1985)

IUCN Red List Status: Not Evaluated

255. *Syngnathoides biaculeatus* (Alligator pipefish)

Distribution: Indo-Pacific: Red Sea and Knysna, South Africa ([3]Dawson, 1986) to Samoa, north to southern Japan, south to New South Wales

IUCN Red List Status: Data Deficient (DD)

256. *Syngnathus abaster* (Black-striped pipefish)

Distribution: Eastern Atlantic: southern Biscay to Gibraltar and also the Mediterranean and Black Sea. Introduced and became established in the former USSR ([2]Dawson, 1986).

IUCN Red List Status: Least Concern (LC)

257. *Syngnathus acus* (Greater pipefish)

Distribution: Eastern Atlantic: Norway, Faroes and British Isles to Western Shara, Senegambia, and from Namibia to Cape of Good Hope and northward to the coast Zululand in the Western Indian Ocean ([1]Dawson, 1986). Also throughout the Mediterranean, Aegean and Black seas. Despite literature records, occurrence in the Indo-Pacific outside South Africa waters lacks conclusive evidence ([3]Dawson, 1986)

IUCN Red List Status: Not Evaluated

258. *Syngnathus affinis* (Texas pipefish)

Distribution: Western Atlantic: Reported from USA (Robins *et al.*, 1991)

IUCN Red List Status: Not Evaluated

259. *Sygnathus auliscus* (Barred pipefish)

Distribution: Eastern Pacific: Santa Barbara Channel in southern California, USA to northern Peru (Eschmeyer *et al.*, 1983)

IUCN Red List Status: Least Concern (LC)

260. *Syngnathus californiensis* (Kelp Pipefish)

Distribution: Eastern Pacific: Bodega Bay in northern California, USA to southern Baja California, Mexico (Eschmeyer *et al.*, 1983)

IUCN Red List Status: Not Evaluated

261. *Syngnathus caribbaeus* (Caribbean pipefish)

Distribution: Western Atlantic: Belize to Suriname; also from the Greater and Losser Antilles (Randall, 1996)

IUCN Red List Status: Not Evaluated

262. *Syngnathus carinatus*

Distribution: Eastern Central Pacific: endemic to the northern Gulf of California (Dawson, 1985)

IUCN Red List Status: Data Deficient (DD)

263. *Syngnathus dawsoni*

Distribution: Western Central Atlantic: US Virgin Islands (Eschmeyer, 2003)

IUCN Red List Status: Not Evaluated

264. *Syngnathus euchrous* (Chocolate pipefish)

Distribution: Eastern Pacific: Redondo Beach in southern California, USA to central Baja California, Mexico (Eschmeyer *et al.*, 1983)

IUCN Red List Status: Not Evaluated

265. *Syngnathus exilis* (Barcheek pipefish)

Distribution: Eastern Pacific: Half Moon Bay in central California, USA to south central Baja California and Guadalupe Island off north central Baja California, Mexico (Eschmeyer)

IUCN Red List Status: Not Evaluated

266. *Syngnathus floridae* (Dusky pipefish)

Distribution: Western Atlantic: disjunct distribution from Bermuda and Chesapeake Bay (USA), including the northern Gulf of Mexico, the Bahamas and the western Caribbean Sea (McEachran and Fechhelm, 1998) to Panama

IUCN Red List Status: Least Concern (LC)

267. *Syngnathus folletti* (Southern pipefish)

Distribution: Southwest Atlantic: Uruguay (Eschmeyer, 2003)

IUCN Red List Status: Not Evaluated

268. *Syngnathus fuscus* (Northern pipefish)

Distribution: Western Atlantic: Gulf of St. Lawrence in Canada to northeastern Florida in USA; also in northwestern Gulf of Mexico (Lazzari and Able, 1990)

IUCN Red List Status: Not Evaluated

269. *Syngnathus insulae* (Guadalupe pipefish)

Distribution: Eastern Central Pacific: endemic to Guadalupe Island, Mexico (Dawson, 1985)

IUCN Red List Status: Not Evaluated

270. *Syngnathus leptorhynchus* (Bay pipefish)

Distribution: Eastern Pacific: Sitka, Alaska to southern Baja California in Mexico; the northern population ranges from Alaska to Monterey Bay, southern population from Morro Bay southward (Eschmeyer *et al.,* 1983)

IUCN Red List Status: Not Evaluated

271. *Syngnathus louisianae* (Chain pipefish)

Distribution: Western Atlantic: Virginia in USA, Bermuda and northern Gulf of Mexico in USA to Campeche in Mexico and Jamaica; about from Bahamas (Robins and Ray, 1986)

IUCN Red List Status: Not Evaluated

272. *Syngnathus macrobrachium*

Distribution: Southeast Pacific: Tumbes, Peru to Puerto Montt, Chile (Dawson, 1985)

IUCN Red List Status: Data Deficient (DD)

273. *Syngnathus macrophthalmus*

Distribution: Western Indian Ocean: known only from Suez and around Hurghada (Al Ghardaqa) in the northwestern Red Sea (Dawson, 1985)

IUCN Red List Status: Not Evaluated

274. *Syngnathus makaxi* (Yucatan pipefish)

Distribution: Western Central Atlantic: Mexico (Herald and Dawson, 1972)

IUCN Red List Status: Not Evaluated

275. *Syngnathus pelagicus* (Sargassum pipefish)

Distribution: Western Atlantic: Maine (USA), Bermuda and northern Gulf of Mexico to Argentina. Nova Scotia, Antilles, western Caribbean from Yucatan to Colombia (Smith, 1997). Taxonomic status of the eastern Atlantic population (from Mauritania to Gabon) needs further study and records from Cape of Good Hope, South Africa and the Indo-Pacific area require confirmation.

IUCN Red List Status: Not Evaluated

276. *Syngnathus phlegon*

Distribution: Eastern Atlantic: throughout the Mediterranean Sea and adjacent Atlantic coasts ([2]Dawson, 1986)

IUCN Red List Status: Not Evaluated

277. *Syngnathus rostellatus* (Nilsson's pipefish)

Distribution: Northeast Atlantic: Bergen, Norway and southern portions British Isles to Bay of Biscay ([2]Dawson, 1986)

IUCN Red List Status: Not Evaluated

278. *Syngnathus safina*

Distribution: Western Indian Ocean: Gulf of Aqaba (Eschmeyer, 2003)

IUCN Red List Status: Not Evaluated

279. *Syngnathus schlegeli* (Seaweed pipefish)

Distribution: Northwest Pacific: around Vladivostok, Russia Federation southward to the Gulf of Tonkin (Dawson, 1985)

IUCN Red List Status: Not Evaluated

280. *Syngnathus schmidti*

Distribution: Europe: Black Sea and Sea of Azov ([2]Dawson, 1986)

IUCN Red List Status: Not Evaluated

281. *Syngnathus scovelli* (Gulf pipefish)

Distribution: Western Atlantic: northeastern Florida and northern Gulf of Mexico to the Caribbean coast of Central and South America; absent from West Indies (Robins and Ray, 1986)

IUCN Red List Status: Not Evaluated

282. *Syngnathus springeri* (Bull pipefish)

Distribution: Western Atlantic: Canada (Scott and Scott, 1988) to North Carolina in USA, Bahamas and northeastern Gulf of Mexico to Panama

IUCN Red List Status: Not Evaluated

283. *Syngnathus taenionotus*

Distribution: Mediterranean Sea: northwestern Adriatic Sea (Wheeler, 1979)

IUCN Red List Status: Not Evaluated

284. *Syngnathu tenuirostris* (Narrow-snouted pipefish)

Distribution: Mediterranean Sea: southern Adriatic Sea, Tyrrhenian Sea. Also known from the Black Sea and Sea of Azov ([2]Dawson, 1986)

IUCN Red List Status: Not Evaluated

285. *Syngnathus typhle* (Broadnosed pipefish)

Distribution: Eastern Atlantic: Vardo, Norway, Baltic Sea and the British Isles to Morocco. Also throughout the Mediterranean, Black Sea and Sea of Azov. Record off Ghana is still questionable ([2]Dawson, 1990)

IUCN Red List Status: Not Evaluated

286. *Syngnathus variegatus*

Distribution: Eurasia: Black Sea and Sea of Azov (Dawson, 1986)

IUCN Red List Status: Not Evaluated

287. *Syngnathus watermeyeri* (River pipefish)

Distribution: Africa: known only from areas of tidal influence within the Kariega, Kasouga and Bushmans Rivers, South Africa (Dawson, 1986)

IUCN Red List Status: Critically Endangered (CR) (B1ab(i,ii,iii);C2a(i))

288. *Trachyrhamphus bicoarctatus* (Double-ended pipefish)

Distribution: Indo-West Pacific: Red Sea and East Africa (Dawson, 1986) to New Caledonia, north to southern Japan; Mariana Islands in Micronesia

IUCN Red List Status: Not Evaluated

289. *Trachyrhamphus longirostris* (Straightstick pipefish)

Distribution: Indo-West Pacific: Red Sea and western Indian Ocean eastward to Japan and eastern Australia (Paxton *et al.,* 1989)

IUCN Red List Status: Not Evaluated

290. *Trachyrhamphus serratus* (Saw pipefish)

Distribution: Indo-West Pacific: southern India and Sri-Lanka to Korea and Japan (Dawson, 1985)

IUCN Red List Status: Not Evaluated

291. *Urocampus carinirostris* (Hairy pipefish)

Distribution: Western Pacific: Australia and Papua New Guinea (Paxton *et al.*, 1989)

IUCN Red List Status: Not Evaluated

292. *Urocampus nanus* (Barbed pipefish)

Distribution: Northwest Pacific: restricte to southern Japan and adjacent coasts

IUCN Red List Status: Not Evaluated (Dawson, 1985)

293. *Vanacampus margaritifer* (Mother-of-pearl pipefish)

Distribution: Western Pacific: Southport, southern Queensland to Port Phillip Bay, Victoria and off Rottnest Island, Western Australia (Dawson, 1985)

IUCN Red List Status: Not Evaluated

294. *Vanacampus phillipi* (Port Phillip pipefish)

Distribution: Eastern Indian Ocean: southern Australia, from Albany and Cottesloe, western Australia to Moruya Point, southern New South Wales (Dawson, 1985)

IUCN Red List Status: Not Evaluated

295. *Vanacampus poecilolaemus* (Australian long-nosed pipefish)

Distribution: Eastern Indian Ocean: southern Australia (Dawson, 1985)

IUCN Red List Status: Not Evaluated

296. *Vanacampus vercoi* (Verco,s pipefish)

Distribution: Eastern Indian Ocean: known only from South Australia (Dawson, 1985)

IUCN Red List Status: Not Evaluated

Discussion

Among all the members of Syngnathidae family, seahorse and pipefish are the most popular. Seahorses are distributed throughout the world, in both tropical and temperate zones, ranging from about 45 degrees north to 45 degrees south. They favour saltwater habitats or brackish waters of estuaries, containing mangroves, coral reefs and sea grass beds. The range varies by gender. Females may wander as far as 100 square yards, centring around, but reducing feeding competition with their oft-pregnant partner who confines himself to a single square yard (Vincent, 1996). Because prey is abundant in areas with good water exchange, they tend to colonize only edges of suitable habitats (Breder, 1942). Pipefishes are cosmopolitan in distribution between 71°N and 56°S latitudes (Dawson, 1985). Recent reports revealed a northward shift in distribution by 15° latitude and attributed this to increase in the northern sea surface temperature (Fleischer *et al.*, 2007). About 15 genera (29 per cent) are represented in the Atlantic ocean and 47 (92 per cent) in the Indo-Pacific region. Most species inhabit estuarine and coastal marine areas, some in freshwaters and very few in hyper saline environments (Simmons, 1957). We have reported the taxonomy and distribution of *Microphis deocata* from the north eastern Himalayan rivers (Saikia and Goswami, 2012; Goswami *et al.*, 2007). A series of threats of natural and anthropogenic hazards are being simultaneously experienced by the species.

Syngnathids has been the subject of large scale target fishing which has resulted in rapid decline of the wild populations. This was still worsened by indirect habitat destruction and loss (Vincent, 1995). Substantial quantities of syngnathids are also trapped as by catch in various fishing operations throughout the world. There are already reports on extinction of the riverine pipefish, *Syngnathus watermeyeri* in South Africa (Whitfield, 1996) and local extermination of *Doryichthys cuncalus* in India (Chhapgar and Pande, 1986). In Australia, *Vanacampus vercoi* was identified as vulnerable to habitat loss because of its restricted geographical range and preference to shallow waters (Pogonoski *et al.*, 2002). Considering the need for conservation, all seahorses, 6 pipefishes, 2 seadragons and 5 pipehorses are listed in 2007 IUCN Red List of Threatened Animals. Of the 6 pipefishes, *Doryrhamphus dactyliophorus, Microphis caudocarinatus, M. spinachioides, Syngnathoides biaculeatus* and *Syngnathus abaster* are classified as Data Deficient and *S. watermeyeri* as Critically Endangered. For the present review, we have referred the publications of various authors and also use the database of Fish Base.

References

Allen, G.R. and Adrim, M., 2003 Coral reef fishes of Indonesia. *Zool. Stud.*, 42(1): 1–72

Allen, G.R. and Erdmann, M.V., 2008. *Corythoichthys benedetto*, a new pipefish (Pisces: Syngnathidae) from Indonesia and Papua New Guinea. *Aqua Int. J. Ichthyol.*, 13(3–4): 121–126.

Allen, G.R. and Steene, R.C., 1988. Fishes of Christmas Island Indian Ocean. Christmas Island Natural History Association, Christmas Island, Indian Ocean, 6798, Australia, p. 197.

Allen, G.R. and Kuiter, R.H., 1995. *Bhanotia pauciradiata*, a new species of pipefish (Syngnathidae) from Flores, Indonesia. *Rev. Fr. Aquariol.*, 21(3–4): 75–76.

Allen, G.R. and Kuiter, R.H., 2004. *Dunckeroncampus naia*, a new pipefish (Pisces: Syngnathidae) from the Western Pacific. *Aqua, J. Ichthyol. Aquat. Biol.*, 9(1): 1–6.

Allen, G.R., 1991. Field guide to the freshwater fishes of New Guinea. Christensen Research Institute, Madang, Papua New Guinea.

Anderson, R.C., Randall, J.E. and Kuiter, R.H., 1998. New records of fishes from the Maldives Islands, with notes on other species. *Ichthyol. Bull.*, 67(2): 20–36.

Anonymous, 1997. Fish registrations within the museum database of the Vertebrate Section of the Royal Museum for Central Africa. MRAC, Tervuren, Belgium.

Bell-Cross, G. and Minshull, J.L., 1988. The fishes of Zimbabwe. National Museums and Monuments of Zimbabwe, Harare, Zimbabwe, p. 294.

Blot, J., 1980. La faune ichthyologique des gisements du Monte Bolca (Province de Verone, Italie). *Bulletin du Museum d'Histoire Naturelle de Paris*, 4 (1980): 339–396.

Bohlke, J.E. and Chaplin, C.C.G., 1993. *Fishes of the Bahamas and Adjacent Tropical Waters,* 2nd edition. University of Texas Press, Austin.

Breder, C.M. and Rosen, D.E., 1966. *Modes of Reproduction in Fishes.* T.F.H. Publications, Neptune City, New Jersey, p. 941.

Breder, C.M. Jr. and Edgarton, H.E., 1942. An analysis of the locomotion of the seahorse, hippocampus, by means of high speed cinematography. *Annals of the New York Academy of Sciences,* 43: 145–172.

Browne, R.K. and Smith, K., 2007. A New pipefish, *Stigmatopora narinosa* (Syngnathidae) From South Australia. *Memoirs of Museum Victoria,* 64: 1–6.

Chhapgar, B.F. and Pande, J.M., 1986. Pise Dam: An ecological disaster for the freshwater pipefish *Doryichthys cuncalus* (Ham.Buch). *J. Bom. Nat. Hist. Soc.,* 83(1): 232–235.

Dawson, C.E., 1983. Synopsis of the Indo-Pacific pipefish genus *Siokunichthys* (Syngnathidae), with description of *S. nigrolineatus* n. sp. *Pac. Sci.,* 37: 49–63.

Dawson, C.E., 1983. Western Atlantic occurrence of the genus *Minyichthys* (Pisces: Syngnathidae), with description of *M. inusitatus* n. sp. *Copeia,* (3): 774–776.

Dawson, C.E., 1984. Revision of the genus *Microphis* Kaup (Pisces: Syngnathidae). *Bull. Mar. Sci.,* 35(2): 117–181.

Dawson, C.E., 1985. Indo-Pacific pipefishes (Red Sea to the Americas). The Gulf Coast Research Laboratory Ocean Springs, Mississippi, USA.

[1]Dawson, C.E., 1986. Syngnathidae. p. 281–287. In J. Daget, J.–P. Gosse and D.F.E. Thys van den Audenaerde (eds.) *Checklist of the freshwater fishes of Africa (CLOFFA).* ISNB, Brussels, MRAC, Tervuren, and ORSTOM, Paris, Vol. 2.

[2]Dawson, C.E., 1986. Syngnathidae. p. 628–639. In P.J.P. Whitehead, M.–L. Bauchot, J.–C. Hureau, J. Nielsen and E. Tortonese (eds.) *Fishes of the North-Eastern Atlantic and the Mediterranean.* UNESCO, Paris. Vol. 2.

[3]Dawson, C.E., 1986. Syngnathidae. In: *Smiths' Sea Fishes,* (Eds.) M.M. Smith and P.C. Heemstra. Springer-Verlag, Berlin, pp. 445–458.

Dawson, C.E., 1990. Syngnathidae. pp. 658–664. In J.C. Quero, J.C. Hureau, C. Karrer, A. Post and L. Saldanha (eds.) *Checklist of the Fishes of the Eastern Tropical Atlantic (CLOFETA).* JNICT, Lisbon, SEI, Paris, and UNESCO, Paris, Vol. 2.

Eschmeyer, W.N. (ed.), 2003. *Catalog of Fishes.* Updated database version of March 2003. Catalog databases as made available to FishBase in March 2003.

Eschmeyer, W.N. (ed.), 2004. *Catalog of Fishes.* Updated database version of January 2004. Catalog databases as made available to FishBase in January 2004.

Eschmeyer, W.N. (ed.), 2006. *Catalog of Fishes.* Updated database version of April 2006. Catalog databases as made available to FishBase in April 2006.

Eschmeyer, W.N., Herald, E.S. and Hammann, H., 1983. *A Field Guide to Pacific Coast Fishes of North America.* Houghton Mifflin Company, Boston, U.S.A., p. 336.

Fleischer, D., Schaber, M. and Pipenburg, D., 2007. Atlantic snake pipefish (*Entelurus aequoreus*) extends its northward distribution range to Svaibard (Arctic Ocean). *Poi.Biol.,* doi. 10. 1007/s 00300–007–0322–y.

Floeter, S.R., Gasparini, J.L., Rocha, L.A., Ferreira, C.E.L., Rangel, C.A. and Feitoza, B.M., 2003. Brazilian reef fish fauna: Checklist and Remarks (updated Jan., 2003). Brazilian Reef Fish Project: www.brazilianreeffish.cjb.net

Foster, S.J. and Vincent, A.C.J., 2004. Life history and ecology of seahorse: Implications for conservation and management. *J. Fish. Biol.*, 65: 1–61.

Foster, R. and Gomon, M.F., 2010. A new seahorse (Teleostei: Syngnathidae: Hippocampus) from south-western Australia. *Zootaxa*, 2613: 61–68.

Francis, M.P., 1993. Checklist of the coastal fishes of Lord Howe, Norfolk, and Kermadec Islands, Southwest Pacific Ocean. *Pac. Sci.*, 47(2): 136–170.

Fricke, R., 1999. Fishes of the Mascarene Islands (Réunion, Mauritius, Rodriguez): an annotated checklist, with descriptions of new species. Koeltz Scientific Books, Koenigstein. *Theses Zoologicae*, 31: 759.

Fricke, R., 2004. Review of the pipefishes and seahorses (Teleostei: Syngnathidae) of New Caledonia, with descriptions of five new species. *Stuttgarter Beitr. Naturkd., Ser. A Nr.*, 668: 1–67.

Gommon, M.F. and Kuiter, R.H., 2009. Two new pygmy seahorses (Teleostei: Syngnathidae: Hippocampus) from the Indo-West Pacific. *Aqua*, 15(1): 37–44.

Gomon, M.F., 2007. A new genus and miniature species of pipehorse (Syngnathidae) from Indonesia. *Aqua Int. J. Ichthyol.*, 13(1): 25–30.

Goswami, U.C., Viswanath, W., Jayaram, K.C. and Dutta, J.S., 2007. Fish fauna of Northeast India: Natural and anthropogenic hazards. In: *Natural and Anthropogenic Hazards on Fish and Fisheries*, (Ed.) U.C. Goswami. NPH, Delhi, pp. 21–46.

Greenfield, D.W. and Thomerson, J.E., 1997. *Fishes of the Continental Waters of Belize.* University Press of Florida, Florida, p. 311.

Heemstra, P.C., 1995. Additions and corrections for the 1995 impression. p. v–xv. In: *Revised Edition of Smiths' Sea Fishes*, (Eds.) M.M. Smith and P.C. Heemstra. Springer-Verlag, Berlin.

Herald, E.S. and Dawson, C.E., 1972. A new subspecies of the gulf pipefish, *Syngnathus scovelli makaxi* (Pisces: Syngnathidae). *Copeia*, 1972(4): 781–784.

Hoese, D.F., Bray, D.J., Paxton, J.R. and Allen, G.R., 2006. Fishes. In: *Zoological Catalogue of Australia*, (Eds.) O.L. Beasley and A. Wells. Volume 35. ABRS and CSIRO Publishing: Australia Part 1, pp. xxiv 1–670, Part 2, pp. xxi 671–1472, Part 3, pp. xxi 1473–2178.

Jeyaseelan, M.J.P., 1998. Manual of fish eggs and larvae from Asian mangrove waters. United Nations Educational, Scientific and Cultural Organization. Paris, p. 193.

Kami, H.T., 1971. Check–list of Guam fishes, supplement I. *Micronesica*, 7(1–2): 215–228.

Kottelat, M., 1993. Technical report on the fishes from fresh and brackish waters of Leyte, Philippines. Technical Report prepared for the Deutsche Gesellschaft für Technische Zusammenarbeit (GTZ) GmbH and ViSCA–GTZ Ecology Program, Visayan State College of Agriculture, Philippines. Route de Fregiêcourt 96c, Case postale 57, CH–2952 Cornol, Switzerland, p. 54.

Kuiter, R.H., 1993. *Coastal Fishes of South-eastern Australia.* University of Hawaii Press. Honolulu, Hawaii, p. 437.

Kuiter, R.H., 1998. Pipefishes of the syngnathid genus *Dunckerocampus* (Sygnathiformes: Sygnathidae), with a description of a new species from the Indian Ocean. *Aqua.,* 3(2): 81–84.

Kuiter, R.H., 2001.Revision of the Australian seahorses of the genus Hippocampus (Syngnathiformes: Syngnathidae) with descriptions of nine new species. *Rec. Aus. Mus.* 53: 293–340.

Kuiter,R.H. and Debelius, H., 1994. *Southeast Asia Tropical Fish Guide*. IKAN–Unterwasserarchiv, Frankfurt, Germany, p. 321.

Kulbicki, M., Randall, J.E. and Rivaton, J., 1994. Checklist of the fishes of the Chesterfield Islands (Coral Sea). *Micronesica*, 27(1/2): 1–43.

Lazzari, M.A. and Able, K.W., 1990. Northern pipefish, Syngnathus fuscus, occurrences over the Mid-Atlantic Bight continental shelf: Evidence of seasonal migration. *Environ. Biol. Fish.*, 27: 177–185.

Lieske, E. and Myers, R., 1994. *Collins Pocket Guide. Coral Reef Fishes: Indo-Pacific and Caribbean including the Red Sea.* Haper Collins Publishers, p. 400.

Lourie, S.A. and Randall, J.E., 2003. A new pygmy seahorse, *Hippocampus denise* (Teleostei: Syngnathidae), from the Indo-Pacific. *Zool. Studies*, 42(2): 284–291.

Lourie, S.A. and Kuiter, R.H., 2008. Three new pygmy seahorse species from Indonesia (Teleostei: Syngnathidae: Hippocampus). *Zootaxa,* 1963: 54–68.

Lourie, S.A., Vincent, A.C.J. and Hall, H.J., 1999. *Seahorses: An Identification Guide to the World's Species and their Conservation.* Project Seahorse, London, p. 214.

Masuda, H., Amaoka, K., Araga, C., Uyeno, T. and Yoshino, T., 1984. *The Fishes of the Japanese Archipelago*, Vol. 1. Tokai University Press, Tokyo, Japan. p. 437 (text).

May, J.L. and Maxwell, J.G.H., 1986. *Trawl Fish from Temperate Waters of Australia.* CSIRO Division of Fisheries Research, Tasmania. p. 492.

McEachran, J.D. and Fechhelm, J.D., 1998. *Fishes of the Gulf of Mexico, Vol. 1: Myxiniformes to Gasterosteiformes.* University of Texas Press, Austin, p. 1112.

Myers, R.F., 1991. *Micronesian Reef Fishes*, Second Edn. Coral Graphics, Barrigada, Guam, p. 298.

Nielsen, E., 1960. A new Eocene teleost from Denmark. *Meddelelser fra Dansk Geologisk Forening*, 14: 247–252.

Orr, J.W., 1995. Phylogenetic relationships of Gasterosteiform fishes (Teleostei: Acanthomorpha). *Ph.D. Thesis*, University of Washington, Seattle, WA, USA., 1995).

Paulin, C., Stewart, A., Roberts, C. and McMillan, P., 1989. *New Zealand Fish: A Complete Guide*. National Museum of New Zealand Miscellaneous Series, 19: 279.

Paxton, J.R., Hoese, D.F., Allen, G.R. and Hanley, J.E., 1989. Pisces. Petromyzontidae to Carangidae. Zoological Catalogue of Australia, Vol. 7. Australian Government Publishing Service, Canberra, p. 665.

Pietsch, T.W., 1995. Historical Portrait of the Progress of Ichthyology From its Origin to Our Own Time (translated from G.Cuvier's original by A. J.Simpson). Baltimore, MD: Johns Hopkins University Press.

Pogonoski, J.J., Pollard, D.A. and Paxton, J.R., 2002. Conservation overview and action plan for threatened and potentially threatened marine and estuarine fishes. Environment Australia, Department of Environment and Heritage, Canberra, Australia.

Rainboth, W.J., 1996. *Fishes of the Cambodian Mekong*. FAO Species Identification Field Guide for Fishery Purposes. FAO, Rome, p. 265.

Randall J.E. and Lourie, S., 2009. Hippocampus tyro, a new seahorse (Gasterosteiformes: Syngnathidae) from the Seychelles. *Smithiana Bulletin*, 10: 19–21.

Randall, J.E. and Anderson, C., 1993. Annotated checklist of the epipelagic and shore fishes of the Maldives Islands. *Ichthyol. Bull. of the J.L.B. Smith Inst. of Ichthyol.* 59: 47.

Randall, J.E. and Earle, J.L., 1994. *Doryrhamphus aurolineatus*, a new pipefish (Syngnathidae) from Masirah Island, Oman. *Fauna of Saudi Arabia*, 14: 282–286.

Randall, J.E., 1996. *Caribbean Reef Fishes*, Third Edn. Revised and Enlarged. T.F.H. Publications, Inc. Ltd., Hong Kong, p. 368.

Randall, J.E., Williams, J.T., Smith, D.G., Kulbicki, M., Tham, G.M., Labrosse, P., Kronen, M., Clua, E. and Mann, B.S., 2003. Checklist of the shore and epipelagic fishes of Tonga. *Atoll Res. Bull. Nos.*, pp. 497–508.

Robins, C.R. and Ray, G.C., 1986. *A Field Guide to Atlantic Coast Fishes of North America*. Houghton Mifflin Company, Boston, U.S.A., p. 354.

Robins, C.R., Bailey, R.M., Bond, C.E., Brooker, J.R., Lachner, E.A., Lea, R.N. and Scott, W.B., 1991. Common and scientific names of fishes from the United States and Canada. *Am. Fish. Soc. Spec. Pub.* 20: 183.

Russell, B.C. and Houston, W., 1989. Offshore fishes of the Arafura Sea. *Beagle*, 6(1): 69–84.

Saikia, B. and Goswami, U.C., 2012. Distribution and Conservation status of freshwater species of pipefishes from the family Syngnathidae. *Proc. Zool. Soc. India* (In Press).

Scott, W.B. and Scott, M.G., 1988. Atlantic fishes of Canada. Can. *Bull. Fish. Aquat. Sci.*, 219: 731.

Simmons, E.G., 1957. An ecological survey of the upper Laguna Madre of Texas. *Publ. Inst. Mar. Sci. Unio. Texas,* 4: 156–200.

Skelton, P.H., 1993. A complete guide to the freshwater fishes of southern Africa. Southern Book Publishers, p. 388.

Smith, C.L., 1997. National Audubon Society field guide to tropical marine fishes of the Caribbean, the Gulf of Mexico, Florida, the Bahamas, and Bermuda. Alfred A. Knopf, Inc., New York, p. 720.

Talwar, P.K. and Jhingran, A.G., 1991. *Inland Fishes of India and Adjacent Countries.* Volume 2. A.A. Balkema, Rotterdam.

Teugels, G.G., Leveque, C., Paugy, D. and Traore, K., 1988. État des connaissances sur la faune ichtyologique des bassins côtiers de Côte d'Ivoire et de l'ouest du Ghana. *Rev. Hydrobiol. Trop.*, 21(3): 221–237.

Vianna, M. and Bockmann, F.A., 1995. Note about *Cosmocampus albirostris* (Kaup, 1856) (Gasterosteiformes, Syngnathidae) in the Brazilean southeast coast. *Arq. Biol. Technol.*, 38(3): 893–898.

Vincent, A.C.J., 1995. A role of daily greetings in maintaining seahorse pair bonds. *Aim. Behav.*, 49: 258–260.

Vincent, A.C.J. and Sadler, L.M., 1995. Faithful pair bonds in wild seahorses, *Hippocampus whitei. Anim. Behav.*, 50: 1557–1569.

Vincent, A.C.J., 1994. Seahorses exhibit conventional sex roles in mating competition, despite male pregnancy. *Behaviour*, 128: 131–157.

Vincent, A.C.J., 1996. *International Trade in Seahorses.* Cambridge, UK. TRAFFIC International.

Wheeler, A., 1979. Syngnathidae. In: *Checklist of the Fishes of the North-Eastern Atlantic and of the Mediterranean (CLOFNAM),* (Eds.) J.C. Hureau and Th. Monod. UNESCO, Paris, 1: 274–279.

Whitfield, A.K. and Bruton, M.N., 1996. Extinction of the river pipefish *Syngnathus watermeyeri* in the Eastern Cape Province, South Africa. *S. Afr. J. Sci.*, 92: 59–60.

Woods, C.M.C., 2000. Preliminary observations on breeding and rearing the seahorse Hippocampus abdominalis (Teleostei: Syngnathidae) in captivity. *N.Z.J. Mar. Freshwater Res.*, 34: 475–485.

Yoshino, T. and Yoshigou, H., 1998. First records of two freshwater pipefishes of the genus Microphis (Syngnathiformes: Syngnathidae) from Japan. *Ichthyol. Res.*, 45(2): 201–204.

2013, Animal Diversity, Natural History and Conservation, Vol. 3 *Pages* **177–187**
Editors: **V.K. Gupta and Anil K. Verma**
Published by: **DAYA PUBLISHING HOUSE, NEW DELHI**

Chapter 11

Review on Freshwater Riverine Fin-Fish Diversity in Northern Part of West Bengal, India

Amal Kumar Patra*
Department of Zoology,
Ananda Chandra College, Jalpaiguri – 735 101, West Bengal, India

ABSTRACT

The increasing population of the world has threatened mankind in several ways amongst which the most important one is the food problem. Fish being a valuable and easily accessible source of food, rich in protein contents can be used to overcome the food problem to a certain extent. From pre-historic period these aquatic animals have been used as protein rich diet for human being. Flesh of a fish is constituted by approximately 60-80 per cent water, 13–20 per cent protein small amount of fat, minerals like calcium, phosphorus and iron and vitamins(A, D and B-complex).They provide almost 75 per cent of required proteins for functioning of human body. Omnivorous people of North Bengal prefer the riverine fin-fish than cultured and as well as transported fish from neighbouring states. The knowledge of diversity of fish is one of the pre-requisite for adopting the scientific conservation strategies. In these aspects fish is one of the important research resources in our country. This review focuses on the research works of ichthyofauna diversity studied until now and provides a ready reference for researchers involved and interested in freshwater ichthyofaunal diversity.

Keywords: *Fin-fish, Ichthyofauna, North Bengal.*

---

* *Corresponding author.* E-mail: amalpatra@yahoo.co.in

Introduction

Northern part of West Bengal, popularly mention as North Bengal, is endowed with numerous freshwater rivers. The major riverine fishery resources of Northern West Bengal is provided by the Teesta, Torsha, Jaldhaka, Mahananda, Raidak, Sankosh, Kaljani, Korotoa, Punarbhaba and Atrai river systems (Figure 11.1). After origin from diverse north eastern hill stations they finally blend with Yamuna in

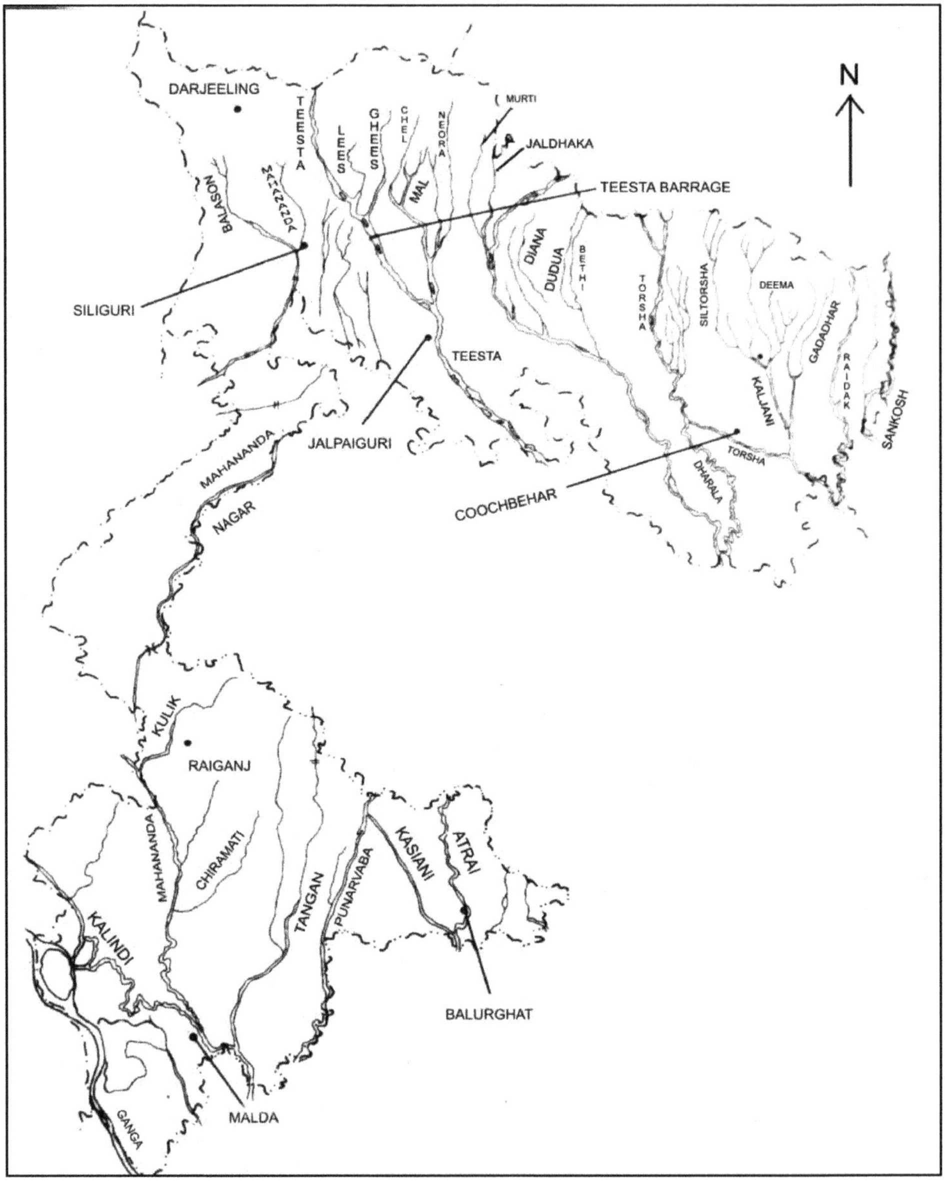

Figure 11.1: Rivers of Northern Part of West Bengal.

Bangladesh. Chief stream of these rivers are connected by numerous tributaries. Their basins sustain the life and livelihood of tea gardeners, farmers, fishermen and slum dwellers. It has long been recognized that the ichthyofauna of the Eastern Himalayas and Assam hills is very rich for this area forms a meeting place of the Indo-Chinese, Indo-Malayan and Indian elements of the Oriental Realm (Hora and Gupta, 1940).The study of the fish of northern part of West Bengal is, therefore, of special interest from a zoogeographical point of view. Comprehensive account of the fish fauna of north Bengal was reported by different ichthyologists at different times. The Pioneer worker, Hamilton (1822), described a fair number of fish that were obtained from North Bengal and it was published in "*An account of fishes found in the River Ganges and its branches*". Ichthyofaunal diversity of North Bengal is reviewed here in two different sub-captions.

Study Before 21st Century

In 1937, M.G.E. Shaw and E.O.Shebbeare published a research article on "The fishes of North Bengal". They studied the fish of rivers, streams and ponds in the hills and plains of the Darjeeling District and the adjoining Doors. They listed 131 species of fishes and a few of which are stated to be exotic as they had been purchased only from the markets at Siliguri and similar other places served by railway. They followed the literature of Dr. C.T. Regan for taxonomy. Among the 131 species, six were found under the order Isospondyli, 101 species were found under the order Ostariophysi, 17 species were found under the order Percomorphi, 3 species were found under the order Opisthomi and a single species each was found under the order Symbranchii, Solenichthyes, Plectognathi and Synentognathi. According to the literature of Jayaram, 1999 these 131 species belong to nine different orders and members of order cypriniformes were more dominated than other eight orders (Figure 11.2).

In November-December, 1938, S.L. Hora and his party of the ZSI (Zoological Survey of India) visited Kalimpong Doors (Mongpong, Chunbhati, Ghish and Burrikhola) and the portion of the Terai in the Siliguri Sub-Division (Siliguri, Sevoke, Kalijhora, Naksalbari and Kharibari) of Darjeeling District. They collected the fish fauna resource from a large number of small streams, ponds and ditches of this region. After visit Hora and Gupta, 1940 published a research article on a collection of fish from Kalimpong Doors and Siliguri Terai, North Bengal. In the article they reported 58 species belonging to five orders and sixteen families. Among the 58 species, 41 species were found under the order Cypriniformes, 8 in Siluriformes, 7 in Perciformes and a single species was found under the order Beloniformes and Synbranchiformes (Figure 11.3).They divided the fish fauna into two principal ecological associations on the basis of the types of locality visited. The first association comprised pool dwelling forms which can be further divided into two categories- those that live in clear water pools in the course of the streams and those that live in stagnant pieces of water. The second association comprised those species that were captured from shallow swift currents. They also added and altered in a list of 131 species of fish of northern Bengal published by Shaw and Shebbeare, 1937.

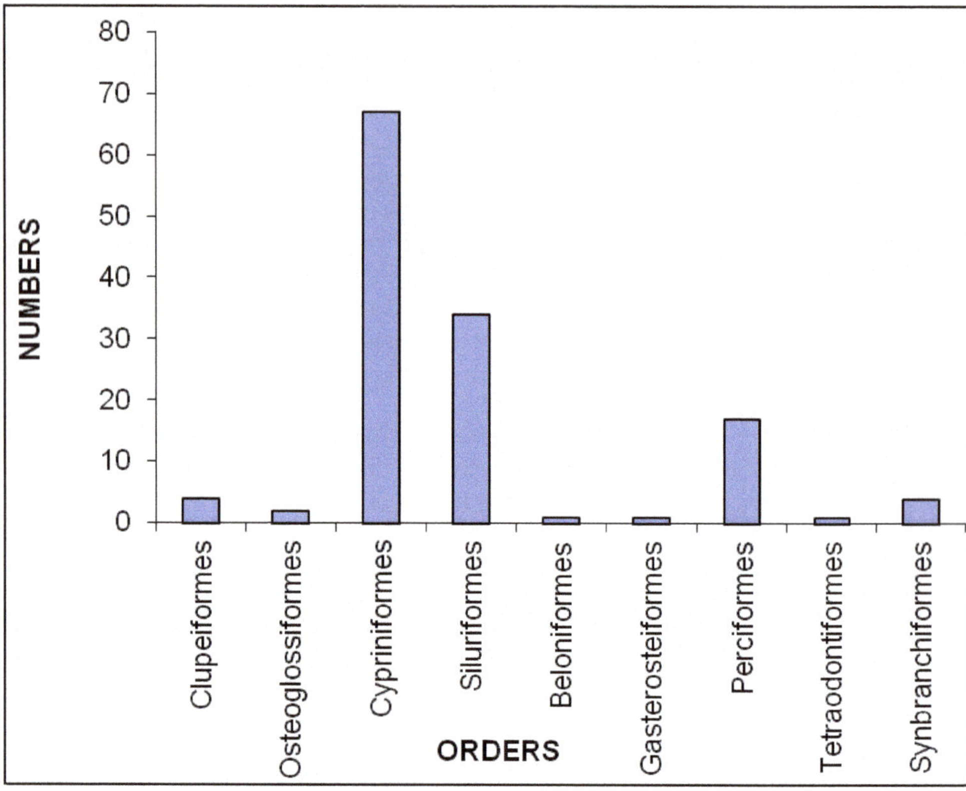

Figure 11.2: Ichthyofauna Diversity in North Bengal (Shaw and Shebbeare, 1937).

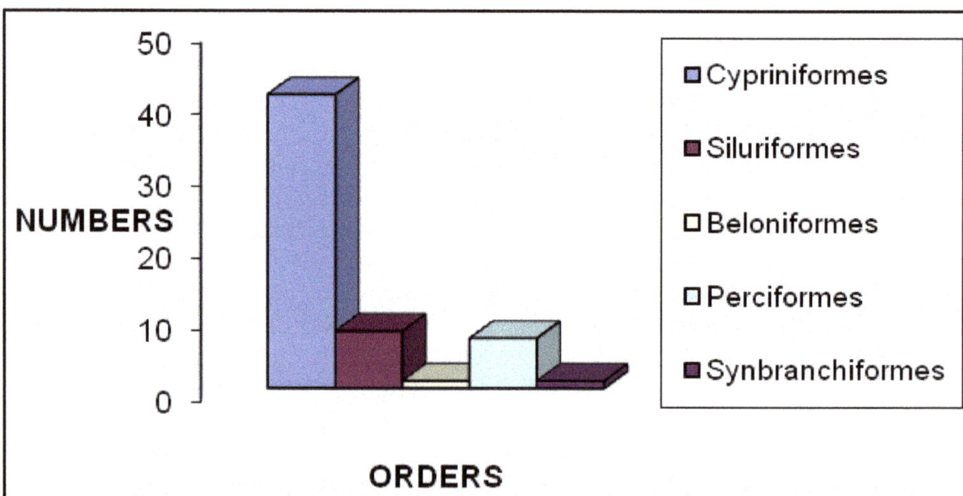

Figure 11.3: Ichthyofauna Diversity at Kalimpong Doors and Siligury Terai of North Bengal (Hora and Gupta, 1940).

With a view mainly to collect specimens of two elusive bagarid species, Jayaram and Singh undertook a survey of North Bengal Rivers in February, 1970. During 1977 they published a research article on a collection of fish from North Bengal. They studied the fish fauna of Mahananda and Teesta with many tributaries. This report listed 96 species, of which 17 was not reported by Shaw and Shebbeare (1937) and Hora and Gupta (1940). Among the 96 species 47 species were found under the order Cypriniformes, 26 in Siluriformes, 12 in Perciformes, 3 in Clupiformes and Synbranchiformes and a single species was found under the order Beloniformes, Osteoglossiformes, Tetraodontiformes, Cyprinodontiformes and Muguliformes (Figure 11.4).

Study Throughout 21st Century

A review of the freshwater fish fauna of West Bengal, India with suggestions for conservation of threatened and endemic species was published by R.P.Barman, 2007.The main aim of this review was to update the list of the freshwater fish of West Bengal in view of the changes in the nomenclature of the fish from this state by Sen, 1992. The work had listed 239 species of fish. This list also included seven Indian endemic fish species. The distribution list of the seven Indian endemic species found in this state revealed that six species had been recorded from North Bengal along with other states including two species, *Batasio batasio* (Hamilton) and *Hora horai* Misra which are exclusively restricted in North Bengal only throughout India. The

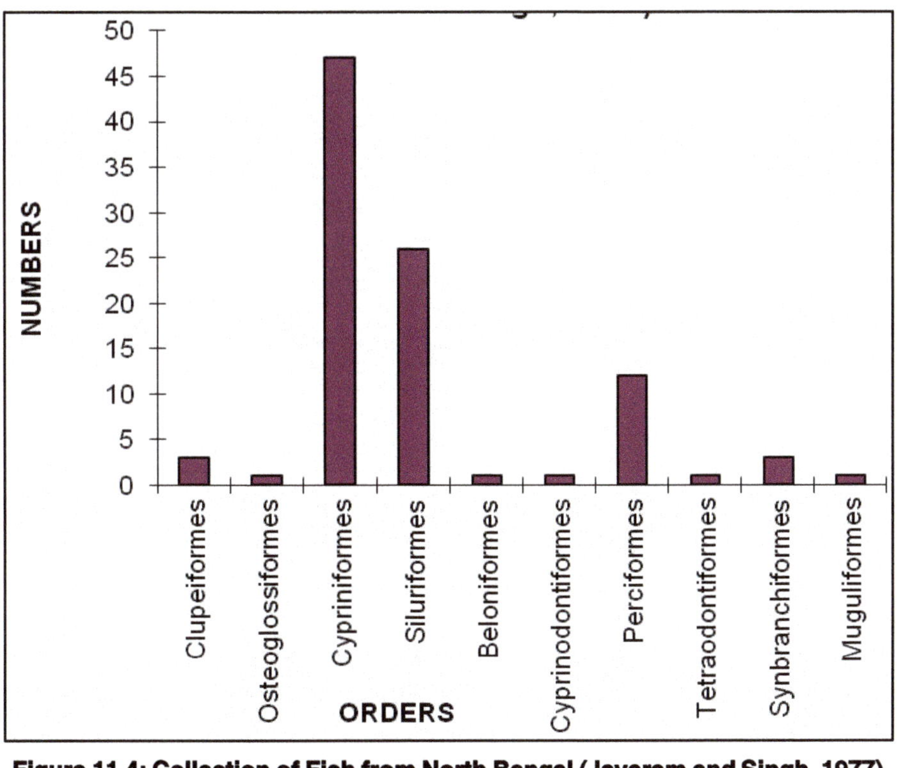

Figure 11.4: Collection of Fish from North Bengal (Jayaram and Singh, 1977).

author concluded that, since North Bengal contains several endemic, threatened and some unique species, this region may be considered as the 'Hot spot' of bio-diversity in West Bengal.

The ichthyofaunal diversity in the freshwater river of South Dinajpur District of West Bengal, India was reported by Chakraborty and Bhattacherjee, 2008. The main objective of the study was to estimate the present status of the annual fish catch of South Dinajpur district, West Bengal, with reference to the species composition of the rivers and kharis during pre-monsoon, monsoon and post-monsoon seasons.A two-year survey (2004-06) was undertaken in major rivers, like Atrai, Punarbhaba, Tangan, Jamuna, and in minor rivers, like Ichhamoti (Figure 11.5). They visited 46 fields in major rivers with the assistance of local fishermen during the entire study. Fish were collected and immediately preserved in neutral formalin (10 per cent).The fish were identified by the literature of Shaw and Shebbeare (1937), Menon (1974), Jhingran (1991), Talwar and Jhingran (1991), Menon (1999) and Jayaram (1999).The study reported 49 species which belong to 5 orders, 6 suborders, 17 families, 11 subfamilies and 40 genera. Among the 49 species, 23 species were found under the order Cypriniformes, 16 in Siluriformes, 8 in Perciformes and a single species was found under the order Cyprinodontiformes (Beloniformes) and Tetraodontiformes (Figure 11.6).

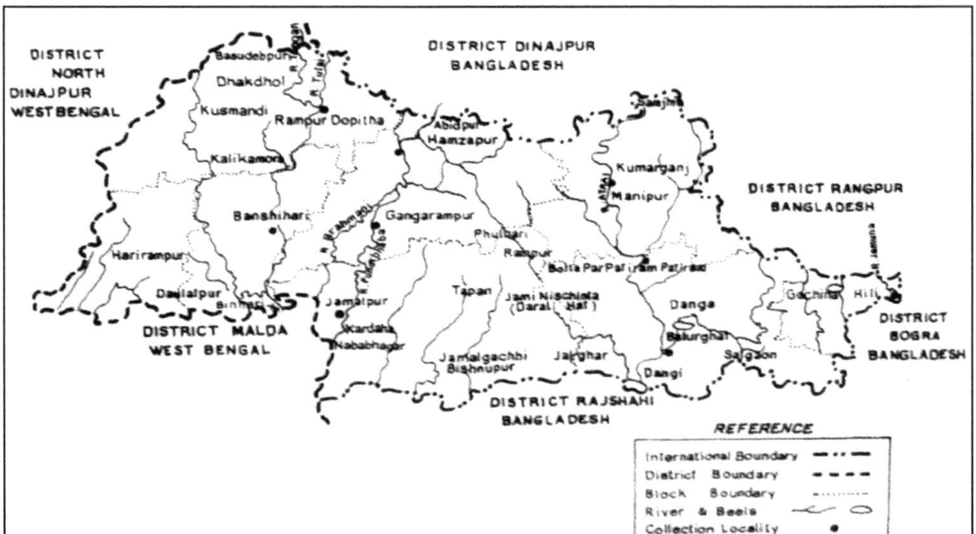

Figure 11.5: Rivers of South Dinajpur.

Impact of fishing methods on conservation of ichthyofauna diversity of river Relli (latitude 26°51' to 27°12'N and longitude 88°28' to 88°53'E) in Darjeeling Himalaya of West Bengal was published by Acharjee and Barat, 2010.They reported fourteen species of fish which were captured by diversion of river channel, cast netting, scoop netting, angling, spearing, rock-striking or hammering, electro-fishing, river poisoning, kurthunnu (local device), tip (a trap), thokre and thali.

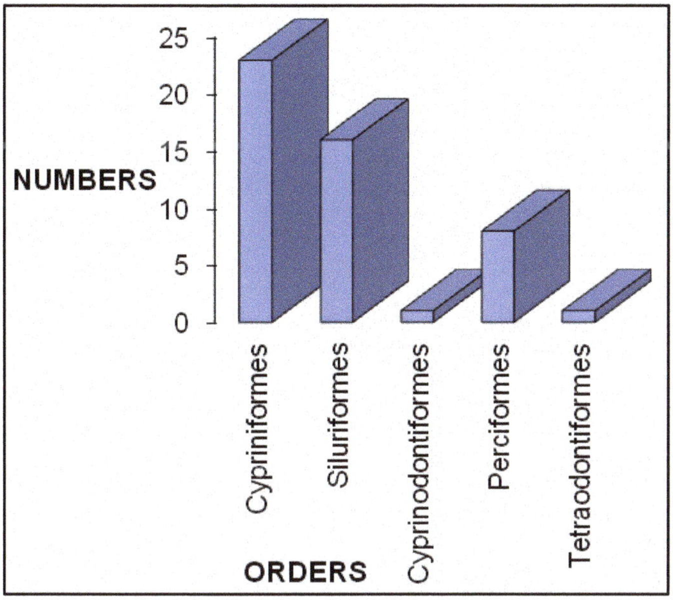

Figure 11.6: Ichthyofaunal Diversity in the Freshwater Rivers of South Dinajpur (Chakraborty and Bhattacherjee, 2008).

The ichthyofauna diversity of the Karala River (latitude 26°47′13" N to 26°28′42" N, longitude 88°32′17" E to-88°44′27"E and elevation from mean sea level 402 ft. to 268 ft.), a tributary of Teesta River at Jalpaiguri District of West Bengal was studied from the period of February, 2009 to July, 2010.The main objective was to study the magnitude and diversity of finfish resource in this river. Three different stations (Figure 11.7) were selected in that river for ichthyofauna diversity. The station-I (Aquiduct) is near the origin of the river. The station-II (Hakimpara) is located in the district town and behind the hospital and market. The station-III is located at the confluence of Teesta and Karala near Daspara. The fishes were captured at three stations by different nets. At station-I the fishes were captured by cast net (mesh size 6mm. X 6 mm.) and naphi jal (local contrivance, mesh size 5 mm × 5 mm). At station II and III the fishes were captured by vessel net or khara jal (local contrivance, mesh size 6 mm. X 6 mm.), gill net (variable mesh sizes), and cast net (5 mm × 5 mm mesh size). After capture the color, color patterns, spots etc. were noted immediately and their picture was taken by a powerful camera. The fishes were killed by formalin solution containing one part commercial formalin (37-40 per cent HCHO) + nine part distilled water and 7 gm Borax/liter (Jayaram, 1981). All the fishes were kept in this buffer formalin solution for 4-5 hours for proper fixation. The fishes were identified by the literature of Talwar and Jhingran, (1991) and Jayaram (1999, 2006). The diversity and evenness indices were calculated by the references of Shannon and Wiener (1949) and Pielou (1975).

A total of fifty five species belonging to eight orders and twenty families were identified. Among the 55 species, thirty one species were found under the order Cypriniformes, ten species under the order Perciformes, seven species were found

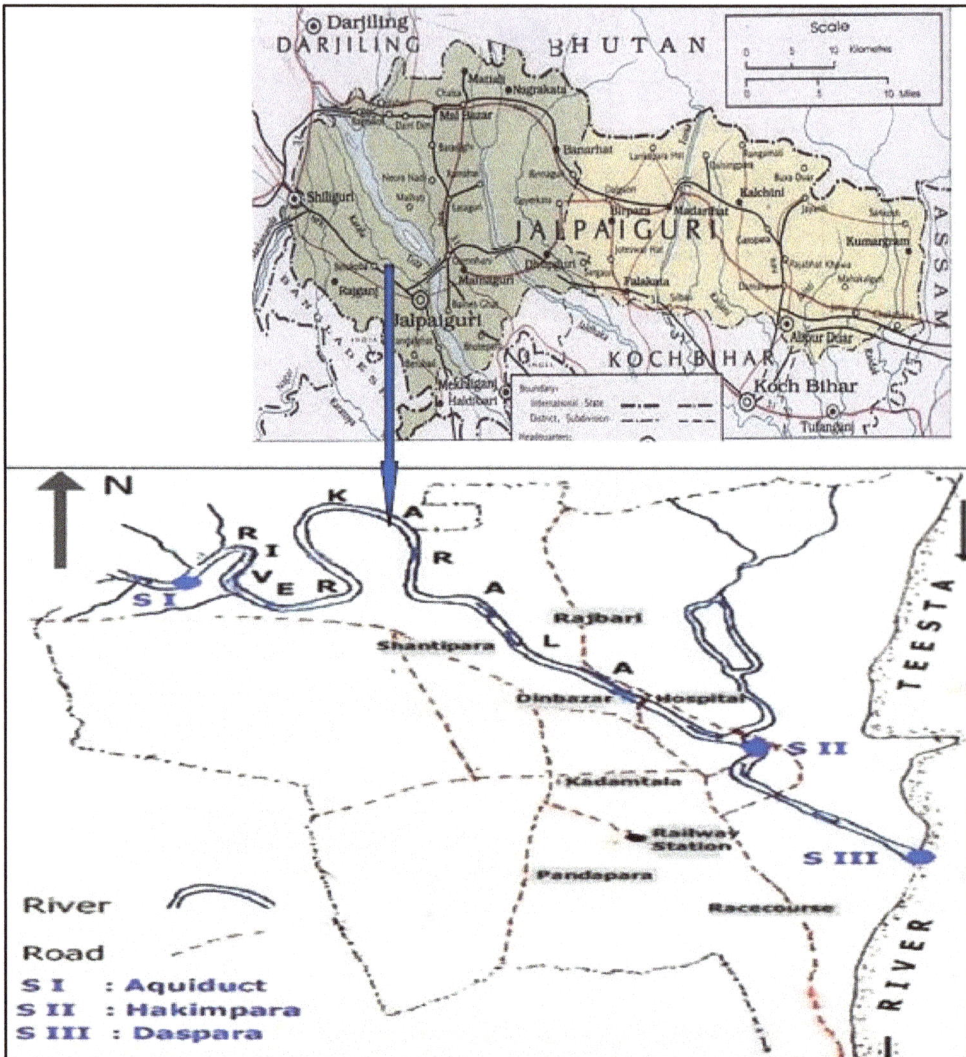

Figure 11.7: Karala River and Study Sites.

under the order Siluriformes, three species under the order Synbranchiformes and a single species each under the order Osteoglossiformes, Anguilliformes, Beloniformes and Tetraodontiformes each (Patra *et al.*, 2011). Of these eight orders, cypriniformes is considered as dominant order (Figure 11.8). On the basis of catch frequency (The frequency of occurrence of each species was calculated based on the number of occasions the species was collected during the samplings by Tamang *et al.*, 2007) 20 per cent were common, 13 per cent were abundant, 11 per cent were frequent, 16 per cent were occasional, 22 per cent were sporadic, 14 per cent were rare and 4 per cent were extremely rare. According to IUCN, 4 per cent was endangered, 16 per cent was vulnerable, 36 per cent was near threatened and rest (44 per cent) were indeterminate

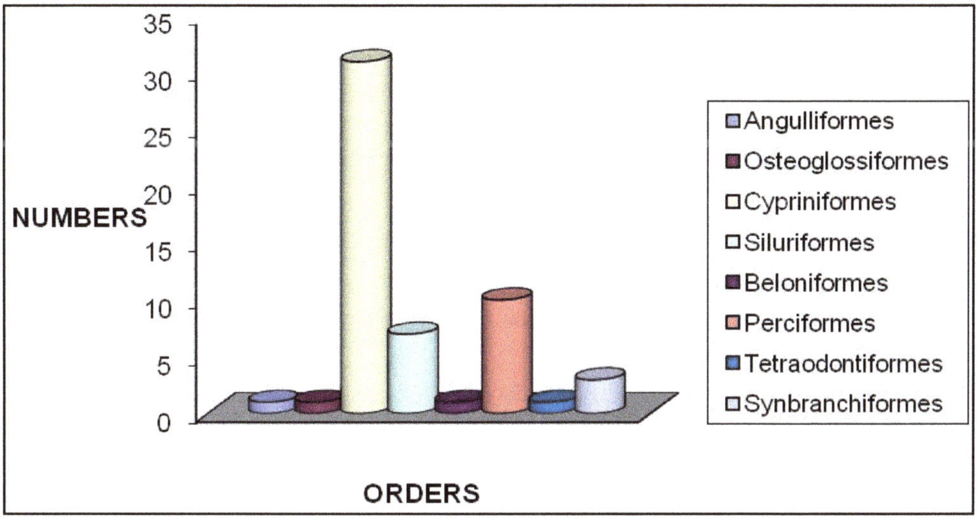

Figure 11.8: Ichthyofauna Diversity in Karala River (Patra *et al.*, 2011).

or out of danger. The species richness in three sampling sites of this river showed considerable variations and higher richness was recorded in the mid stream to down stream. Maximum species richness was recorded from SII (Hakimpara, total number=54) and SIII (Daspara, total species=52) while lower species richness was recorded from SI (Aquiduct, total number=36).The species diversity index of different sampling sites ranged from 3.18 to 3.75. In this study maximum fish diversity index was recorded higher in SIII (H2 = 3.75) as compared with SII (H2 =3.63) and SI (H2 =3.18).The evenness index of three sampling stations (SI=0.887, SII=0.91and SIII =0.949) indicates uneven distribution of fish in this tributary.

Conclusions

West Bengal is an important state of irrigated India. Northern part of this state is flooded by plentiful freshwater rivers which are filled up through rain water as well as glacier melting water. Water resource of these rivers not only sustains the tea gardener and farmers but helps to generate electricity also. One of the important biotic resources of these rivers is fish. The knowledge of the diversity of fish is one of the prerequisite for adopting the proper conservation strategies of fish fauna. Out of 39,900 identified vertebrates; fish contains 54.44 per cent of which 38.72 per cent is freshwater species (Jayaram, 1999).Till date more than 140 species have been reported from the rivers of North Bengal and they contribute 58.58 per cent of Bengal freshwater fish fauna. Hot spot, a term borrowed from physical sciences, is declared on the basis of endemic species richness (Myers, 1998). Northern West Bengal contributes six endemic fish of total seven Indian endemic fish fauna (Barman, 2007). Two species of catfish are exclusively restricted in this region. Since North Bengal contains numerous endemic and some distinctive fish fauna, this region may be consider as the "Hot Spot" of biodiversity in West Bengal.

Acknowledgements

I am thankful to my beloved colleagues Dr. Tanmay Datta, Department of Zoology; Bipul Chandra Sarkar, Department of Geography; Dr. Suman Sengupta, Department of Chemistry; Santanu Ghosh Dastidar, Department of Computer Science and my dear student Debarshi Sarkar, Research fellow in BHU for encouragement of the review work.

References

Acharjee, M.L. and Barat, S., 2010. Impact of fishing methods on conservation of ichthyofauna of river Relli in Darjeeling Himalaya of West Bengal. *J. Env. Biol.*, 31: 431– 434.

Barman, R.P., 2007. A review of the freshwater fish fauna of West Bengal, India with suggestions for conservation of the threatened and endemic species. *Records of the Zoological Survey of India, Occasional Paper*, 263: 1– 48.

Chakraborty, T. and Bhattacherjee, S., 2008. The Ichthyofaunal diversity in the freshwater rivers of South Dinajpur District of West Bengal, India. *J. Bombay Nat. Hist. Soc.*, 105(3): 292–298.

Hamilton, F., 1822. An account of fishes found in the River Ganges and its branches. Edinburg and Richardson and Co., London, VII +405 pp. + 39 pls.

Hora, S.L. and Gupta, J.C., 1940. On a collection of fish from Kalimpong, Duars and Siliguri Terai, North Bengal. *J.Roy. Asiat. Soc. Bengal, Science*, 6(8): 77–83.

Jayaram, K.C. and Singh, K.P., 1977. On the collection of fish from North Bengal. *Rec. Zool. Sur. Ind.*, 72(1–4): 243–275.

Jayaram, K.C., 1981. Methods of preservation of fishes, p. 5. In: Director, ZSI (Ed.). *The Freshwater Fishes of India, Pakistan, Bangladesh, Burma and Srilanka: A Handbook.* Calcutta Laser Graphics (P) Ltd., Calcutta, iii+475 pp. + 13 pls.

Jayaram, K.C., 1999. *The Freshwater Fishes of the Indian Region*. Narendra Publishing House, New Delhi, xxvii + 551 pp. + 18 pls.

Jayaram, K.C., 2006. Catfishes of India. Narendra Publishing House, New Delhi, xxii+383+11 pls.

Jhingran, V.G., 1991. *Fishes and fisheries of India.* Hindustan Publishing Company, New Delhi, p. 727.

Menon, A.G.K., 1974. A checklist of the fishes of the Himalayan and Indo-Gangetic Plains. *Inland Fish. Soc., India* (Barrackpore), pp. viii +136.

Menon, A.G.K., 1999. Freshwater fishes of India. *Records of the Zoological Survey of India, Occasional Paper*, 175: 1–336.

Myers, N.1998. Global biodiversity priorities and expanded conservation policies In: *Conservation in a Changing World*, (Eds.) G.M. Mace, A. Balmford and B.R. Ginsberg. Cambridge Univ. Press, Cambridge, pp. 273–285.

Patra, A.K., Sengupta, S. and Datta, T., 2011. Physico–chemical properties and Ichthyofauna diversity in Karala River,a tributary of Teesta River at Jalpaiguri District of West Bengal, India. *Int. J. App. Biol.Phar.Tech.,* 2(3): 47–58.

Pielou, E.C., 1975. *Ecological Diversity.* John Wiley and Sons, New York, NY, p. 165.

Sen, T.K., 1992. Freshwater fish. Fauna of West Bengal. *State fauna series,* 3 (2): 102–242.

Shannon, C.E. and Weaner, W., 1949. *The Mathematical Theory of Communication.* University of Illinois Press, Urbana, p.163.

Shaw, G.E. and Shebbeare, E.O., 1937. The fishes of North Bengal. *Journal of the Royal Asiatic Society of Bengal* (*Science*), 3(1): 1–137.

Talwar, P.K. and Jhingran A.G., 1991. *Inland Fishes of India and Adjacent Countries,* Oxford and IBH Publishing Company, New Delhi, pp. xix +1158.

Tamang, L., Chaudhry, S. and Choudhury, D., 2007. Ichthyofaunal contribution to the state and comparison of habitat contiguity on taxonomic diversity in Senkhi stream, Arunachal Pradesh, India. *J. Bombay Nat. Hist. Soc.,* 104(2): 170–177.

2013, Animal Diversity, Natural History and Conservation, Vol. 3 *Pages* **189–195**
Editors: **V.K. Gupta and Anil K. Verma**
Published by: **DAYA PUBLISHING HOUSE, NEW DELHI**

Chapter 12

Length-Weight Relationships of Eight Fish Species from the Bulgarian Black Sea Waters

Maria Yankova[1]*, Danail Pavlov[2] and Vesselina Mihneva[1]
[1]Institute of Fishing Resources
[2]Society of Innovative Ecologists in Bulgaria

ABSTRACT

Length-weight relationships of 9755 specimens of fish caught in the Bulgarian Black Sea waters were studied. Spiny dogfish (*Squalus acanthias* L.), thornback ray (*Raja clavata* L.), pontic shad (*Alosa immaculata* Eichwald), whiting (*Merlangius merlangus euxinus* Pallas), horse mackerel (*Trachurus mediterraneus* Aleev), bonito (*Sarda sarda* Bloch), round goby (*Neogobius melanostomus* Pallas) and turbot (*Psetta maxima* L.) were collected from May 2006 to December 2008. The fish were caught by different fishing techniques such as gill net, trawl net and trap nets. The slope or allometric coefficient (*b*) of the functional regression between length and weight values varied between 2.301 and 3.839. The length-weight relationships for *S. acanthias* and *R. clavata* have been reported for the first time for the Bulgarian Black Sea waters.

Keywords: *Length-weight relationships, Fish species, Black Sea waters, Bulgaria.*

Introduction

The knowledge of the biology of commercially important fish in economic terms (size values, *i.e.*, minimum, maximum, mean, and size relationships, *i.e.*, length-weight)

---

* *Corresponding author.* E-mail: maria_y@abv.bg

helps on sustainable exploitation of the Black Sea natural resources. Length and weight measurements in conjunction with age data give information on the stock composition, age at maturity, life span, mortality, growth and production (Bolger and Connoly,1989; Diaz *et al.,* 2000). Length-weight relationships (LWR) allow fisheries scientists to convert growth in length equations to growth in weight in stock assessment models (Morato *et al.,* 2001; Stergiou and Moutopoulos, 2001), estimate biomass from length frequency distribution (Petrakis and Stergiou, 1995; Dulèiæ and Kraljeviæ, 1996) and calculate fish condition (Petrakis and Stergiou, 1995). However, the information about the length-weight relationships of fish species in the Western part of the Black Sea is very scarce and incomplete. Data about previous investigations of LWR for fish species from the Mediterranean, Black and Azov Seas are presented in Table 12.1.

Table 12.1: Some of the Results of LWR Parameters Estimated in different Localities for the Similar Fish Species Targeted in the Present Study

	L_{min}-L_{max}	*a*	*b*	*Area*	*Sources*
M. merlangus	9-24	0.0039	3.24	Mid Black Sea	Samsun and Erkoyuncu, 1998
	5-40	0.0052	3.14	Eastern Black Sea	Genç *et al.,* 1999
	5-32.5	0.0042	3.24	Black Sea	Ismen, 2002
		0.0048	3.11	Bulgaria	Maximov et al.2009
S. sarda		0.0149	2.984	Adriatic Sea	Zorica and Sinovèiæ, 2008
A. immaculata	9-36	0.0629	2.55	Bulgaria	Kolarov, 1991
	11.6-31.6	0.00212	3.39	Mid Black Sea	Samsun, 1995
T. mediterraneus	10.5–17	0.035	3.30	Bulgaria	Yankova *et al.,* 2010
		0.0051	3.17	Eastern Black Sea	Genç *et al.,* 1999
S. acanthias	27.00-70.50	0.0031	3.10	Aegean Sea	Filiz and Mater, 2002
		0.0040	2.95	Southeastern Black Sea	Avsar, 1996
R. clavata	20.50 -99.00	0.0016	3.29	Aegean Sea	Filiz and Mater, 2002
P. maxima		0.0074	3.22	Black Sea	Samsun et al.2006
N. melanostomus		0.0972	2.431	Sea of Azov	Froese and Pauly, 2009

In this study, we report the length-weight relationship parameters of eight species that are of economic and ecological relevance in the commercial fisheries off the Bulgarian Black Sea Coast.

Materials and Methods

The investigated area includes the Bulgarian Black Sea waters in front of Cape Kaliakra, Varna, Cape Emine, Bourgas, and Sozopol (Figure 12.1) and the study was conducted from May 2006 to December 2008. A total of 9755 specimens from eight fish species were collected by different fishing techniques (gill net with mesh size 2a

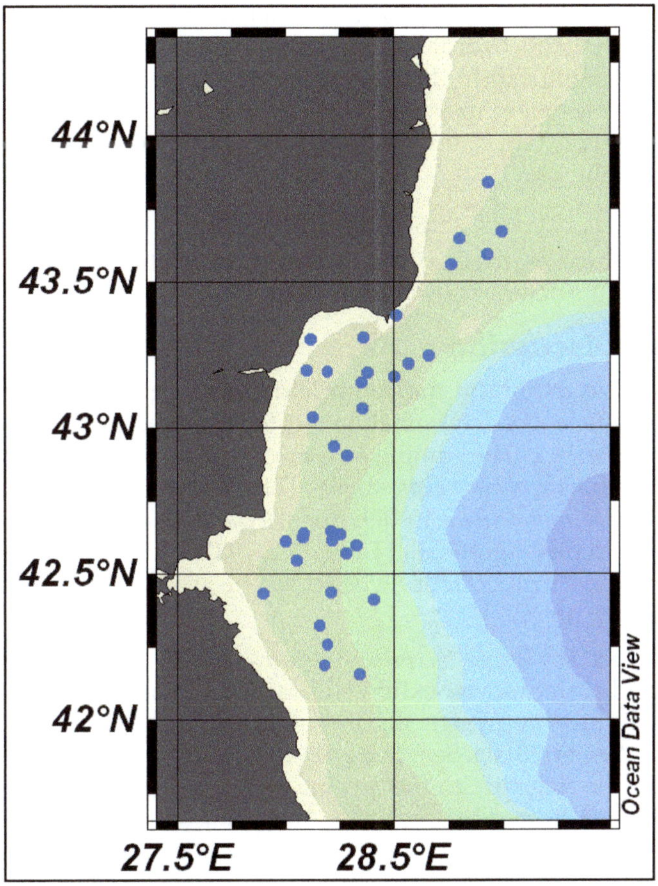

Figure 12.1: Scheme of Area Investigated

= 400 mm; trawl net with an "Effective" part of the mouth of 16 m, vertical opening of 4 m, mesh size of the cod end of 6.5 mm; and trap nets). The samples were transported to the research laboratory in polythene bags containing ice blocks to prevent spoilage and then stored in a deep freezer (-30°C) to avert deterioration. Total length of each fish (measured to the nearest 0.1 cm) was taken from the tip of the snout to the extended tip of the caudal fin using a measuring board. Body weight was measured to the nearest gram using a balance. Fîur species: *Squalus acanthias, Psetta maxima, Raja clavata*, and *Merlangius merlangus euxinus* were collected by the research vessels "Elis" and "RK-3" (boat lengths 13.7 and 17 m, engine horsepower 1.7 and 1.8 hp). After hauling, the catch was removed and analyses were carried out on the deck of the research vessel and later in the laboratory. The mathematical function suggested by Le Cren (1951) was used in estimating length –weight relationship.

$$W = aL^b$$

where,

W is total body weight (g), L the total length (cm), a and b are the coefficients of the functional regression between W and L. Allometric coefficient b value larger or smaller than 3.0 shows an allometric growth or isometric growth when it is equal to 3.0 (Bagenal and Tesch, 1978). In order to confirm whether *b* values obtained in the linear regressions were significantly different from the isometric value (b=3), t-test for independent samples were used. Data concerning t-test are represented in Table 12.2. The values of constant *a* and *b* were estimated from the log transformed values of length and weight, *i.e.*,

$$logW = loga + b\ logL$$

via least square linear regression. The degree of association between the variables was computed by the determination coefficient, r^2.

Results and Discussion

The sample size, minimum, maximum and mean lengths and weights, the values of *a* and *b*, and coefficient of determination r^2 of ten fish species belonging to 8 families are shown in Table 12.2. The sample size ranged from 22 individuals for *Squalus acanthias* to 3910 for *Neogobius melanostomus*. The coefficient of determination ranged from 0.78 for *Alosa immaculata* to 0.99 for *Merlangius merlangus euxinus* and all regressions were highly significant (Table 12.2, $p<0.001$). The *b* value ranged from 2.3019 for *Raja clavata* to 3.8390 for *Sarda sarda*. The growth was negative allometric for *R. clavata* and *A. pontica.*

S. acanthias, M. merlangus euxinus, Trachurus mediterraneus, S. sarda, Neogobius melanostomus and *Psetta maxima* exhibited positive allometric growth. The findings reported in this study represent the first data on the LWRs of *S. acanthias* and *R. clavata* in the Western Black Sea. The growth was negative allometric (Table 12.2) for *R. clavata*. In the Aegean Sea the *b* value for the same species denoted positive allometric growth (Filiz and Mater, 2002). Àccording to Moutopoulos and Stergiou (2002), such differences in the *b* values could be ascribed to differences in the number of specimens examined, as well as to the area and season effects. The latter implies distinctions in the observed length ranges of the specimens caught, to which duration of sample collection can be added as well.

The horse mackerel showed positive allometric growth (Table 12.2), with result similar to the data reported by Yankova *et al.* (2010) for Bulgarian waters, but different of Genç *et al.* (1999) for eastern Black Sea. This discrepancy may be caused by the specific habitats conditions and morphological differences. According to Aleev (1959) the species is represented by four local sub-populations, each one with its own biological characteristics.

The first length-weight relationship of *A. immaculata* in Bulgarian waters (Kolarov, 1991) displayed negative allometry similar to those established in the present study (Table 12.2), but different from the population data in the central Black Sea, that has exhibited a positive allometry (Samsun, 1995). The differences of allometric growth (between the values of *b*) might arose from several ecological factors variations, such as temperature, specific spawning and feeding conditions, and biotope characteristics as well (Ricker, 1975).

M. merlangius and *S. sarda* exhibited positive allometric growth (Table 12.2). The prevoius data about waiting LWRs in the Black Sea are presented at Table 12.1.

Table 12.2: Mean Total Length, Weight, Parameters of the LWR, and Statistical Results from the t-test for 8 Species from the Bulgarian Black Sea Coast

	Length Characteristics			Weight Characteristics				Parameters of Relationship					
Species	*N*	*MTL*	*± SE*	TL_{min}–TL_{max}	M_W	*± SE*	W_{min}–W_{max}	*a*	*b*	r^2	*SE(b)*	*t-test*	*P*
Squalidae													
Squalus acanthias	22	125	0.02	112–144	10.35	0.68	6.2–14.2	0.001	3.153	0.81	0.18	0.16	0.89
Rajidae													
Raja clavata	24	68.81	1.38	56–79	2.89	0.24	1.2–5.5	0.001	2.301	0.86	0.16	1.71	0.11
Clupeidae													
Alosa immaculata	191	29.11	0.14	24.2–37.7	318.66	4.34	175–515	0.071	2.488	0.78	0.24	2.70	0.03
Gadidae													
Merlangius merlangus euxinus	3715	15.23	0.05	5.5–22.5	29.02	0.29	1.05–80.9	0.004	3.151	0.99	0.15	0.88	0.43
Carangidae													
Trachurus mediterraneus	1432	13.38	0.04	7–18.4	20.19	0.21	4.5–55	0.005	3.168	0.92	0.35	0.97	0.38
Cybiidae													
Sarda sarda	411	33.35	0.08	29–37.6	540.3	5.26	300–880	0.001	3.839	0.89	0.22	2.34	0.08
Gobiidae													
Neogobius melanostomus	3910	15.45	0.01	13.6–19.2	58.11	0.16	37.5–113	0.006	3.346	0.98	0.21	0.28	0.80
Bothidae													
Psetta maxima	50	53.05	0.67	44.7–71.7	2.72	0.12	1.39–5.96	0.001	3.278	0.84	0.25	0.90	0.42

During the present study, the *b* values were generally in agreement with the former results. The allometric coefficient of Bonito given by Zorica and Sinovèiæ (2008) for the Adriatic Sea, is smaller than those obtained in the present study. This might be due to differences in sampling periods and to variations in the food availability or water temperatures between the both basins.

The growth were positive allometric for *P. maxima*, *S. acanthias* and *N. melanostomus*. Samsun *et al.* (2007) also reported positive allometric growth for *P. maxima* in the Black Sea. The length-weight relationship constants of spur dog fish from Aegean Sea are close to the data given in this study, but different from the *b* values given by Avsar (1996) for southeastern Black Sea. This difference might be due to interpopulation variations and to probably the differences in sampling time.

Acknowledgments

The authors would like to thank the crews of fishing vessels "Elis" and "RK-3" for their help and collaboration.

References

Aleev, Yu. G., 1959. On the reproduction of the southern population of the horse mackerel in the western areas of Black Sea. *Tr. Sevastopol. Biol. St.,* 12: 270–285. (in Russian).

Avsar, D., 1996. Sex, Age and Growth of the Spur dog (*Squalus acanthias*, L., 1758) in the Southeastern Black Sea. *Yugoslav Jour. of Operations Research*, 6(2): 295–304.

Bagenal, T. and Tesch, F.W., 1978. Age and growth. In: *Methods for Assessment of Fish Production in Freshwater*. IBP Handbook No. 3, (Ed.) T. Bagenal. Blackwell Scientific Publications, Oxford, pp. 101–136.

Bolger, T. and Connoly, P.L., 1989. The selection of suitable indices for the measurement and analysis of fish condition. *Journal of Fish Biology*, 34: 171–182.

Diaz, L.S. Roa, A., Garcia, C.B. Acero, A. and Navas, G., 2000. Length-weight relationships of demersal fishes from the upper continental slope off Colombia. *The ICLARM Quarterly*, 23(3): 23–25.

Filiz, H. and Mater, S., 2002. A preliminary study on length–weight relationships for seven elasmobranch species from North Aegean Sea, Turkey. *Journal of Fisheries and Aquatic Sciences,* Issue (3–4): 401–409.

Froese, R. and Pauly, D. (Eds.), 2009. FishBase.World Wide Web electronic publication. www.fishbase.org, version (02/2009).

Genç, Y., Zengin, M. Baþar, S. Tabak, Ý. Ceylan, N. Çiftci, Y. Ustündað, C. Akbulut, B. and Sahin T., 1999. The Research Project of economical Marine Products. TKB, Central Fisheries Research institute, Trabzon, p. 157 (in Turkish).

Ismen, A., 2002. A preliminary study on the population dynamics parameters of whiting, (*Merlangus merlangus euxinus*) in the Turkish Black Sea coastal waters. *Turkish Journal of Zoology,* 26: 157–166.

Kolarov, P., 1991. *Alosa pontica pontica* (Eichwald, 1938) In: H. Hoestlandt (Ed.), the freshwater fishes of Europe. Vol. 2. Clupeidae, Anguillidae, http: // www.fishbase.org/,last accessed: 28 March 2006). AULA– Verlag Wiesbaden: 337–368.

Le Cren, E.D., 1951. The length–weight relationship and seasonal cycle in gonad weight and condition in the perch (*Perca fluviatilis*). *J. of Anim. Ecol.,* 20: 201–219.

Maximov, V., Raykov, V., Yankova, M. and Zaharia, T., 2009. Whiting (*Merlangius merlangus euxinus*) population parameters in the Romanian and Bulgarian Black Sea Coast during 2000–007. *4th International Environmental Conference of Katerini–Pieria: "Life Quality and Capacity Building in the frame of a Safe Environment"*. Abstracts, p. 61.

Morato, T.P., Afonso, P., Lourinho, P., Barreiros, J.P., Santos, R.S. and Nash, R.D.M., 2001. Length-weight relationships for 21 coastal fish species of the Azores, north-eastern Atlantic. *Fishery Research,* 50: 297–302.

Moutopoulos, D.K. and Stergiou, K.I., 2002. Length-weight and length-length relationships of fish species from the Aegean Sea (Greece). *Journal Applied Ichthyology,* 18: 200–203.

Petrakis, G. and Stergiou K.I., 1995. Weight-length relationships for 33 fish species in Greek waters. *Fishery Research,* 21: 465–469.

Samsun, O., 1995. The weight-length relationship of the shads (*Alosa pontica* Eichw., 1938) in the mid of the Turkish Black Sea. *E. Û. Su. Ûrnleri Dergisi,* 12(1–2): 15–21(in Turkish).

Samsun, N. and Erkoyuncu, I., 1998.The research on the estimation of some parameters of whiting (*Gadus merlangus euxinus* Nordmann, 1840) caught by the bottom trawlers from the viewpoint of fishery biology. *E. Û.Su. Ûrnleri Dergisi,* 15(1–2): 19–31(in Turkish).

Samsun, N., Kalayci, F. and Samsun, O., 2007. Seasonal Variation in Length, Weight, and Sex Distribution of Turbot (Scophthalmus maeoticus Pallas, 1811) in the Sinop Region (Black Sea) of Turkey. *Turk. J. Zool.,* 371–378.

Stergiou, K.I. and Moutopoulos, D.K., 2001. A review of length-weight relationships of fishes from Greek marine waters. *Naga, ICLARM Q.,* 24(1–2): 23–39.

Yankova, M., Raykov, V., Gerdzhikov, D. and Bogomilova, P., 2010. Growth and Length–Weight Relationships of the Horse Mackerel, *Trachurus mediterraneus ponticus* (Aleev, 1956), in the Bulgarian Black Sea Coast. *Turk. J. Zool.,* 34(1): 1–8.

Zorica, B. and Sinovèiæ, G., 2008. Biometry, length-length and length-weight relationships of juveniles and adults of Atlantic bonito, *Sarda sarda,* in the eastern Middle Adriatic Sea ISSN: 0001–5113. *Acta Adriatica.,* 49 (1): 65–72.

2013, Animal Diversity, Natural History and Conservation, Vol. 3 *Pages* **197–214**
Editors: **V.K. Gupta and Anil K. Verma**
Published by: **DAYA PUBLISHING HOUSE, NEW DELHI**

Chapter 13

Exotic Ornamental Fishes in India: Perspectives and Problems

A.K. Singh, Dinesh Kumar, Sharad C. Srivastava and Abubakar Ansari*

National Bureau of Fish Genetic Resources

Canal Ring Road, P.O. Dilkusha, Lucknow – 226 002, Uttar Pradesh, India

ABSTRACT

Ornamental fish comprising of more than 4000 freshwater and 1400 marine species are traded internationally each year, making it one of the most important components of the global fish trade. The ornamental fish trade in India is dominated mainly by the presence of over 600 alien fish species varieties such as goldfish, angel, guppy, swordtail, oscar, platy, pacu and may others. They have been introduced from different parts of world but mainly from Asia. Majority of them are freshwater while marine aquaria fishes are also becoming popular in recent years. In India, there is a huge demand for exotic ornamental fishes. The domestic ornamental fish market is worth around INR 500 million and the demand is increasing at 20 per cent annually. The export from country is mainly confined to freshwater varieties from the natural sources predominantly from the Eastern and Western ghats (85 per cent) and a few locally bred varieties of exotic species (15 per cent). Ornamental fish breeding and culture is confined in and around the metro-cities *viz.* Mumbai, Kolkata and Chennai, Kerala and there are about 212 fish varieties being bred locally. As per recent information, Andhra Pradesh, Orissa, Kerala, Assam, Bihar and many other states are producing ornamental fishes locally. The introduction and spread of alien ornamental species in various parts of

* *Corresponding author.* E-mail: aksingh56@rediffmail.com

the country is regarded as a major threat to fish biodiversity causing concern for ecological safety. In this paper, the details of ornamental fish varieties in Indian ornamental market are described with their importance and ecological concerns.

Keywords: *Ornamental fish, Introductions, Ecological concern, Aquarium trade, Risk assessment.*

Introduction

Fish keeping is an age old hobby and Chinese are the pioneer in ornamental fish keeping. In ancient time during 1000 AD in Sung Dynasty, Chinese used a variety of containers for fish keeping such as dishes, bowls and small tanks that permitted viewing of he fish from the top. First public aquarium was opened in Regent's Park, London on May 21, 1853 with the inspiration of Philip Henry Gosse, Inventor of institutional aquarium. The other cities that quickly followed by London were Paris (1859), New York, Boston (1859), Hamburg (1864), Berlin (1869) Brighton (1872), Washington (1873), San Francisco (1894) and Chennai (1909). Today fish keeping is a very popular hobby in the world including Asia resulting into a burgeoning ornamental fish trade. The live import-export trade of faunal aquatic resources dealing with ornamental aquatic species is consequently increasing in India also day by day (Ponniah *et al.,* 2007). Some of these species are cultured while others are imported and traded. The trade of such ornamental species are for freshwater, brackish as well as marine species. The live ornamental species trade consists of both finfish and invertebrate species.

Ornamental fishes are considered as a sign of peace and beauty because of their silence movement, colours and they take very little place of an area rather than other pet animals and birds. While keeping of these fishes is one of the most accepted and fashionable hobby in the developed and developing countries and the trade of this market is now growing very rapidly, the rising attention in aquarium fishes has resulted in steady increase in aquarium fish trade globally. According to FAO (2004), export earnings from ornamental fish trade is US$ 251 million and more than 60 per cent of the production comes from the households of developing countries. The wholesale value of the global ornamental fish trade is estimated to be over 1 billion US$. The entire industry, including accessories and fish feed, is estimated to be worth more than 14 billion US$. The largest demand (99 per cent) for ornamental fish is from home hobbyists, and the remaining 1 per cent is from public aquaria and research institutes (Bassleer, 1995).

More than 4000 freshwater and 1400 marine species are traded internationally each year, making it one of the most important components of the global fish trade (Wittington and Chong, 2007). The ornamental fish trade in India is dominated mainly by the presence of alien fish species/variety such as goldfish, angel, guppy, swordtail, oscar, platy and they have been introduced from different parts of the world but mainly from Asia (Kapoor and Abidi, 2004). Majority of them are freshwater while marine aquaria fishes are also becoming popular in recent years. However, there is no record available to document the period and place from where huge number of such aquarium fishes have been introduced to India. These fishes have been mainly introduced by private aqua-traders. There are more than 600 species/variety of exotic ornamental fish species in the country. More than 200 species of these introduced species are bred in different parts of the country (Ghosh *et al.,* 2003). Some of the commonly available exotic ornamental fishes in the Indian markets have been listed (Table 13.1). Most of these freshwater exotic aquarium fishes are bred in India by aquaculturists and aquarium shop keepers in their indoor small cement cisterns or earthen ponds and tanks for their supply in the domestic markets or for export purpose. Ornamental fish species imported from countries like Singapore, Malaysia, and Sri Lanka are in high demand in Indian fish markets. An estimate carried out by the Marine Products Exports Development Authority of India, (MPEDA), shows that there are approximately 1.0 million ornamental fish hobbyist in India (Felix *et al.,* 2012). The market in Chennai alone is turning out to be one of the centers of breeding and export-import of ornamental fishes. In Chennai, many fish breeders use in their backyards for fish breeding and selling the bred stock to the bigger companies, which are engaged in export business. Even 600 species/variety of introduced ornamental fish are not enough for the customers and every day entrepreneurs are expecting some more new varieties. Every year, 10-15 new species are important to India which is being attempted to breed by local breeders. Local fish breeders are enlarging business and also supplying ornamental fish species to some other parts of country.

In India, there is a huge demand for exotic ornamental fishes. The domestic ornamental fish market is worth over Rs. 500 million and the demand is increasing at 20 per cent annually. India's share in global ornamental fish trade is negligible and it contributes about 0.008 per cent only to the global trade. In addition to trade of exotic varieties, the country is also exporting freshwater varieties of local ornamental fishes from the natural sources predominantly from the north-eastern States (85 per cent) and a few local bred varieties of exotic species (15 per cent). Ornamental fish breeding and culture is confined in and around the metro cities in India viz. Mumbai, Kolkata and Chennai. Recent reports are available that Andhra Pradesh, Orissa, Kerala, Assam, Bihar and many other states have started producing ornamental fishes locally. At present the ornamental fish export from India is dominated by the wild caught local species.

Commonly Cultivable Ornamental Fish Species

In India, both exotic (Figure 13.1) and native ornamental fishes are marketed. 291 exotic species varieties of ornamental fish species have been recorded (NBFGR Annual Report, 2011), over 212 species are bred in different parts of India while only about 52 species are among native fish species earmarked as aquarium fishes. Among exotic ornamental fishes, there are two categories of fishes:

1. Live bearers (Viviparous) like guppy, molly, platy and sword tail which release young ones in batches.
2. Egg layers which lay eggs on the walls of the container or on the surface of aquatic plants. Among egg layers, gold fish, koi, gourami, zebra fish, etc, are common. Out of these, the gold fish, *Carassius auratus* (Linn.) and Koi, *Cyprinus carpio* belonging to family Cyprinidae are preferred by most of hobbyists because of its attractive coloration and peculiar morphological

Figure 13.1: Some Commonly Available Exotic Ornamental Fishes in India

characteristics (colour, tail shapes etc.). Both of these categories of fishes are good for community tanks due to compatibility and non-aggressiveness. The other ornamental fishes include honey gourami, rosy barb, glass fish, etc. (Kour and Dhawan, 2009). The commonly available alien ornamental fish species are listed in Table 13.1.

Table 13.1: List of Exotic Aquarium Fish Species having Potential for Indian Markets as well as Export

Sl.No.	*Species Name*	*Common Name*	*Family*	*Risks of Introduction*
1.	*Abramites hypselonotus*	Brown head stander or Marbled head stander	Anostomidae	NA
2.	*Acanthodoras spinosissimus*	Talking catfish	Doradidae	NA
3.	*Acanthophthalmus semicinctus*	Kuhlii or Coolie loach	Cobitidae	NA
4.	*Acipenser ruthenus*	Starlet Sturgeon	Acipenseridae	NA
5.	*Aequidens dorsiger*	Red breasted dwarf cichlid	Cichlidae	NA
6.	*Aequidens maroni* or *Cleithracara maronii*	Keyhole cichlid	Cichlidae	NA
7.	*Aequidens pulcher* or *Andinoacara pulcher*	Blue acara	Cichlidae	NA
8.	*Aequidens rivulatus* or *Andinoacara rivulatus*	Green terror	Cichlidae	NA
9.	*Ancistrus hoplogenys*	Bristte nose catfish	Loricariidae	NA
10.	*Ancistrus dolichopterus*	Bushy mouth catfish	Loricariidae	NA
11.	*Anostomus anostomus*	Striped head stander	Anastomidae	NA
12.	*Aphyocharax anisitsi*	Blue blood fin	Characidae	NA
13.	*Aphyocharax paraguayensis*	Three spotted or Paraguay tetra	Characidae	
14.	*Aphyocharax rathbuni*	Neon blood fin or Green fire tetra	Characidae	NA
15.	*Aphyosemion arnoldi*	Arnold's killi fish	Aplochelidae	NA
16.	*Aphyosemion australe*	Lyretail panchax	Nothobranchiidae	NA
17.	*Aphyosemion bivittatum*	Two stripe lyretail	Nothobranchiidae	NA
18.	*Aphyosemion striatum*	Killi fish	Nothobranchiidae	NA
19.	*Aphyosemion walkeri*	Killi fish	Nothobranchiidae	NA
20.	*Apistogramma agassizi*	Agassiz's dwarf cichlids	Cichlidae	NA
21.	*Apistogramma ramirezi*	Ram cichlids	Cichlidae	NA
22.	*Apistogramma reitzigi*	Dwarf cichlids	Cichlidae	NA
23.	*Apistogramma trifasciata*	Dwarf cichlids or three strip apisto fish	Cichlidae	NA

Contd...

Table 13.1–*Contd...*

Sl.No.	*Species Name*	*Common Name*	*Family*	*Risks of Introduction*
24.	*Apistogramma caetei*	Blue apistogramma	Cichlidae	NA
25.	*Apistogramma nijsseni*	2 Spotted dwarf cichlid	Cichlidae	NA
26.	*Aplocheilus dayi*	Ceylon killi Fish	Aplochelidae	NA
27.	*Apteronotus albifrons*	Black ghost or knifefish	Apteronotidae	LR
28.	*Astronotus ocellatus*	Oscar	Cichlidae	NA
29.	*Astyanax fasciatu* or *Astyanax mexicanus (Anoptichthys jordani)*	Banded astyanax	Characidae	NA
30.	*Atractosteus spatula*	Aligator gar fish	Lepisosteidae	NA
31.	*Balantiocheilos melanopterus*	Silver shark	Cyprinidae	LR
32.	*Barbodes everetti* or *Puntius evertti*	Clown barbs	Cyprinidae	LR
33.	*Aulonocara nyassae*	Pecock blue cichlid	Cichlidae	NA
34.	*Barbodes hexazona*	Six banded barbs	Cyprinidae	NA
35.	*Barbodes lateristriga*	Spanner barbs	Cyprinidae	
36.	*Barbodes pentazoa* or *Puntius pentozoa*	Five banded barbs	Cyprinidae	NA
37.	*Barbodes schwanenfeldi* or *Barbonymus schwanenfeldi*	Tin foil barbs	Cyprinidae	NA
38.	*Barbus cumingi*	Two spot barbs	Cyprinidae	NA
39.	*Barbus eugrammus*	Zebra striped barbs	Cyprinidae	NA
40.	*Bedotia geayi*	Madagascar rainbow	Bedotiidae	NA
41.	*Belontia signata*	Combtail paradise	Osphronemidae	NA
42.	*Betta splendens*	Fighter at least or Siamese fighting fish	Osphronemidae	LR
43.	*Boehlkea fredcochui*	Blue tetra	Characidae	NA
44.	*Botia horae*	Skunk loach	Cobitidae	NA
45.	*Botia macracantha* or *Chromobotia yasuhikotakia*	Clown loach	Cobitidae	LR
46.	*Botia modesta*	Blue loach	Cobitidae	LR
47.	*Botia sidthimunki*	Dwarf loach	Cobitidae	LR
48.	*Brachydanio albolineatus* or *Danio albolineatus*	Pearl danio	Cyprinidae	NA
49.	*Brachydanio frankei* or *Danio rerio*	Leopard danio or Zebra fish	Cyprinidae	NA
50.	*Brachydanio kerri* or *Danio kerri*	Kerri danio	Cyprinidae	NA
51.	*Brochis splendens*	Emerald catfish	Callichthyidae	NA
52.	*Brycinus longipinnis*	Long finned tetra	Characidae	NA

Contd...

Table 13.1–*Contd...*

Sl.No.	*Species Name*	*Common Name*	*Family*	*Risks of Introduction*
53.	*Capoeta oligolepis* or *Puntius oligipinnis*	Checkerad barbs	Cyprinidae	NA
54.	*Capoeta semifasciolatus*	Schuberti barbs	Cyprinidae	NA
55.	*Capoeta titteya*	Chery barbs	Cyprinidae	NA
56.	*Capoeta tetrazona*	Tiger barbs	Cyprinidae	LR
57.	*Carassius auratus* (at least 10 varieties)	Goldfish	Cyprinidae	LR
58.	*Carnegiella strigata*	Marble hatchet	Gasteropelecidae	NA
59.	*Cheirodon axelrodi* or *Paracheirodon axelrodi*	Cardinal tetra	Characidae	NA
60.	*Chilodus punctatus*	Spotted head stander	Chilodontidae	NA
61.	*Cichlasoma/ Herichthys cyanoguttatum*	Texas cichlid	Cichlidae	NA
62.	*Cichlasoma festivum*	Flag cichlid	Cichlidae	NA
63.	*Cichlasoma labiatum* or *Amphilophus labiatum*	Red devil	Cichlidae	NA
64.	*Cichlasoma managuense*	Juguar cichlid	Cichlidae	NA
65.	*Cichlasoma meeki* or *Thorichthys meeki*	Fire mouth	Cichlidae	NA
66.	*Cichlasoma nigrofasciatum* or *Amatitlania nigrofasciata*	Convict cichlids	Cichlidae	NA
67.	*Cichlasoma octofasciatum* or *Rocio octofasciata*	Jack dempsy	Cichlidae	NA
68.	*Cichlasoma salvini* or *Amphilophus salvini*	Salvini cichlid	Cichlidae	NA
69.	*Cichlasoma severum*	Green severum	Cichlidae	NA
70.	*Cichlasoma urophthalmus*	Mayan cichlid	Cichlidae	HR
71.	*Colisa labiosa*	Thick lipped gourami	Osphronemidae	NA
72.	*Colossoma macropomum*	Cachama	Serrasalmidae	MR
73.	*Copella nattereri*	Spotted tetra	Lebiasinidae	NA
74.	*Corydoras aeneus*	Bronze corydoras	Callichthyidae	NA
75.	*Corydoras julii*	Julii corydoras	Callichthyidae	NA
76.	*Corydoras paleatus*	Mottled corydoras	Callichthyidae	NA
77.	*Corydoras melini*	False banditcory	Characidae	NA
78.	*Corydoras sterbai*	Sterbai corydoras	Callichthyidae	NA
79.	*Corynopoma riisei*	Swordtail caracin	Characidae	NA
80.	*Ctenopoma acutirostre*	Lepord ctenopoma	Anabantidae	NA
81.	*Cyphotilapia frontosa*	Frantosa cichlid	Cichlidae	NA

Contd...

Table 13.1–*Contd...*

Sl.No.	*Species Name*	*Common Name*	*Family*	*Risks of Introduction*
82.	*Cyprinus carpio*	Koi carp	Cyprinidae	LR (-KHV)
83.	*Dermogenys pusillus*	Half beak	Hemiramphidae	NA
84.	*Distichodus sexfasciatus*	Six banded distichodus	Distichodontidae	NA
85.	*Epalzeorhynchos kalopterus*	Flying fox	Cyprinidae	NA
86.	*Epiplatys dageti* or *Epiplatys monroviae*	Redchin panchax	Nothobranchiidae	NA
87.	*Fundulpanchax gardneri*	Blue lyretail	Nothobranchiidae	NA
88.	*Glossolepis incisus*	Red rainbow	Melanotaeniidae	NA
89.	*Gnathonemus petersi*	Elephant nose	Mormyridae	NA
90.	*Gymnocorymbus ternetzi*	Widow tetra	Characidae	NA
91.	*Gyrinocheilus aymonieri*	Chinees algae eater	Gyprinocheilidae	NA
92.	*Haplochromis moori* or *Cyrtocara moori*	Blue hump head	Cichlidae	NA
93.	*Helostoma temmincki*	Kissing gourami	Helostomatidae	NA
94.	*Hemichromis bimaculatus*	Jewel fish	Cichlidae	NA
95.	*Hemigrammus caudovittatus* or *Hyphesso bryconanititsi*	Bunos aires tetra	Characidae	NA
96.	*Hemigrammus erythrozonus*	Glowlight tetra	Characidae	NA
97.	*Hemigrammus hyanuary*	January tetra	Characidae	NA
98.	*Hemigrammus ocellifer*	Head & tall light tetra	Characidae	NA
99.	*Hemigrammus pulcher*	Garnet tetra	Characidae	NA
100.	*Hemigrammus rhodostomus* or *Bleheri rhodostomus*	Rummy nose tetra	Characidae	NA
101.	*Hyphessobrycon bentosi*	Ornate tetra	Characidae	NA
102.	*Hyphessobrycon bifasciatus*	Yellow tetra	Characidae	NA
103.	*Hyphessobrycon callistus*	Serpae tetra	Characidae	NA
104.	*Hyphessobrycon erythrostigma*	Bleeding heart tetra	Characidae	NA
105.	*Hyphessobrycon flammeus*	Flame tetra	Characidae	NA
106.	*Hyphessobrycon griemi*	Griem's tetra	Characidae	NA
107.	*Hyphessobrycon herbertaxelrodi*	Black neon tetra	Characidae	NA
108.	*Hyphessobrycon pulchripinnis*	Lemon tetra	Characidae	NA
109.	*Hyphessobrycon scholzei*	Black line tetra	Characidae	NA
110.	*Hyphessobrycon simulans*	Green neon tetra	Characidae	NA
111.	*Hypostomus plecostomus*	Sucker mouth catfish	Loricariidae	NA
112.	*Inpaichthys kerri*	Purple emperor	Characidae	NA
113.	*Iriatherina werneri*	Thread fin rainbow	Melanotaeniidae	NA
114.	*Jordanella floridae*	Flag fish	Cyprinodontidae	NA

Contd...

Table 13.1–*Contd...*

Sl.No.	*Species Name*	*Common Name*	*Family*	*Risks of Introduction*
115.	*Julidochromis marlieri*	Tanganyika lake dwarf cichlids	Cichlidae	NA
116.	*Julidochromis regani*	Convict julie	Cichlidae	NA
117.	*Julidochromis gombi*	Tanganyika lake dwarf cichlids	Cichlidae	NA
118.	*Kryptopterus bicirrhis*	Glass catfish	Siluridae	NA
119.	*Labeo bicolor*	Red tail black shark	Cyprinidae	NA
120.	*Labeo erythrurus*	Red tail shark	Cyprinidae	NA
121.	*Lamprologus brichardi*	Lake tanganyika dwarf cichlids	Cichlidae	NA
122.	*Lamprologus leleupi*	Lake tanganyika dwarf cichlids	Cichlidae	NA
123.	*Lamprologus tetracanthus*	Four spine cichlids	Cichlidae	NA
124.	*Leporinus fasciatus*	Banded leporinus	Anastomidae	NA
125.	*Macropodus opercularis*	Chinees paradise fish	Osphronemidae	LR
126.	*Megalamphodus megalopterus*	Black phantom tetra	Characidae	NA
127.	*Megalamphodus sweglesi*	Red Phantom tetra	Characidae	NA
128.	*Melanotaenia boesemani*	Boeseman's rainbow fish	Melanotaeniidae	NA
129.	*Melanotaenia lacustris*	Blue rainbow	Melanotaeniidae	NA
130.	*Melanotaenia axelrodi*	Axelrodi rainbow	Melanotaeniidae	NA
131.	*Melanotaenia splendida australis*	Red tailed blue rainbow	Melanotaeniidae	NA
132.	*Melanotaenia maccullochi*	Pigmy or dwarf rainbow	Melanotaeniidae	NA
133.	*Melanotaenia splendida*	Splendid rainbow	Melanotaeniidae	NA
134.	*Metynnis hypsauchen*	Silver doller	Characidae	NA
135.	*Moenkhausia pittieri*	Diamond tetra	Characidae	NA
136.	*Moenkhausia sanctaefilomenae*	Red eye tetra	Characidae	NA
137.	*Monodactylus sebae*	African moony	Monodactylidae	NA
138.	*Nannacara anomala*	Dwarf cichlid	Cichlidae	NA
139.	*Nannostomus beckfordi*	Golden pencil fish	Labiasinidae	NA
140.	*Nannostomus eques*	Rocket pencil fish	Lebiasinidae	NA
141.	*Nannostomus marginatus*	Dwarf pencil fish	Lebiasinidae	NA
142.	*Nannostomus trifasciatus*	Three striped pencil Fish	Lebiasinidae	NA
143.	*Nannostomus unifasciatus*	One Lined Pencil fish	Lebiasinidae	NA
144.	*Nematobrycon palmeri*	Emperor tetra	Characidae	NA
145.	*Nomorhamphus celebensis*	Half beak	Hemiramphidae	NA
146.	*Nothobranchius guentheri*	Red tail notho	Nothobranchiidae	NA
147.	*Nothobranchius korthausae*	Killi fish	Nothobranchiidae	NA

Contd...

Table 13.1–*Contd...*

Sl.No.	*Species Name*	*Common Name*	*Family*	*Risks of Introduction*
148.	*Nothobranchius palmqvisti*	Killi fish	Nothobranchiidae	NA
149.	*Osphronemus gorami*	Gaint gourami	Osphronemidae	LR
150.	*Otocinclus arnoldi*	Dwarf sucking cat fish	Loricariidae	NA
151.	*Panchax playfairi*	Playfairi panchax	Aplocheilidae	NA
152.	*Panaque nigrolineatus*	Royal pleco	Loricaiidae	NS
153.	*Pangasius hypophthalmus*	Striped shark	Pangasiidae	MR
154.	*Pantodon buchholzi*	Butterfly fish	Pantodontidae	NA
155.	*Paracheirodon innesi*	Neon tetra	Characidae	NA
156.	*Parosphromenus deissneri*	Licorice gourami	Osphronemidae	NA
157.	*Peckoltia pulcher*	Dwarf sucker catfish	Loricariidae	NA
158.	*Peckoltia vittata*	Dwarf sucker catfish	Loricariidae	NA
159.	*Pelmatochromis thomasi*	African butterfly cichlid	Cichlidae	NA
160.	*Pelvicachromis pulcher*	Rainbow krib	Cichlidae	NA
161.	*Phenacogrammus interruptus*	Congo tetra	Alestidae	NS
162.	*Phractocephalus hemioliopterus*	Red tail catfish	Pimelodidae	NA
163.	*Platydoras costatus*	Raphaed catfish	Doradidae	NA
164.	*Poecilia latipinna*	Sail fin molly fish	Poeciliidae	MR
165.	*Poecilia reticulata*	Guppy	Pociliidae	MR
166.	*Poecilia sphenops*	Marble molly	Poeciliidae	NA
167.	*Poecilia velifera*	Sail fin molly	Poeciliidae	NA
168.	*Priapella intermedia*	Isthmian priapella	Poeciliidae	NA
169.	*Prionobrama Filigera*	Glass blood fin	Characidae	NA
170.	*Pristella maxillaris*	X-Ray tetra	Characidae	NA
171.	*Pseudepiplatys annulatus* or *Fundulus diaphnus*	Banded killi fish	Cyprinodontidae	NA
172.	*Pseudotropheus auratus*	Lake Malavi cichlids	Cichlidae	NA
173.	*Pseudotropheus lombardoi*	Lake Malavi cichlids	Cichlidae	NA
174.	*Pseudotropheus tropheops*	Lake Malavi Cichlids	Cichlidae	NA
175.	*Pseudotropheus zebra*	Lake Malavi cichlids	Cichlidae	NA
176.	*Pterolebias longipinnis*	Long fin killie	Rivulidae	NA
177.	*Pterophyllum scalare*	Angel fish	Cichlidae	LR
178.	*Pterygopichthys* spp. (*disjunctivus* and *perdalis*)	Sucker mouth catfish	Loicariidae	HR
179.	*Puntius lineatus*	Longfin killie fish	Cyprinidae	NA
180.	*Puntius nigrofasciatus*	Ruby or Nigro barb	Cyprinidae	LR
181.	*Puntius sachsi*	Schuberti barbs	Cyprinidae	NA

Contd...

Table 13.1–*Contd...*

Sl.No.	*Species Name*	*Common Name*	*Family*	*Risks of Introduction*
182.	*Puntius somphongsi*	Tinfoil barbs	Cyprinidae	NA
183.	*Pyrrhulina vittata*	Banded pyrrhulina	Lebiasinidae	NA
184.	*Rasbora caudimaculata*	Giant scissortail rasbora	Cyprinidae	NA
185.	*Rasbora heteromorpha*	Harlequin	Cyprinidae	LR
186.	*Rasbora kalochroma*	Clown rasbora	Cyprinidae	NA
187.	*Rasbora maculata*	Pigmy rasbora	Cyprinidae	NA
188.	*Rasbora trilineata*	Three lined rasbora	Cyprinidae	NA
189.	*Scleropages formosus*	Asiatic arowana (Feng Shui Fish)	Osteoglossidae	NA
190.	*Pygocentrus nattereri*	Red bellied piranha	Serrasalmidae	HR
191.	*Schilbe mystus*	African butter Catfish	Schilbeidae	NA
192.	*Sorubim lima*	Shovel nose catfish	Pimelodidae	NA
193.	*Sorubimichthys planiceps*	Firewood catfish	Pimelodidae	NA
194.	*Sphaerichthys osphromenoides*	Chocolate gourami	Osphronemidae	NA
195.	*Sturosoma panamanase*	Royal twig catfish	Loricariidae	NA
196.	*Symphysodon aequifasciata*	Common discus	Cichlidae	LR
197.	*Symphysodon discus*	Red discus	Cichlidae	LR
198.	*Synodontis contractus*	Bugeye squeaker	Mochokidae	NA
199.	*Synodontis nigriventris*	Blotched upside down catfish	Mochokidae	NA
200.	*Tanichthys albonubes*	White cloud mountain minnow	Cyprinidae	NA
201.	*Telmatherina ladigesi*	Celebese rainbow	Telmatherinidae	NA
202.	*Thayeria boehlkei*	Hockey stick	Characidae	NA
203.	*Trichogaster leeri*	Pearl gourami	Osphronemidae	LR
204.	*Trichogaster microlepis*	Moonlight gourami	Osphronemidae	NA
205.	*Trichogaster pectoralis*	Snake skin gourami	Osphronemidae	NA
206.	*Trichogaster trichopterus*	Three spot gourami	Osphronemidae	MR
207.	*Trichopsis pumilus*	Pygmy gourami	Osphronemidae	NA
208.	*Trichopsis vittatus*	Crocking gourami	Osphronemidae	NA
209.	*Tropheus duboisi*	Maswa or Doboisi cichlid	Cichlidae	NA
210.	*Uaru amphiacanthoides*	Uaru	Cichlidae	NA
211.	*Xiphophorus helleri*	Sward tails fish	Poeciliidae	MR
212.	*Xiphophorus maculates*	Platy fish or moon fish	Poceiliidae	MR

Abbreviations: NA: Not assessed; LR: Low risk; MR: Moderate risk; HR: High risk.

Rearing of Ornamental Fishes in India

Cement cisterns, glass aquaria, earthen ponds and earthen pots are used as culture tanks. Cement tanks are not only easy to maintain, but are also durable. The spawning tank (after the removal of parents) can be used as rearing tank. Compatible species (two or three species) can be reared together; otherwise, single species rearing is preferred. Well aerated tube well water is utilized for rearing and fishes reach marketable size in about 4 to 6 month period. By following proper rearing and management practices, eight to ten crops are easily taken in a year.

Feeding of Ornamental Fishes

In general, most of the ornamental fish species require crude protein level in the range of 30-45 per cent, crude fat (lipid) 4-8 per cent and carbohydrates between 30-50 per cent. Well balanced manufactured feeds are used for aquarium fishes while some times, different feed ingredients are selected and mixed to get the appropriate protein, lipid and energy level. Commonly available ingredients such as agricultural by products (groundnut cake, rice polish, wheat bran, maize bran, soybean meal) and animal protein sources (fish meal, meat meal) are mixed. The feed is provided in the form of dough or pelleted form.

Ecological Concern of Market Pressure on Local Species

The increased demand of local ornamental fish species has caused increased collection pressure on some fish species from wild. This situation has adversely impacted on the population of these local species in wild inviting attention to protecting them under Convention on International Trade of Endangered Species (CITES). For example *Puntius danisoni,* commonly called ''Kerala queen" is a native fish of Indian west coast and it is high in demand by traders. Increased pressure on wild catch has been declining fast this fish.

Ecological Concern of Import

The introduction and spread of alien ornamental species in various parts of the country is regarded as a major threat to fish biodiversity and hence ecological safety. In India, introductions of species such as the sucker mouth catfish, piranha, red belly pacu, oscar, are among the higher profile biological invaders. However, only a small number of people are aware that small, freshwater ornamental fish species in ponds and aquaria can also cause damage to aquatic environments and local fish diversity (Singh and Lakra, 2011). It is therefore, awareness and educational programmes is imperative.

Risk Assessment

The process of Risk Analysis comprises of three components – Hazard Identification, Risk management and Risk communication. Since transboundary boundary movements of alien ornamental fish are increasing, it should be made mandatory to undertake risk analysis prior to import so as to counteract the possible risk of introduction (Lakra *et al.,* 2006; 2008). The risk analysis method used to address the elements of risks are described as follows:

(A) Hazard Identification

Hazard identification involved determining which, of all the microbial diseases and parasites known from ornamental fishes, may be introduced through import, becomes established, and which may have an adverse effect on native biota. The initial step in a risk analysis is to identify the diseases and parasites of potential concern in the commodity under consideration. This process of hazard identification, must be relevant to the fish species that are permitted entry into a country. The freshwater fishes are more at risk because of their confinement by lakes and catchments and their closer association with humans, than the marine fishes in open waters which have less contact with land-based humans. It is widely recognized that ornamental fish have caused the spread of common parasites and diseases, such as ectoparasitic lice, anchor worms, white spot ("Ich"), *Cryptobia,* etc, that are well known to ornamental fish specialists (Thilakaratne *et al.,* 2003). The potential to spread such serious parasites and diseases, may have a devastating impact on the native fishes of the importing country. However there are many examples of disease transfer in the past (McCann *et al.*, 1996, Langdon 1990, Humphrey and Ashburner, 1993), and the evidence that ornamental fishes and the water used to transport them (Trust and Bartlett, 1974) can act as vectors for viruses and other pathogens of national and international significance continues to increase. In examples from Australia, atypical strains of the bacterial fish pathogen *Aeromonas salmonicida* were introduced through importation of infected goldfish *Carassius auratus* (Humphrey and Ashburner, 1993), and subsequently spread to other species including trout, carp and silver perch. A detail of the introductions and transfers of parasites and other disease agents by finfish was reported by Ganzhorn *et al.* (1992). Not only fishes but eggs may also be infected. The water and the containers of ornamental fishes may also be contaminated and serve as vehicles for the introduction of pathogens. The risk of establishment of these parasites and disease agents depends on their biology, life cycle, the fate of the shipment, and the presence or absence of appropriate hosts in the receiving country (Ganzhorn *et al.,* 1992). A recent case of disease transmission through trade is the spread of koi herpes virus (KHV). Although known in several parts of the world since 1998, it was first reported in Indonesia in 2002 (NACA/ACIAR, 2002). Since then it has spread to several countries in the Asian region, in most cases associated with movement of live fish. KHV outbreaks in the region have significant trade implications for the high-value ornamental koi carp industry and the important food fish, the common carp. Active trade in ornamental fish poses a potential risk for the further spread of this disease in the Asian region. Recognizing the significance and responding to confirmed outbreaks in Indonesia and Japan, "infection with koi herpes virus" was added in 2003 to the list of diseases prevalent in the region in the quarterly disease reports from Asian countries (NACA/FAO, 2004). Since its listing, several countries in the region have increased their surveillance, testing and quarantine programmes.

(B) Risk Management

There is a growing recognition among policy-makers, the public and the business community that the invasive alien species problem needs to be addressed. Preventing the introduction of invasive alien ornamental species is generally recognized as the most effective way to mitigate the threat of invasive. To be effective, efforts should be made to combat unintentional introductions, while prevention of unintentional introductions can be key primarily to manage invasive species. Prevention is generally far more cost effective and environmentally desirable than measures taken following introduction of an alien invasive species. Priority should be given to prevention of entry of alien invasive species. Invasive alien species prevention measures can be classified in many ways and do not always fall into neatly defined, discrete categories. At the international level, control and prevention of invasive alien species depends upon a complex intersection of environmental, trade and transportation agreements. In developing prevention measures, it is advocated to generally establish a series of requirements governing how to import ornamental fish species and the same are detailed as follows:

Pre-Border Measures

Pre-border measures are intended to reduce the likelihood that invasive alien species will ever reach new countries (or ecosystems). They are inherently the best form of prevention; if effective, they virtually eliminate the possibility of accidental or intentional introduction. Regulatory officials are thereby in a better position to address potential concerns as opposed to dealing with a new species at border control points, which can never be 100 per cent escape proof.

Border Measures

Border measures are intended to intercept invasive alien species at a political boundary, to complement pre-border measures, or where pre-border measures are not in place or are not fully effective. Border measures prohibit or restrict importation of potentially invasive species or regulate pathways of potentially invasive species. Customs rules, inspections and quarantine procedures are longstanding types of border measures that can be of significant value in addressing invasive alien species.

Post-Entry Measures

Even after a product has been allowed through the border, there are still a range of regulatory tools that can be used to prevent invasive alien species. For example, monitoring for high risk unintentionally introduced invasive alien species is common at nurseries and around high volume ports. Some alien species introduced into a country may be benign in one area, but potentially invasive when transported to a different part of that country, with different ecological conditions. Measures applicable only to domestic commerce can prevent the further spread of an alien species that has become invasive in only one part of the country. In many instances, these measures may be quite similar to pre-border and at-border measures. In other situations, they may reflect the domestic character of post-entry prevention. Internal prevention measures are particularly important in archipelagic states and states with islands to prevent the spread of invasive alien species between islands or between islands and the mainland.

In addition to various above mentioned approached, NBFGR, Lucknow has developed a detailed National Plan and Strategic Guidelines on Aquatic exotic species which need to be followed strictly.

(C) Risk Communication

A frequently used approach to communicate the risk of invasive alien species is through listing of species after risk assessment whether a particular species can be imported. The success of such a listing system is inherently related to its adaptability and flexibility, particularly with regard to processing new submissions and proposals for movement from one list to another. There are three types of lists referred to as black, white and grey lists to denote the type of risks associated with import of ornamental fish species.

(i) Black Lists

Black lists (also referred to as "dirty lists") identify those species whose introduction is prohibited due to their potential adverse effects on the environment or human, animal or plant health. Such lists can be a significant component of an invasive alien species prevention regime since they clearly state which species should be banned from import. Black lists are the most common type of communication mechanism and are found in a range of countries, including the United States, Australia, New Zealand, and Poland. Black lists and their bans on species can be fully consistent with international trade rules, and are extremely effective when countries have the flexibility to add and remove species as available information and necessary precaution dictate.

(ii) White Lists

While black lists identify species that are denied access or import into a country, white lists (also referred to as "clean lists") identify species that are low-risk and approved for introduction. The key idea behind a white list approach is that intentional introduction of an organism should be authorized only if the species is on this list of safe organisms. As with black lists, white lists are used in conjunction with risk assessments to determine whether a species may be intentionally introduced. The presence of a given species on a white list reflects that a risk assessment has determined the species to be safe (or, in some cases, white listed species may already be established and beyond hope of control). White lists can be used on their own or in tandem with black lists as a means to increase transparency for potential importers and, ultimately, to facilitate trade in those species. White lists are commonly used in New Zealand, Argentina as well as in the Australian states of the Northern Territory and Western Australia.

(iii) Grey Lists

There may be instances in which the risk associated with an alien species cannot be determined conclusively, and hence the species cannot yet be listed on either a white or a black list. These species can be put on an intermediate "grey" or "pied" list, subject to further review. Sophisticated versions of grey lists place the species within a spectrum or matrix according to the likelihood of invasiveness based on presently available information. Grey lists are frequently developed as potential importers submit requests to a government to see whether their product or organism is eligible for import. The list serves as a virtual holding area as the appropriate regulatory agency makes the relevant determination. Grey lists thus serve a precautionary function by temporarily preventing the import of a species until further assessment can be conducted to determine whether a species is safe or potentially invasive. In some senses, grey lists function as provisional black lists.

Conclusion

Interest from private and government sector in promoting the culture and trade of ornamental aquatic species, particularly ornamentals, has been spurred by their growing potential for increasing rural employment and generating income among small rural and even urban families. By 2000, the global total wholesale value of live ornamental fishes both freshwater and marine (live animals for aquarium only) was estimated at US$900 million with an estimated retail value of US$3 billion. Asia provides more than 50 per cent of the global total ornamental fish supply (FAO, 2000). Estimates place the annual value of the marine ornamental trade at US$200–330 million; the overall value of the marine fish trade accounts for about 10 per cent of the international ornamental fish trade (marine and freshwater included). Ornamental fish are also produced for regional and international markets in Cameroon, Kenya, Uganda, Malawi, South Africa and Zambia, though no accurate figures are available, except for South Africa. Southeast Asia is the hub of the ornamental fish trade, supplying up to 85 percent of the aquarium trade.

Ornamental fish culture in India involves breeding of exotic species and wild native fish collection from northeast and southern part of India. This activity is a small segment in the fisheries Sector. India is endowed with a suitable climate, water resources and large manpower base. The potential manpower can be easily trained into ornamental fish breeding technology and fast growing domestic market. Moreover, the unit value of ornamental fish is higher than the food fish. Hence this sector offers good opportunity for rural and urban households to augment income and link them to the International trade. Almost all of the tank bred ornamental fish in India comes from the small-scale or back yard type of breeding units. But due to the lack of adequate infrastructure and key inputs like appropriate feed, quality brood stock etc these breeding units are not in a position to produce varieties which are in demand in the international market. In India about 90 per cent ornamental fishes are bred and supplied throughout India from West Bengal, Tamil Nadu and Maharashtra. Organized trade in ornamental fish depends on assured and adequate supply of demand, which is possible only by mass breeding of local and exotic verities and institutional strengthening of the sector on scientific basis following precautionary measures.

Acknowledgments

Authors are grateful to Dr. J. K. Jena, Director NBFGR, Lucknow for his consistent support and guidance. Financial support of UP State Biodiversity Board, Lucknow is also thankfully acknowledged.

References

Bassleer, G., 1995. European ornamental fish statistics, with special emphasis on the German market. *Ornamental Fish International Journal*, 10, February, 2-3.

FAO., 2000. The state of world fisheries and aquaculture. FAO Fisheries Department. Rome, p. 142.

Felix, N., Jameson, J. D. and Brindo, A. 2012. Ornamental fish farming - a successful cottage industry in rural and urban India. *Aquaculture Asia,* 17(3): 14-18.

Ganzhorn, J., Rohovec, J. S. and Fryer, J. L. 1992. Dissemination of microbial pathogens through introductions and transfers of finfish. Pages 175-192 in A. Rosenfield and R. Mann, editors. Dispersal of Living Organisms Into Aquatic Ecosystems. Maryland Sea Grant, College Park, p. 471.

Humphrey, J. D. and Ashburner, L.D., 1993. Spread of the bacterial fish pathogen *Aeromonas salmonicida* after importation of infected goldfish, *Carassius auratus,* into Australia. *Australian Veterinary Journal,* 70: 453-454.

Kapoor, D, and Abidi, R. 2004. Lucrative alien ornamental fish species for aquarium trade of India. Publisher National Bureau of Fish genetic Resources, Lucknow, p. 140.

Kour, V. I. and Dhawan Asha. 2009. Ornamental fish culture and breeding a lucrative and upcoming business. *Science Tech Entrepreneur,* April 2009.

Lakra, W.S., Rehana, A., Singh, A.K., Rathore, G., Sood, N., and Rajaswaminathan, T. (eds). 2006 Fish Introductions and Quarantine: Indian Perspectives. National Bureau of Fish Genetic Resources, Lucknow.

Lakra, W.S., Singh, A.K., and Ayyappan, S. (eds). 2008 Fish Introduction in India: Status, Potential and Challenges. Narendra Publishers, New Delhi.

Langdon, J. S. 1990. Disease risks of fish introductions and translocations. In: Introduced and translocated fishes and their ecological effects. DA Pollard (ed.) Bureau of Rural Resources Proceedings No. 8, Australian Government Publishing Service, Canberra, pp. 98-107.

McCann, J. A., Arkin, L. N. and Williams, J. D. 1996. Nonindigenous Aquatic and Selected Terrestrial Species of Florida. Status, Pathway and Time of Introduction, Present Distribution, and Significant Ecological and Economic Effects.

NACA/ACIAR. 2002. Report of the Emergency Disease Control Task Force on a serious disease of koi and common carp in Indonesia. Network of Aquaculture Centers in Asia-Pacific (NACA). Bangkok, Thailand, p. 24.

NACA/FAO. 2004. Emerging trends and experiences in Asia-Pacific Aquaculture 2003. Bangkok, p. 150.

NBFGR. 2011. Annual Report. National Bureau of Fish Genetic Resources, Lucknow.

Ponniah, A.G., Ravisankar, T. and Ksilasarn, M. 2007. Ornamental fish resources and utilization potential, problems and prospects of ornamental fisheries in India. proceedings of national workshop on ornamental fish culture development in India. 22-23 February 2007, Chennai.

Singh, A.K. and Lakra, W.S. 2011. Risk and benefit assessment of alien fish species of the aquaculture and aquarium trade into India. *Reviews in Aquaculture,* 3: 3–18.

Singh, A.K. and Lakra, W.S. 2006. Alien fish species in India: Impact and emerging scenario. *Journal of Ecophysiology and Occupational Health,* 6 (3–4): 165–174.

Thilakaratne, I. D., Rajapaksha, G., Hewakopara, A., Rajapakse, R. P. and Faizal, A. C. 2003. Parasitic infections in freshwater ornamental fish in Sri Lanka. *Disease of Aquatic Organism,* 54: 157-162.

Trust, T.J. and Bartlett, K.H. 1974. Occurrence of potential pathogens in water containing ornamental fishes. *Applied Microbiol*ogy, 28: 35–40.

Whittington R.J. and Chong, R. 2007. Global trade in ornamental fish from an Australian perspective: The case for revised import risk analysis and management strategies. *Preventive Veterinary Medicine,* 81(1–3): 92–116.

2013, Animal Diversity, Natural History and Conservation, Vol. 3 *Pages* **215–236**
Editors: **V.K. Gupta and Anil K. Verma**
Published by: **DAYA PUBLISHING HOUSE, NEW DELHI**

Chapter 14

Diversity of Freshwater Turtles and Tortoises in Northeast India

D.K. Sharma* and Chittaranjan Baruah
Turtle Conservation and Research Program (TCRP) and Bioinformatics Centre, Department of Zoology, Gauhati University, Guwahati – 781 014, Assam, India

ABSTRACT

The chapter reports 22 turtle species so far recorded from northeast India, belonging to 3 families namely Geoemydidae (13 species), Trionychidae (7 species) and Testudinidae (2 species). The freshwater turtle population is declining due to wanton killing, destruction of eggs or destruction of their natural environment. Out of 22 species found in Northeast, 6 species namely *Pangshura sylhetensis*, *Cuora mouhotii*, *Nilssonia nigricans*, *Chitra indicia*, *Manouria emys* and *Indotestudo elongata* are listed by IUCN as most endangered species of Asia. *Chitra indica* is on the new list of the 25 most endangered turtle species of the world. Lack of recent field information has created a gap in conservation activities and status evaluation of almost all turtle species of northeast India. Immediate adoption of conservation measures is essential for the conservation of turtle diversity of this region.

Keywords: *Captive breeding, Conservation, Tortoises, Turtles.*

Introduction

The Chelonians are the oldest living reptiles that evolved 200 million years ago, extending valuable service as scavengers in the sea, rivers and other water bodies.

* *Corresponding author.* E-mail: dksgu@yahoo.co.uk

Turtle refers to all species included in the order Testudines, covering turtles, tortoises and terrapins. They bear scales, lay eggs and are ectothermic in nature. The common names generally signify the habitat of the species and living pattern. Turtles bear webbed feet for swimming and spend most of their life in water. Freshwater turtles live in freshwater like rivers, ponds and lakes. In addition to swimming, they also climb to the banks, logs or rocks to bask in the sunshine especially during low temperature. Tortoises are land-dweller and they do not possess webbed feet and their feet are round and stumpy suitable for walking on land. Terrapins are often found in brackish and swampy areas, where they spend most of the time both in land and water. Turtles play a major role in cleaning the river and tanks and thus significantly controlling pollution in the freshwater system with its wide range of diversity (Moll and Moll, 2004).

Turtles are among the world's most endangered vertebrates and almost half of their 328 recognized species (452 taxa) have been facing threat with extinction (Turtle conservation coalition, 2011). Presently, eight species (10 taxa) of tortoise and freshwater turtles have become extinct since 1500 AD and out of this 320 species (442 taxa) of living turtles and tortoise, 7 species are marine and 313 species (435 taxa) are from the living freshwater turtle group (Turtle Taxonomy Working Group, 2010). According to the draft Red List of IUCN (Tortoise and Freshwater Turtle Specialist Group) status, 156 turtle species (47.6 per cent of total species) are recognized as globally threatened, while 90 (27.4 per cent) are Critically Endangered or Endangered species (Turtle Taxonomy Working Group, 2010).

The present study has been conducted in the north-eastern and northern (Brahmaputra-Ganges region) parts of India. The Brahmaputra-Ganges region of India and Bangladesh exhibits the greatest range overlap for turtles in a single area in the world, with 19 species known from each of 4 connected hydrologic unit compartments [(HUCs) (Figure 14.1) (Iverson, 1992)]. The lower Gangetic Plain is Moist Deciduous Forest Ecoregion, representing the third-most species-rich turtle Ecoregion in the world (18 species), accounted for 12 species. Other Ecoregions that include the same high turtle species richness (but are within the Indo-Burma and Himalaya biodiversity hotspots) include Meghalaya Subtropical Forests, Brahmaputra Valley Evergreen Forests, Mizoram–Manipur–Kachin Rainforests, and Terai–Duar Savannas and Grasslands. In addition, observations on climate change suggest that glacial melt in the Himalaya will probably affects water flow in the Ganges-Brahmaputra Basin (Xu *et al.,* 2009), thus adding another conservation concern to this priority area.

The north-eastern region of India covering an area of 2, 62,179 km^2 (7.6 per cent total area of the country) is a biologically highly diverse area and one of the most important biodiversity hot spots of the world. The region is at the conjunction of the Himalaya and Indo-Burma biodiversity hotspots, rich in turtle diversity (Das, 1990). However, the region has been poorly explored scientifically with regard to surveys, conservation and monitoring of turtle fauna. The turtle and tortoise diversity of the northeast India is the highest in the country and 22 of the 29 species are found in this region (Das, 1995; Ernst *et al.,* 2000; Das, 2001; Praschag and Gemel, 2002). However, the lack of recent field information has created a gap in status, conservation and

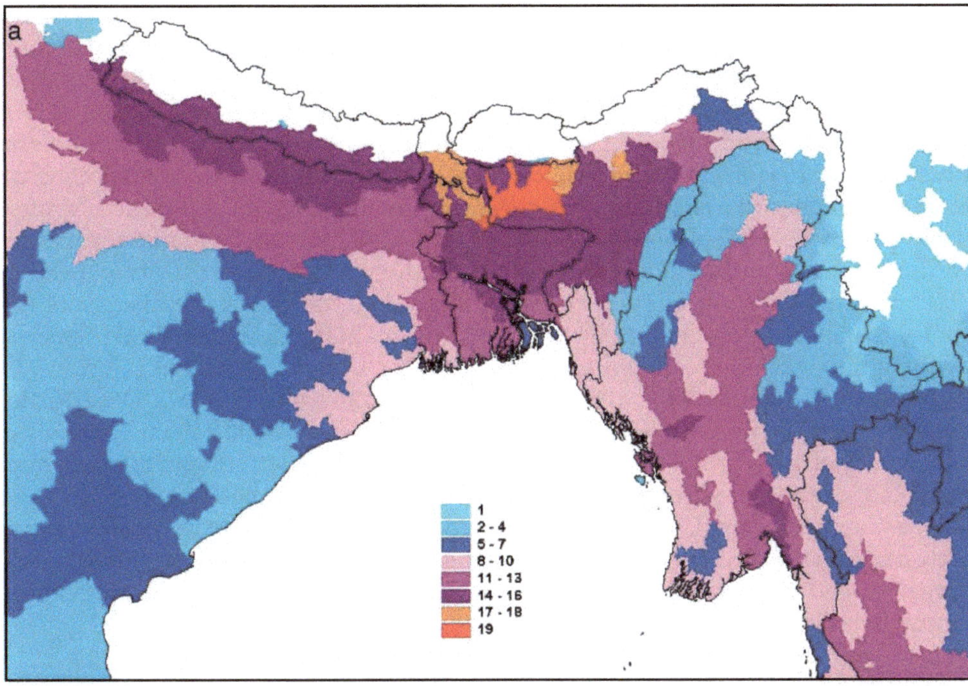

Figure 14.1: The Study Area Belongs to World's Greatest Turtle Richness Area, Based on the Co-occurrence of Species in Hydrologic Unit Compartment in the Ganges-Brahmaputra River Basin Drainage of India and Bangladesh in South Asia (Buhlmann *et al.,* 2009)

management of turtle species in the region. This chapter reports a checklist of turtles and tortoises from the northeastern states which are recorded in the field work during the last four years.

The northeastern region of India at the conjunction of the Himalaya and Indo-Burma biodiversity hotspots, enriched with a diversity of 22 turtle species has earned the destination of being included in the priority of turtle conservation area (Choudhury, 1995, 1998; Das, 2001; Ahmed *et al.,* 2009; Buhlmann *et al.,* 2009). Despite the fact that freshwater turtles and tortoises have been the focus of recent and intense conservation attention (Van Dijk *et al.,* 2000), there are still a little work carried out for most of the Critically Endangered or data deficient turtle taxa. The necessary role that conservation genetic studies have in turtle conservation is apparent and suggests areas demanding future research.

Materials and Methods

The study was carried out during May 2005 till January 2011 with the aim to record the distribution and conservation status of freshwater turtle and tortoise species in Northeast India. Field surveys were carried out randomly throughout the states and at selected observation points in the Brahmaputra and Barak rivers, using three different methods including interviews of peoples using photo sheets of turtles, trapping using nets with the help of fishermen and visual encounter surveys (VES). Binoculars (20 × 50) are also used to observe turtles in some areas, where it is not possible to go physically. During the study period few individuals were caught and released themselves without having any damages. In some forested areas day and night surveys were carried out to find out turtle species. Hideouts like bushes, leaf litters, gaps in large rock boulders and gaps in roots of large trees were searched for hiding or hibernating turtles or tortoises. Hill stream beds and river banks were surveyed to search for turtle tracks, following which sometimes yielded the specimens. Besides, the fishermen and local communities were interviewed and the freshwater turtle species were identified from the by catch of fishermen. Sporadic records of seizure of turtles and market survey were conducted during the study period to assess the threat to some of the species due to trading. Basking turtles were spotted from a distance by binoculars from boat or by walking along the bank. Species identification was done following Pritchard (1979), Daniel (1983), Das (1985, 1995, 2002), Prachag and Gemel (2002), Praschag *et al.* (2007a and b, 2009). Dead turtle specimens and turtle carapace were identified following Smith (1933).Taxonomy and nomenclature are followed after various measurements *viz.* straight line carapace length (CL), carapace width (CW), plastron length (PL) and shell height (SH) were taken using the Vernier Callipers. Specimens were weighed (M) to the nearest gram using a spring balance. Scute and bone terminology of the shell were followed after Zangerl (1969). Subsequently the live specimens were released in their respective natural habitat.

Observations

A total of 22 turtle species, belonging to 3 families namely Geoemydidae (13 species), Trionychidae (7 species) and Testudinidae (2 species) have been recorded from northeast India (Table 14.1). The recorded chelonians were photographed and the sex of the specimens were attempted to identify though it is difficult morphologically. Population of the chelonians outside the protected areas is very thin or nonexistent as habitat is altered or disturbed and very often large individuals are killed for meat. It has been observed that generally the females are being caught and killed due to their size and the small sized males are released into the water and a sex ratio imbalance has been appeared as major threat. During the present investigation, the Kaziranga National Park has been recorded to be the habitat of 17 freshwater turtle species along with one tortoise species. A total of fifteen species of turtles, belonging to 2 families namely Geoemydidae (10 species), Trionychidae (5 species) have been recorded from the Rajiv Gandhi Orang National Park.

A Brief Account on the Turtle Species

Family: Geoemydidae (Plate 14.1)

Pangshura tecta (Gray, 1831)

It is locally known as 'Futuki Salika Dura' in Assam. Females do not exceed about 24 cm in length and the shell is strikingly elevated, the first two vertebral scutes being somewhat keeled and the third rising to a sharp point. This species is common in the northeast India. It is also an inhabitant of ponds in Kamakhya and Hajo. Two clutches of 3-12 eggs are laid during the winter season in sandy or muddy banks of the water bodies.

Table 14.1: Showing the Diversity and Distribution of Freshwater Turtles and Tortoises in Northeast India

Sl.No.	*Family*	*Common Name*	*Scientific Name*	*AP*	*AS*	*NL*	*MN*	*MZ*	*ML*	*TR*	*SK*	*WLPA Schedule*
		Critically Endangered										
1.	Trionychidae	Black soft-shell turtle	*Nilssonia nigricans*	+	+	–	–	–	–	+	–	IV
		Endangered										
2.	Geoemydidae	Assam roofed turtle	*Pangshura sylhetensis*	+	+	+	+	+	+	–	–	I
3.	Geoemydidae	Keeled box turtle	*Cuora mouhotii*	+	+	–	+	+	+	–	–	Not listed
4.	Geoemydidae	Three-striped roof turtle	*Batagur dhongoka*	–	+	–	–	–	–	–	–	Not listed
5.	Trionychidae	Narrow-headed soft-shell turtle	*Chitra indica*	+	+	–	–	–	–	–	–	II
6.	Testudinidae	Elongated tortoise	*Indotestudo elongata*	–	+	–	–	+	+	–	–	IV
7.	Testudinidae	Asian brown tortoise	*Manouria emys phayrei*	+	+	+	+	+	+	–	–	I
		Vulnerable										
8.	Geoemydidae	Tricarinate hill turtle	*Melanochelys tricarinata*	+	+	+	–	+	+	–	+	IV
9.	Geoemydidae	Indian eyed turtle	*Morenia petersi*	–	+	–	+	–	–	–	–	Not listed
10.	Trionychidae	Asian giant softshell turtle	*Pelochelys cantorii*	–	–	–	+	–	–	–	–	I
11.	Geoemydidae	Crowned river turtle	*Hardella thurjii*	+	+	–	–	–	+	–	–	Not listed
12.	Geoemydidae	spotted pond turtle	*Geoclemys hamiltonii*	+	+	–	–	–	+	–	–	IV
13.	Geoemydidae	South Asian box turtle	*Cuora amboinensis*	+	+	+	+	–	–	–	–	Not listed
14.	Trionychidae	Indian peacock soft-shell turtle	*Nilssonia hurum*	+	+	–	+	–	+	+	–	I
15.	Trionychidae	Gangetic soft-shell turtle	*Nilssonia gangatica*	+	+	–	+	–	–	+	–	I
16.	Trionychidae	Asiatic soft-shell turtle	*Amyda cartilaginea*	–	–	–	+	+	–	–	–	IV

Contd...

Table 14.1–*Contd...*

Sl.No.	*Family*	*Common Name*	*Scientific Name*	*AP*	*AS*	*NL*	*MN*	*MZ*	*ML*	*TR*	*SK*	*WLPA Schedule*
		Near Threatened										
17.	Geoemydidae	Brown roofed turtle	*Pangshura smithii*	+	+		–	–	+	–	–	Not listed
18.	Geoemydidae	Indian black turtle	*Melanochelys trijuga*	–	+	+	–	+	+	–	+	Not listed
19.	Geoemydidae	Assam leaf turtle	*Cyclemys gemeli*	+	+	–	+	+	+	–	–	I
		Least Concern										
20.	Geoemydidae	Indian roofed turtle	*Pangshura tecta*	+	+	–	–	+	+	–	–	I
21.	Geoemydidae	Indian tent turtle	*Pangshura tentoria*	+	+	–	+	+	+	+	–	Not listed
22.	Trionychidae	Indian flap-shelled turtle	*Lissemys punctata andersoni*	+	+	–	+	–	+	+	+	I

+: Present; –: Absent; AP: Arunachal Pradesh; AS: Assam; NL: Nagaland; MN: Manipur; MZ: Mizoram; ML: Meghalaya; TR: Tripura; SK: Sikkim.

Plate 14.1: Family–Geoemydidae

A: *Pangshura tentoria;* B: *Pangshura smithii;* C: *Pangshura tecta;* D: *Pangshura sylhetensis;*E: *Melanochelys tricarinata;* F: *Cuora amboinensis;* G: *Geoclemys hamiltonii;* H: *Hardella thurjii;* I: *Morenia petersi;* J: *Cyclemys gemeli;* K: *Batagur dhongoka;* L: *Cuora mouhotii;* M: *Melanochelys trijuga.*

Contd...

Plate 14.1–*Contd...*

Pangshura tentoria (Gray, 1834)

It is locally known as 'Tambu Dura' in Assam. The species grows to 26.5 cm straight carapace length. Three subspecies have been described: *tentoria* (Gray, 1834) from peninsular India, *circumdata* (Mertens, 1969) from the western tributaries of the Ganga as well as Brahmaputra and the flaviventer (Gunther, 1864) from the northern tributaries of the Ganga. This species is represented by the subspecies *P. tentoria tentoria* and *P. tentoria flaviventer* in Assam. It is found throughout the Brahmaputra and Barak rivers. This species is the most common by-catch during fishing activities in the river Brahmaputra. Nesting takes place during the months of September to

November in the sandy riverine chars. In ponds and marsh lands the species is seen to nest on the mud bank. A clutch of 3-8 eggs are usually laid by digging nearly 30 cm pit. The eggs are consumed by the people and other predators like dogs and Jackals.

***Pangshura smithii* (Gray, 1863)**

It is locally known as 'Muga Dura' in Assam. The females may reach a length of 22.7 cm and the males are just 10.8 cm in the carapace length. The species occupies an elongate, narrow band of territory in the northern and northeastern parts of the Indian subcontinent. It is a common species in the river Brahmaputra in Assam. The species basks in community on the river banks. It lays 6-12 eggs per clutch.

Two subspecies are recognized. In India, *P. smithii smithii* is has been reported from Assam (Manas National Park, Kaziranga National Park and Rajiv Gandhi Orang National Park) and the other subspecies *P. smithii pallidipes* is reported from Bihar and Uttar Pradesh (Gangra River system).

***Pangshura sylhetensis* (Jerdon, 1870)**

It is locally known as 'Asomi Dura' and 'Phulen Dura' in Assam. It is a small (carapace length to 20.5 cm) freshwater turtle with thirteen pairs of marginal scutes. The head is moderately small, snout pointed, shorter than the orbit and projecting much beyond the lower jaw. It has narrowest distribution range compared to other three Pangshura species. It lays 6-12 eggs per clutch. The species is omnivorous and mostly nocturnal in nature (Baruah *et al.,* 2010b).

***Batagur dhongoka* (Gray, 1834)**

It is locally known as 'Tinipatia Dura' in Assam. This species attains a length of 48 cm and has a modest vertebral kill, which reaches its highest development on the second and third vertebral scutes. The species has been reported from Assam. Carapaces of the dead individuals have been observed in Kaziranga National Park and Chandubi beel area of Assam. Clutch size is 20 to 35 eggs, which are laid in March to April, along the edges of the rivers or in midriver sandbanks. The species is mostly herbivorous in nature.

***Cuora amboinensis* (Daudin, 1802)**

The Southeast Asian Box Turtle, *Cuora amboinensis* is locally known as 'Jap Dura' in Assam. It is a small (carapace length to 25 cm), semi-aquatic turtle. It is distributed throughout northeast India and other parts of Southeast Asia. The species is very common in most of its range, perhaps due to its ability to adapt to a variety of habitats. They are found in rivers, lakes, marshes, ponds, channels, creeks, swamps and paddy fields in and around lowland forests. This species also exists in Manas National Park, Orang National Park, Kaziranga National Park and Deihing Patkai Wildlife Sanctuary. The species lays 1-4 eggs per clutch. Diet includes both plant and animal matter. This species is omnivorous but primarily vegetarian diet.

***Cuora mouhotii* (Gray, 1862)**

It is locally known as 'Siria Jap Dura' in Assam. The carapace is elongated and does not exceed 18 cm. It is distributed throughout northeast India. The species inhabits in evergreen, semi-evergreen and subtropical broadleaf forests. It lays 1-5 elongate eggs per clutch. It is an omnivorous species.

Melanochelys trijuga (Schweigger, 1812)

It is locally known as 'Kola Sil Dura'. It is a medium sized (carapace length to 38.3 cm) forest dwelling species living in dry deciduous, moist deciduous and semi evergreen forests near water bodies or swampy grasslands. It has been observed in Assam and Meghalaya. Breeding takes place during monsoon season and egg lying was observed during October to November. A clutch of 8-16 elongated eggs was recorded from a female of carapace length 33 cm in captivity. It is omnivorous in dietary habits.

Melanochelys tricarinata (Blyth, 1856)

It is locally known as 'Bamuni Dura' in Assam. It is a small (carapace length up to 17.4 cm), entirely terrestrial turtle from the deciduous and evergreen forests and riverine grasslands. During this study we recorded the species from Manas National Park, Kaziranga National Park, Rajiv Gandhi Orang National Park, Burachapari Wildlife Sanctuary and forested areas of Kamrup district. The species was earlier reported from a relatively narrow sub-Himalayan belt (Das, 2009a). It is mostly active during night time; captive individuals feed on vegetables, wild fruits, different species of snails and crabs. Tikader and Sharma (1985) reported that 3 to 6 eggs are laid in a single clutch and incubation period is 60-72 days, hatching taking place during April to May. Nesting occurs during winter months in Assam.

Cyclemys gemeli (Gray, 1831)

It is locally known as 'Sepeta Dura' in Assamese. This is a moderate-sized species (carapace length to 24 cm), with flat and oblong carapace. It is a common species in Assam and Arunachal Pradesh. It inhabits in streams, rivers and pools at lower hills. It is an omnivorous species. It nests four to six times per year, producing 2-4 elongate eggs each time.

Geoclemys hamiltonii (Gray, 1831)

The spotted pond turtle, locally known as 'Nal Dura' and 'Futuki Dura' in Assam, is a medium sized (carapace length to 40.5 cm) freshwater turtle from lentic waterbodies of the northern and northeastern India. The species in observed in Assam and Meghalaya. Significant population exists in the Kaziranga National Park, Rajiv Gandhi Orang National Park and Dibru Saikhowa National Park in Assam. It lays two clutches of 18-30 eggs that hatch within 60-76 days. The species is primarily carnivorous, feeding on snails and insect larvae along with vegetative material.

Hardella thurjii (Gray, 1831)

The crowned river turtle, locally known as 'Bor Dura' in Assam, is a large (carapace length to 65 cm) freshwater turtle, largely confined to lentic waters. It has been observed from Assam, Meghalaya and Arunachal Pradesh. It has been observed in the Kaziranga National Park of Assam. It lays 8-19 ellipsoidal eggs per clutch. The species is herbivorous in nature.

Morenia petersi (Anderson, 1879)

It is locally known as 'Bangla Dura' in Assam. It is a small freshwater turtle (carapace length to 22.2 cm) with fairly restricted distribution in northern, eastern and northeastern India, southern Nepal and Bangladesh. It inhabits both in standing and slow moving water bodies. There are three records from protected area of Assam *viz.* Deepor Beel Ramsar Site (Baruah and Sharma, 2010), Kaziranga National Park (Choudhury, 2004) and Pobitora Wildlife Sanctuary (Sengupta *et al.,* 1998). It has been further observed in the Rajiv Gandhi Orang National Park in Assam. It is exclusively herbivorous in wild, with jaw surfaces highly specialized for folivory habit. It lays a single clutch of 6-10 elongated eggs during winter.

Family: Trionychidae (Plate 14.2)

Nilssonia nigricans (Anderson 1875)

The Black Softshell Turtle, *Nilssonia nigricans,* is a large (carapace length up to 91 cm), invariably aquatic species. It is locally known as 'Bormuria Kacho' in Assam. Its distribution in the wild is limited to the River Brahmaputra and its tributaries. The species was described from a Shrine pond in Nasirabad near Chittagong, Bangladesh. Also found in captivity in the Temple ponds of Tripura and Assam, Northeast India. It lays spherical eggs numbering 10 to 38 per clutch (Baruah *et al.,* 2010a). There are several nestings per year. The incubation period for hatching has been recorded at 270 days, however demands further evaluation.

Nilssonia gangetica (Cuvier, 1825)

This species is locally known as 'Ganga Kaso' in Assam. This is a large (carapace length to 70 cm) freshwater turtle, with a round or somewhat oval carapace. Though widespread, this species is restricted to only few localities in the region. This species is found in the river Brahmaputra, Barak and their tributaries, specifically in the Jia Bharali River, Kulshi River, Chandubi beel, Deepor beel and temple ponds of Assam. Nesting takes place during winter and hatchlings were seen during monsoon period. This species nests in the sandy bank and also in sandy loam soils near ponds and other water-bodies. A clutch of 8-85 spherical eggs are laid by digging a deep pit. Incubation lasts 217 to 287 days. It is omnivorous in feeding habit and individuals were seen feeding on cattle or even carcasses near the river banks. It is mostly exploited for meat and medicinal value of the calipee.

Nilssonia hurum (Gray, 1831)

This species is locally named as 'Bor Kaso' in Assam. It is a relatively abundant large (carapace length to 60 cm) riverine species, found in rivers and reservoirs. The carapace is flat and oval or slightly domed shaped. It is beautifully marked and the background colour is dark green, but in juveniles the disk bears four to six large ocelli. Plastron is light grey. The centre of each ocellus is a black disk surrounded by concentric yellow circles. Head is large with black reticulation; large yellow or orange patch behind eye and one across snout. This species is found in the wetlands of Assam. The juveniles of the species are worshiped due to the prominent eye-spots resembling the eyes of Lord Vishnu. It is a common species in Assam and Tripura.

Plate 14.2: Family-Trionychidae

N: *Nilssonia hurum;* O: *Nilssonia gangetica;* P: *Nilssonia nigricans;* Q: *Lissemys punctata andersoni;* R: *Chitra indica;* S: *Amyda cartilaginea;* T: *Pelochelys cantorii.*

This species occurs in several protected areas within its natural range, including the Pakhui Wildlife Sanctuary (Arunachal Pradesh) and Chandubi Reserve Forest (Assam). This species is worshiped in many of the temple ponds in northeast. It lays 20-38 spherical eggs per clutch during July to September. Incubation lasts for 180 days. The species is primarily nocturnal and omnivorous. Individuals maintained in the temple tanks and village ponds in Assam were seen to feed on puffed rice, bread and materials generated as garbage as a result of religious activities (Noureen and Baruah, 2010).

***Chitra indica* (Gray, 1831)**

The narrow-headed softshell turtle is an extremely large (carapace length to 110 cm), highly aquatic species. The dorsum carries a striking pattern of deep yellow and black closely resembling the colour pattern of tiger, as a result of which many locals describe it as 'Baghia Kacho' (tiger turtle) in Assam. This species is observed in the Brahmaputra River and its tributaries in Assam and very good population size in the Ganges river system (Singh and Baruah, 2010). It inhabits rivers with sandy bottom and remains buried at bottom under sand, keeping only the eyes and snout out to catch prey at lightening speed. It is widely distributed on the Indian subcontinent. This study reports the species from Brahmaputra river at Morigaon district having water depth of more than 25 feet. Several specimens were caught by fishermen and released. It is highly aggressive in disposition and prefers steep river banks for basking. A large clutch of 65-193 eggs are laid during June- August near the sandy bank. Eggs are sometimes predated by people, dogs and jackals. Choudhury (1990) reported it form the Lohit river in upper Brahmaputra and Datta (1997) from Dhubri district in lower Assam.

***Lissemys punctata* (Lacepede, 1788)**

Locally known as 'Benga Kaso' in Assam. It is one of the most common freshwater turtle species in Assam, Meghalaya and Tripura. This species is known to live in rivers, ponds, low-land agricultural fields and swampy areas. It lays 2-15 eggs per clutch. The species aestivates during winter and dry summer. It secretes a foul smelling fluid when disturbed as a reflex action, to protect itself from predators. This species is widely poached for the freshwater turtle trade in Tripura.

***Amyda cartilaginea* (Boddaert, 1770)**

It is locally known as 'Malaya Kaso' in Assam. This is a large (carapace length to 80 cm) turtle with an oval carapace. The species is widespread, occurring from peninsular India to China and Southeast Asia. The range of this species is very wide and also discontinuous. It is known from southern Mizoram, Manipur and Nagaland. It inhabits rivers and lakes. Three or four nestings may occur annually, with 6 to 10 eggs per clutch (maximum 30 in a year). Incubation lasts 60 to 145 days. It is a very rare species. It is Carnivorous in dietary habits and consumes fishes, amphibians, crustaceans, aquatic insects and other invertebrates.

***Pelochelys cantorii* (Gray, 1864)**

The Asian giant softshell turtle, locally known as 'Jata kacho'. It is a large freshwater species which can reach 130 cm in total carapace length. The carapace is very rounded and flat, olive-green to a more or less dark brown and decorated with small dark dots or yellow spots. It occurs in a variety of habitats, including lakes and rivers. Clutch size is 24 to 70 eggs. The species consumes aquatic organisms including fish, crustaceans and molluscs in addition to plant matter.

Family: Testudinidae (Plate 14.3)

Indotestudo elongata (Blyth, 1853)

Locally known as 'Halodhia Kaso' in Assam. Males may reach 33 cm and females about 29 cm. This species is distributed throughout northeast India. This forest dwelling species mostly found in Sal dominated moist deciduous and semi-evergreen forests with big rock boulders. It was observed from Manas National Park, Dehing Patkai Wildlife Sanctuary and Karbi Anglong district of Assam. It lays 2-9 eggs per clutch.

Manouria emys (Schlegel and Muller, 1840)

Locally known as 'Pahari Kaso' in Assam. This is the largest tortoise of Asia, with a length of 60 cm and a weight of 30 kg. The sub-species known from northeastern state of India is *M. Emys phayrei* (Blyth, 1853) in which pectoral scutes on plastron touches each other. It inhabits evergreen forests in the hills close to water. It lays 23-51 eggs per clutch that hatch in 60-75 days. It was observed from a few localities of the northeast region *viz.* Manipur, Mizoram, Dehing Patkai Wildlife Sanctuary and Karbi Anglong district of Assam. Two individuals have been encountered in Manipur during the present survey. It has been earlier reported from Nongkhellym Wildlife Sanctuary (Das, 1990), Mailbong, Langting-Mupa of North Cachar Hills, Loomajooting of Nagaland and Nengpui Wildlife Sanctuary of Mizoram.

Discussion

The diversity and abundance of freshwater turtle species in the northeast India suggests that this region is a turtle priority conservation area with 22 species of turtles, the highest turtle diversity area in India. Out of 22 recorded species, as many as 19 species have been recorded from in and around the protected areas of northeastern states.

There is no information on community ecology of the freshwater turtles of northeast till date. Information on ecological requirements of each turtle species, causes of turtle death are also lacking from this region, limiting management options for turtles and tortoises for the managers and policy-makers.

Threat and Conservation Status

Wild populations have been subjected to local exploitation for food. Major threats for the semi-captive populations are the possible health risk from fungal infestation and the inbreeding depression. These captive populations are thriving in few isolated temple ponds in very congested area with foods, provided by the visiting devotees.

Habitat Destruction

The forests and wetlands are shrinking with the decline of chelonian density. Deforestation and the resultant loss of soil, leading to increased siltation of lakes and

Plate 14.3: Family–Testudinidae
U: *Manouria emys phayrei;* V: *Indotestudo elongata.*

wetlands. The deep pools that favoured habitats of soft-shell turtle species are rapidly becoming shallow and choked with silt, leading to the decline in habitat.

Killing for Flesh

Indigenous inhabitants of this region basically consume meat and flesh and so the turtle meat and eggs. The Ningthouja clan of the Meiteis of Manipur, for example, considered it a taboo to consume turtle or tortoise meat (Gupta and Guha, 2002). Many turtle hunters use spears to gore the hapless creatures in shallow water, refusing to spare even juveniles. Turtles are also killed by planting hooked lines in shallow water in addition to nets, bamboo traps etc. As the turtles move into this area, their paddles get caught in the hooks. In addition to nets, various traps made of bamboo and canes are also used.

Superstitious Beliefs

Hanging a carapace in the cattle-shed is believed to be a good luck and to keep snakes away from the premises; and its hanging is believed to keep away burglars. Some of the religious communities in Assam sacrifice this soft-shell turtles to satisfy the Goddess Kali.

Eggs of Hard-shell Turtle (*Pangshura* sp.), rescued from poachers (Photo by Chittaranjan Baruah)

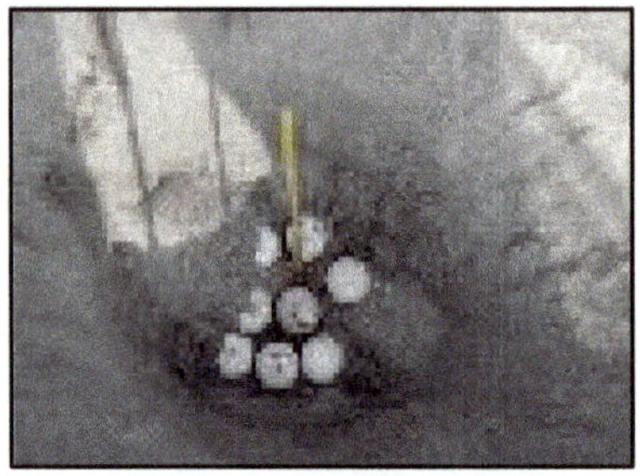

Eggs of Softshell Turtle (*Nilssonia gangetica*) (Photo by Chittaranjan Baruah)

Plate 14.4

Use in Ethnomedicine

Both the flesh and eggs are believed to have remedy for gout and arthritis, while the carapace of soft-shell turtle species is also used as medicine. Due to decline of visible catch and killing, the traders now offer very lucrative prices for flesh and carapace to the tribal hunters and fishermen.

Conservation Efforts

Although poaching is a problem, turtles have historically received community-sanctioned religious protection in many temple tanks in northeast India, notably the Nagshanker pukhuri, Madhab phukhuri, Kamakhya phukhuri in Assam and Tripureswari pukhuri at Udaipur, Tripura. Thus *ex-situ* conservation of chelonians in community and temple tanks and in public gardens could also constitute a useful mechanism for conservation.

Recommendations

Lack of recent field information has created a gap in conservation activities and status evaluation of almost all turtle species. Immediate adoption of conservation measures is essential for the conservation of endangered turtle genetic resource available in this region. As a part of the conservation of genetic diversity, the following studies are proposed:

1. A detail inventory of freshwater turtles inhabiting the river systems of northeast India and their current status is an urgent need of the hour to address the conservation measures of the group.
2. Study should be carried out to examine the possible threats to the turtle species due to habitat loss, forest fire and poaching.
3. Study of reproductive cycle of endangered turtle species is important.
4. Developmental activities and destruction of turtle nesting sites due to rampant sand mines should be stopped or regulated to give protection to the species.

5. Steps to be taken to stamp out the ongoing freshwater turtle trade.
6. Creating awareness programmes among the communities to curb illegal trade of turtles and their parts will be helpful.
7. Community conservation efforts should be properly recognized and management guidelines can be developed for better management of the turtles.
8. To introduce captive breeding programme for the endangered turtles and tortoises of northeast India.
9. To conduct population surveys in temple ponds and potential rivers to discover unknown population northeast India.
10. Temple pond management and protection of nests, prevention of trade and emphasis on conservation genetics.
11. Development of molecular markers for species identification and molecular phylogeny to analyze the polymorphism and evolutionary history.

Table 14.2: Etymology of some Turtle Species Names (Das, 2009b)

Name	*Translation*
GEOEMYDIDAE	
Cuora	From the Malay *Kura*, meaning hard-shelled turtle
amboinensis	Latin implying inhabitant of Amboina (Ambon), Maluku, Indonesia
mouhotii	For Henri Mouhot (1826–61), naturalist and explorer who collected reptiles in Thailand, Cambodia and Laos
Cyclemys	Greek for circle turtle, for the rounded shell
Cyclemys gemeli	For Richard Gemel, Austrian herpetologist
Geoclemys	Latin for earth turtle
hamiltonii	For Dr. Francis Hamilton (1762–1829), botanist and surgeon with the British East India Company based in Bengal and Burma
Melanochelys	Greek for black turtle
tricarinata	Latin for three keeled
trijuga	Latin for three-ridged
thermalis	Latin for heat
Morenia	Greek for sluggishness
petersi	For Wilhelm Carl Hartweg Peters (1815–83) of ZMB
smithii	For Sir Andrew Smith (1797–1872), author of 'Illustrations of the zoology of South Africa'
pallidipes	Latin for pale-footed
sylhetensis	Latin implying an inhabitant of Sylhet
tecta	Latin for roofed
tentoria	Latin for extended, for the elevated shell
circumdata	for surrounding, for the pinkish-orange pleuro-marginal ring
flaviventer	Latin for yellow-bellied

Contd...

Table 14.2–*Contd...*

Name	*Translation*
TESTUDINIDAE	
Indotestudo	Greek for Indian tortoise
elongata	Latin for elongated
Manouria	Greek for rare, domed
emys	Greek for turtle
phayrei	For Lieutenant-Colonel Sir Arthur Purves Phayre (1812–85), British Commissioner at Pegu, Burma and Chief Commissioner of British Burma.
TRIONYCHIDAE	
Amyda	Greek for turtle
cartilaginea	Latin for cartilage
Chitra	After the north Indian vernacular, meaning picture
indica	Latin for pertaining to India
Lissemys	Greek for smooth turtle
punctata	Latin for spotted
andersoni	For John Anderson (1833–1900), Superintendent of the IMRR
Nilssonia	For Sven Nilsson (1787–1883), Professor of Natural History at Lund
gangetica	Latin for pertaining to the River Ganga
hurum	Apparently derived from a north Indian vernacular for the species
nigricans	Latin for blackish
Pelochelys	Greek for mud turtle
cantorii	For Theodore Edward Cantor (1809–1860), Danish surgeon-naturalist with the British East India Company

Ongoing Research

Under the aegis of Turtle Conservation and Research Programme (TCRP), the turtle conservation programme in northeast India has been initiated by the Department of Zoology, Gauhati University with Conservation education, Local awareness programmes and Community participation in the wetland and rivers including *in-situ* egg protection (Baruah *et al.,* 2010b). Most of the field activities are starting in Assam, as outlined in the "Conservation Action Plan for Endangered Turtles and Tortoises of India" (CFH/MCBT, 2006).

Community Conservation Efforts

Many of the large freshwater turtle species are protected by the cultural and religious systems of the local community belief. *Nilssonia nigricans* is protected by the local community. In Assam many temples also serve as rescue centres for species like *Pangshura sylhetensis* and *Geoclemys hamiltonii.* When people find turtles in their agricultural fields during forest fires, they bring them to the temples and later release them into the wild after a few months. The local communities are now protecting the turtles after the initiative taken by the TCRP.

Conclusion

Habitat loss and over exploitation for meat is considered a persistent threat to chelonians of the region. Out of 23 species found in Northeast, 6 species namely *Pangshura sylhetensis, Cuora mouhotii, Nilssonia nigricans, Chitra indicia, Manouria emys* and *Indotestudo elongata* are listed by IUCN (2007) as most endangered species of Asia. *Chitra indica* is on the new list of the 25 most endangered turtle species of the world. The soft-shell turtle *Chitra indica* needs special attention as it is on the new list of the 25 most endangered turtle species of the world. As a part of the conservation of turtle diversity, we emphasize on some important research and conservation initiatives.

Acknowledgments

We gratefully acknowledge Turtle Survival Alliance, the Conservation Leadership Programme (CLP), Rufford Small Grant's Foundation, Mohamed bin Zayed Species Conservation Fund and the British Chelonian Group for supporting endangered turtle conservation program in Northeast India. We thank Shailendra Singh and Kartik Shanker for supporting chelonian research in India. Special thanks go to Peter Praschag, Indraneil Das and Peter CH Pritchard for their encouragement.

References

Ahmed, M. F., Das, A. and Dutta S.K., 2009. *Amphibians and Reptiles of Northeast India: A Photographic Guide.* Aaranyak, Guwahati, India, p. i-xiv and 1-168.

Anderson, J., 1879. Anatomical and Zoological Researches: Comprising an Account of the Zoological Results of the Two Expeditions to Western Yunnan in 1866 and 1875. *Bernard Quaritch, London "1878"*, p. 703–860.

Baruah, C., Ali Reza, A.H.M. and Sharma, D.K., 2010a. Le Trionyx noiratre, la Tortue de Bostami *Nilssonia nigricans* (Anderson 1875) (Ang: Black softshell Turtle, Bostami Turtle), *Chéloniens,* 18: 33–38.

Baruah, C., Sarma, P.K. and Sharma, D.K., 2010b. Status and conservation of Assam roofed turtle Pangshura sylhetensis in the Brahmaputra floodplain, Assam, India. *NeBIO,* 1(3): 42– 47.

Blyth, E. 1853. Notices and descriptions of various reptiles, new or little known. *Journal of the Asiatic Society of Bengal,* 22(7): 623–655.

Blyth, E.1856. Report on the collection presented by Capt. Beddome and Mr. Theobald. *J. Asiatic Soc. Bengal,* 24: 713–720.

Boddaert, P. 1770. Brief van van de Kraakbeenige Schidpad. Epistola testudine cartilaginea. Amsterdam: Kornelis van Tongerlo, p. 39.

Buhlmann, K.A., Akre, T.S.B., Iverson, J.B., Karapatak I S, D., Mittermeler, R.A., Georges, a., Rhodln, A.G.J., VandI J.K., P.P. and G I B Bons, J.W., 2009. A global analysis of tortoise and freshwater turtle distributions with identiication of priority conservation areas. *Chelonian Conservation and Biology,* 8(2): 116–149.

CFH/MCBT., 2006. Conservation Action Plan for Endangered Freshwater Turtles and Tortoises of India. Madras Crocodile Bank Trust, Tamil Nadu, south India, p. 34.

Choudhury, A.U., 1990. Turtles recorded in Dibru–Saikhowa Wildlife Sanctuary, Assam. *Journal of Ecological Society. Pune*, 8: 33.

Choudhury, A.U., 1995. Some freshwater turtles from Lakhimpur district, Assam. *Cheetal,* 34(3 and 4): 18–24.

Choudhury, A.U., 1998. *Pyxidea mouhotii* (Gray) in Southern Assam and Mizoram. *Journal of Bombay Natural History Society*, 95(3): 511.

Choudhury, A. U., 2004. Kaziranga: Wildlife in Assam. Kolkata: Rupa and Co., p. 104.

Cuvier, G., 1825. *Recherches sur les ossemens fossiles de quadropèdes où l'on rètablit les caractères du plusieurs espèces d'animaux que les revolutions du globe paraoissent voir detruites.* Vol. 5. Third edition. Dufour d'Ocagne, Paris.

Daniel, J.C., 1983. The Book of Indian Reptiles. Bombay Natural History Society, Oxford University Press, India p. 141.

Das, I., 1985. Indian turtles: A field guide. World Wildlife Fund–India, Oxford University Press, Calcutta, India p. 119.

Das, I., 1990. Distributional records for Chelonians from Northeastern India. *Journal of Bombay Natural History Society*, 87(1): 90–97.

Das, I., 1995. Turtles and tortoises of India. World Wide Fund for Nature– India. Bombay: Oxford University Press, p. 176.

Das, I., 2001. Die Schildkröten des Indischen Subkontinents. Frankfurt am Main: Chimaira, p. 181.

Das, I., 2002. A photographic guide to snakes and other reptiles of India. New Holland Publishers (UK) Ltd, Garfield House, 86–88 Edgware Road, London w2 2EA, UK, p. 144.

Das, I., 2009a. *Melanochelys tricarinata* (Blyth 1856) – tricarinate hill turtle, three–keeled land turtle. In: Rhodin, A.G.J., Pritchard, P.C.H., van Dijk, P.P., Saumure, R.A., Buhlmann, K.A., Iverson, J.B., and Mittermeier, R.A. (Eds.). Conservation Biology of Freshwater Turtles and Tortoises: A Compilation Project of the IUCN/ SSC Tortoise and Freshwater Turtle Specialist Group. Chelonian Research Monographs, 5: 025.1–025.5.

Das, I., 2009b. The Discovery of Indian Turtles, With Notes on Publications, Type Localities and Type Repositories. Pp. 1–4. In: Vasudevan, K. (Ed.). Freshwater Turtles and Tortoises of India. ENVIS Bulletin: Wildlife and Protected Areas, Vol. 12 (1). WII, Dehradun, India.

Datta, S., 1997. Freshwater turtles and tortoises of Dhubri District. *Zoos' Print*, 12 (6): 1–4.

Daudin, F.M., 1802. *Histoire naturelle, génerale et particulière des reptiles.* Vol. 2. F. Dufart, Paris, p. 432.

Ernst, C.H., Altenburg, R.G.M., Barbour, R.W., 2000. Turtles of the World. World Biodiversity Database, CD–ROM Series, Windows, Version 1.2. Amsterdam: Biodiversity Center of ETI.

Gray, J.E., 1831. *Synopsis Reptilium or short descriptions of the species of reptiles. Part I: Cataphracta, tortoises, crocodiles, and enaliosaurians*. Treuttel, Wurz and Co., London, p. 85.

Gray, J.E., 1834. Characters of several new species of freshwater tortoises (Emys) from India and China. *Proc. Zool. Soc. London*, 1834(18): 53–54.

Gray, J.E., 1862. Notice of a new species of *Cyclemys* from the Lao Mountains, in Siam. *Ann. and Mag. Nat. Hist. Ser.*, 2(10): 157.

Gray, J.E., 1863. Notice of a new species of Batagur from north–western India. *Proc. Zool. Soc. London*, 1863(2): 253.

Gray, J.E., 1864. Revision of the species of Trionychidae found in Asia and Africa, with the descriptions of some new species. *Proc. Zool. Soc. London*, 1864(1): 76–98.

Gunther, A., 1864. The reptiles of British India. Ray Society, London. xxvii + I–XXVI pl. Note: Reprinted around 1982, Oxford and IBH Publishing Co., New Delhi, Bombay and Calcutta, p.452.

Iverson, J.B., 1992. A revised checklist with distribution maps of the turtles of the world. Richmond, Indiana: privately published, p. 374.

Jerdon, T. C., 1870. Notes on Indian Herpetology. *Proceedings of the Asiatic Society of Bengal*. March, pp. 66–85.

Mertens, R., 1969. Eine neue Rasse der Dachschildkröte, Kachuga tecta. *Senckenbergiana Biol.*, 50(1/2), 23–30.

Moll, D. and Moll, E.O., 2004. *Ecology, Exploitation and Conservation of River Turtles*. Oxford University Press, USA, p. 420.

Noureen, U. and Baruah, C., 2010. Le Trionyx a ocelles *Nilssonia hurum* (Gray, 1831) (Ang: Indian peacock softshell turtle). *Chéloniens*, 18: 39–45.

Praschag, P. and Gemel, R., 2002. Identify of Black soft– shell turtle *Aspideretes nigricans* (Anderson, 1875) with remarks on related species (Reptilia: Testudines, Trionychidae). (Translation: Jarmo Perala), *Band 23 Ausgeben*, 5: 87–116.

Praschag, P., Hundsdorfer, A.K. and Fritz, U., 2007a. Phylogeny and taxonomy of endangered South and South–east Asian freshwater turtles elucidated by mtDNA sequence variation (Testudines: Geoemydidae: Batagur, Callagur, Hardella, Kachuga, Pangshura). *Zoologica Scripta*, 36: 429–442.

Praschag, P., Hundsdorfer, A.K., Reza, A.H.M.A. and Fritz, U., 2007b. Genetic evidence for wild–living Aspideretes nigricans and a molecular phylogeny of South Asian softshell turtles (Reptilia: Trionychidae: Aspideretes, Nilssonia). *Zoologica Scripta* 36: 301–310.

Praschag, P., Hundsdorfer, A.K. and Fritz, U., 2009. Further specimens and phylogenetic position of the recently described leaf turtle species *Cyclemys gemeli* (Testudines: Geoemydidae). *Zootaxa*, 2008, 29–37.

Pritchard, P.C.H., 1979. Taxonomy, evolution and zoogeography. In: *Turtles: Perspectives and Research,* (Eds.) M. Harless and H. Morelock. Wiley, New York, pp. 1–42.

Schlegel, H. and Muller, S., 1840. Over Schildpadden van den Indischen Archipel, beschrijving einer nieuwe sort van Sumatra. In: *Verhandelingen over de Natuurlijke Geschiendenis der Nederlandsche Overzeesche Bozittingen, 1839–44. Part 3. Zoologie, Schildpadden,* (Ed.) C.J. Temminck. Leiden: Luchtmans and van der Hoek, plate 4. [in Dutch].

Schweigger, A.F., 1812. *Prodromus monographiae cheloniorum.* Königsberg. *Arch. Naturwiss. Math,* 1: 271–358, 406–458.

Sengupta S., Baruah, M.R., Baruah, M. and Choudhury, N.K., 1998. Report on turtles of Pobitora Wildlife Sanctuary. *Journal NATCON,*10(2): 209–210.

Singh, S. and Baruah, C., 2010. Le Trionyx a rayures *Chitra indica* (Gray, 1831) (Ang: Indian Narrow–headed softshell turtle). *Chéloniens,* 18: 49–53

Smith, M.A., 1933. The fauna of British India, including Ceylon and Burma. Reptilia and Amphibia. Vol. 1 Loricata, Testudines. Today and Tomorrow's Printers and publishers, New Delhi.

Tikader, B.K. and Sharma, R. C., 1985. *Handbook: Indian Testudines.* Zoological Survey of India, Calcutta, p. 156.

Turtle Taxonomy Working Group, [Rhodin, A.G.J., van Dijk,P.P., Iverson, J.B., and Shaffer, H.B.]., 2010. Turtles of the world, 2010 update: annotated checklist of taxonomy, synonymy, distribution, and conservation status. In: Rhodin, A.G.J., Pritchard, P.C.H., van Dijk, P.P., Saumure, R.A., Buhlmann, K.A., Iverson, J.B., and Mittermeier, R.A. (Eds.). Conservation Biology of Freshwater Turtles and Tortoises: A Compilation Project of the IUCN/SSC Tortoise and Freshwater Turtle Specialist Group. Chelonian Research Monographs 5, pp. 000.85–000.164

Van Dijk, P.P., Stuart, B.L. and Rhodin, A.G.J., 2000. Asian Turtle Trade: *Proceedings of a workshop on conservation and trade of freshwater turtles and tortoises in Asia. Chelonian Research Foundation, MTC Printing, Lunenburg, MA.*

Xu, J., Grumbine, R.E., Shrestha, A., Ericsson, M., Yang, X., Wang, Y., Wilkes, A., 2009. The melting Himalayas: cascading effects of climate change on water, biodiversity, and livelihoods. *Conservation Biology,* 23: 520–530.

Zangerl, R., 1969. The turtle shell. In: *Biology of the Reptilian, Vol. 1: Morphology,* (Eds.) A.C. Gans, A d'A. Bellairs, and T.S. Parsons. Academic Press, New York, pp. 311–339.

2013, Animal Diversity, Natural History and Conservation, Vol. 3 *Pages* **237–252**
Editors: **V.K. Gupta and Anil K. Verma**
Published by: **DAYA PUBLISHING HOUSE, NEW DELHI**

Chapter 15

A Glimpse of the Avifaunal Diversity of Jammu Himalayas

Sanjeev Kumar* and D.N. Sahi
Department of Zoology,
University of Jammu, Jammu – 180 006

Abstract

The present work encompasses the extensive avifaunal surveys conducted in the Jammu region of Jammu and Kashmir State for last five years. To carry out the work, the study area was divided into different stations (erstwhile six districts namely, Kathua, Jammu, Udhampur, Doda, Rajouri and Poonch) and periodic visits were carried out to make observations. The study revealed 140 species of birds belonging to 14 orders and 44 families. An effort was made to compile a conservation status of the reported species. Local abundance status of the avian species has also been worked out.

Keywords: *Avifauna, Conservation, Abundance.*

Introduction

Strategically located Jammu and Kashmir State constitutes the northern most extremity of India. Situated between 32°17′ and 36°58′ North latitudes and 80°30′ East longitudes, the total area of the state is 22, 22, 36 sq. km including 78,114 sq. km under the illegal occupation of Pakistan and 42,685 sq. km under that of China. The state is bounded by Pakistan, Afghanistan and China from the West to the East. The state ranks 6th in area and 17th in population among the states and union territories of India.

---

* *Corresponding author.* E-mail: drskpangotra@gmail.com

It has four geographical zones:

1. Sub-mountain and semi-mountain plain known as Kandi or dry belt.
2. The Shivalik ranges.
3. The high mountain zone constituting the Kashmir valley, Pir Panjal ranges and its offshoots including Doda, Poonch and Rajouri districts and part of Kathua and Udhampur districts.
4. The middle run of the Indus river comprising Leh and Kargil.

The state of Jammu and Kashmir is the northernmost state of India comprising three climatic regions *viz.* arctic cold desert areas of Ladakh, temperate Kashmir Valley and sub-tropical region of Jammu. There is a sharp rise of altitude from 1000 ft to 28,250 ft above the sea level within state's four degrees of latitude.

The climate varies from subtropical in Jammu plains to semi-arctic cold in Ladakh with Kashmir and Jammu mountainous tracts having temperate climatic conditions. The annual rainfall also varies from region to region with 9.26 cm in Leh, 65 cm in Srinagar and 111.5 cm in Jammu. A large part of the state forms part of the Himalayan mountains.

The present work is confined to the Jammu division of the state.

Jammu Division

Jammu division is situated in the northwest of India and is bordered by Punjab and Himachal Pradesh in east and Pakistan in west and south while Pir Panjal range of mountains separates from Kashmir division in the north. Due to these extreme variations within a limited distance, there is a marked difference in the climate. Jammu division has been divided into erstwhile six districts, *i.e.* Kathua, Jammu, Udhampur, Doda, Rajouri and Poonch.

Kathua District

The district Kathua of Jammu and Kashmir falls between 32°17′ and 32°55′ North latitudes and 75°17′ to 75°55′ East longitudes. The district is situated in the south east of Jammu district and is bounded in north by Ramnagar and Bhaderwah tehsils of Udhampur and Doda districts respectively. Samba tehsil of Jammu district bounds it in the northwest while Shakargarh tehsil of Pakistan and Gurdaspur of Punjab (India) lie on its south and Chamba District of Himachal Pradesh lies in the east. The district comprises of five tehsils namely, Kathua, Hiranagar, Billawar, Basohli and Bani. The total geographic area of the district is 2651 sq. kms. The temperature rises as high as 46°C in the plains and goes down below 1.5°C in upper hilly areas. The average annual rainfall ranges between 20 cm to 120 cm for different regions.

District Kathua can be divided into three distinct physiographic divisions from north to south. In the north, the Mandhidhar mountain range separates it from Doda district. The Mandhidhar mountain range and its subsidiary mountain ranges have steep slopes cut into precipitous ridges and deep defiles. Most of these mountain ranges are running in the northeast-southwest direction and reach a maximum elevation of 4200 msl.

Towards south of the outer Himalayan hill ranges lies the plain agricultural belt of the district. The most part of the Kathua tehsil and whole of the Hiranagar tehsil fall in this physical division of the district. Minimum elevation of the district (240 msl) lies in this physiographic division.

Jammu District

The district Jammu stretches between 74°24′ and 75°18′ east longitudes and 32°50′ and 33°30′ north latitudes. It was bounded by Udhampur district in the north and northeast, Kathua in the east and southeast, Pakistan in west and POK and Rajouri district in northwest.

The district has an area of 3097sq. km and administratively divided into 5 tehsils and 11 blocks. The average annual rainfall in Jammu district is nearly 113 cm. The hot season lasts from April to June followed by the rainy season from July to September. The intensity of heat in summer is comparable to the plains of the rest of India. Winter lasts from mid-October to March.

Udhampur District

The district is bounded by erstwhile districts *viz.* Rajouri district in the north, in the northeast by Doda district, in the southeast by Kathua district and in the southwest by Jammu district. The district lies between 32°30′ N latitudes and 74°16′ to 75°38′ E longitudes. The district has an area of 4,550 sq. km. Due to the altitudinal variations ranging from 600 m to 3000 m, there is wide variation in temperature in the different parts of the district. The temperature rises as high as 42°C and goes below to 1.5°C in severe winters.

Doda District

Doda district stretches between 32° 53' and 34° 21' north latitude and 75°1' and 76°47' east longitude, with an elevation of about 8,000 ft. to 15,000 ft. The total area of the district Doda is 11,691 sq. km. It is bounded on its eastern side by Ladakh (Zanskar), from northern side it is connected with Anantnag, southern side extends upto Udhampur and southeast side is connected with Himachal Pradesh. Because of its diverse topography and unique ecological features, it is bound to field diverse wildlife. The average annual rainfall is 35mm which is lowest as compared to other districts of the division.

Poonch District

It stretches between 33°45' to 34°08' North latitude to 73°58' to 74°33' East longitude. It comprises of four tehsils namely Haveli, Mendhar, Surankote and Mandi. Distt. Poonch is flanked by distt. Baramula and Distt. Shopian of Kashmir Province in north and eastern side respectively. On western side of it are POK and Distt. Rajouri of Jammu province lies to the Southern side of this dist. The altitude varies from 840 m in some areas of tehsil Mendhar to 4742 m in alpine peaks of tehsil Surankote and Mandi with a geographical area 1674 sq. km. It is totally hilly and mountainous bearing few nallahs which are low lying. Vegetation varies from scrub vegetation to alpine coniferous forests of silver fir *Abies pindrow* depending upon rainfall and altitude.

Rajouri District

It is located between 70° and 74°4′ E longitudes and 32°58′ and 33°35′ in N latitudes. It is flanked by Poonch district in north, Jammu district in south and Udhampur district in east. The district has an area of 2630sq. km and has 7 tehsils and 7 blocks. The district has some peculiar physical features.

Birds are considered as Masters of Air and are among the most familiar animals of our environment and through the years people have observed their activity with keen interest.

The studies on birds have been conducted from time to time in India and Jammu and Kashmir State also. But, so far as the knowledge of present investigator is concerned, there is no detailed study on the birds of Jammu region except a few records namely Ward (1906 -1907), Osmaston (1927), Whistler (1928), Ali and Ripley (1968-74, 1983), Sahi (1985), Alfred *et al.* (2001), Choudhary (2002), Sharma (2003), Ahmed (2004) Wani and Sahi (2005), Kumar (2006) and Kumar and Sahi (2006).

Methodology

For inventorization of aves, Line transect method and Point transect methods were used.

The nomenclature followed in the present work is in accordance with those given in the "Handbook of Birds of India and Pakistan" by Ali and Ripley (1968-74).

For identification and field diagnosis of birds, colourful plates of Ali and Ripley (1968-74), Ali (1996), Grimmett *et al.* (1998) and Grewal *et al.* (2002) have proved quite helpful. Colours are usually the best indicators of identity of a species at a close range or through binoculars. Calls also served as important clue in identifying and locating birds in their habitat.

Classification of birds is in accordance with Grewal *et al.* (2002) followed after *An Annotated Checklist of the Birds of the Oriental Region* by Inskipp *et al.* (1996).

The birds were observed with naked eye and through binoculars (12 × 50 Super Zenith) whenever found necessary. As far as possible, the birds were watched from behind a hide. Casual information was collected from many breeding pairs and only a few were selected for a detailed account.

Photographs were taken with Canon (EOS) fitted with 300 mm Zoom lens.

Observations and Discussion

The present work includes the extensive avifaunal surveys in the erstwhile Jammu region of Jammu and Kashmir state for last five years. To carry out the study, area under investigation was divided into different stations and periodic surveys were undertaken to record the observations. The study revealed 140 species of birds belonging to 14 orders and 44 families (Table 15.1).

Table 15.1: Showing Order-wise Distribution of the Species Recorded from the Study Area

Sl.No.	*Order*	*Order-wise Number of Species*
1	Cicconiformes	25
2	Galliformes	9
3	Gruiformes	5
4	Columbiformes	6
5	Psittaciformes	4
6	Cuculiformes	4
7	Strigiformes	3
8	Coraciformes	7
9	Piciformes	7
10	Upupiformes	1
11	Bucerotiformes	1
12	Passeriformes	60
13	Apodiformes	1
14	Anseriformes	7
	Total	140

The study shows (Table 15.2) that 126 species of birds were reported from Kathua, 119 from Jammu, 116 from Udhampur, 74 from Rajouri, 70 from Poonch and 66 from Doda. The highest number of species, *i.e.* 126 was reported in Kathua district owing to the reason that it encompasses plains (Hiranagar, Kathua) ranging from 240 mt and high mountainous area (Bani, Billawar) upto 4200 mt traversed by Outer to Lesser Himalayas thereby representing diverse habitats. On the other hand, district Doda represents minimum number of species, *i.e.* 66 which is engrossed by only lesser Himalayas thereby giving rise to more or less similar type of habitat.

An effort was also made to work out the conservation status of the birds reported during the present investigation. Out of the 140 species reported 5 species, *i.e.* Monal Pheasant, Chir Pheasant, Peafowl, Indian Shikra and Kestrel are endangered and belong to Schedule I of the Wildlife (Protection) Act of 1972 amended upto 2002. Rest of the species that need attention include White Backed Vulture, Scavenger Vulture, Black Cinerous Vulture and House Sparrow which are also showing decline in their population. House Sparrow which was quite common in houses has declined in number and is now restricted to scrubby and forest areas. Decline in the vulture's population is also due to urbanization as the number of cattle has considerably reduced due to modernization in living standard so they are not able to meet their sustenance due to lack of adequate food for them in the form of carcasses. Also, it has been established that decline in population of vultures is due to excessive use of the drug Diclofenac Sodium which is used to treat the diseased cattle.

Table 15.2: Systematic List of Birds with Station-wise Distribution with Local Abundance

Sl.No.	Name	Station-wise Distribution	Abundance
Class: AVES			
Order: CICCONIFORMES			
Family: PODICIPEDIDAE			
1.	Little Grebe *Podiceps ruficollis capensis* Salvadori	J, K, U	R
Family: PHALACROCORACIIDAE			
2.	Little Cormorant *Phalacrocorax niger* (Vieillot)	J, K, U	R
3.	Large Cormorant *Phalacrocorax carbo*	J, K, U	C
Family: Anhangidae			
4.	Darter *Anhinga rufa melanogaster* (Pennant)	J, K, U	R
Family: ARDEIDAE			
5.	Indian Pond Heron *Ardeola grayii grayii* Sykes	J, K, U, D, P, R	C
6.	Night Heron *Nycticorax nycticorax* (Linnaeus)	J, K	R
7.	Little Egret *Egretta garzetta* (Linnaeus)	J, K, U	R
8.	Cattle Egret *Bubulcus ibis coromandus* Boddaert	J, K, U, D, P, R	C
9.	Grey Heron *Ardea cinerea rectirostris* (Linnaeus)	J, U	R
10.	Black Bittern *Ixobrychus flavicollis flavicollis* (Latham)	J, K, U	R
Family: ACCIPITRIDAE			
11.	Pariah Kite *Milvus migrans govinda* Sykes	J, K, U, D, P, R	C
12.	Black Winged Kite *Elanus caeruleus* (Latham)	J, K, U	O
13.	Indian Shikra *Accipiter badius dussumieri* Temminck	J, K, U, D, P, R	C
14.	Kestrel *Falco tinnunculus*	D, R, P	R
15.	Steppe Eagle *Aquila nipalensis* Hodgson	J, K, U, D, P, R	R
16.	Indian White Backed Vulture *Gyps bengalensis* (Gmelin)	J, K	R
17.	Indian Long Billed Vulture *Gyps indicus* G. R.Gray	J, K, U, D, P, R	O

Contd...

Table 15.2–*Contd...*

Sl.No.	*Name*	*Station-wise Distribution*	*Abundance*
18.	Himalayan Griffon Vulture *Gyps himalayensis* Hume	J, K, U, D, P, R	O
19.	White Scavenger Vulture *Neophron percnopterus* (Linnaeus)	J, K, U, R	O
20.	Black Cinerous Vulture *Aegypius monachus*	J, K, U	R
Family: CHARADRIIDAE			
Sub family: RECURVIROSTRINAE			
21.	Black Winged Stilt *Himantopus himantopus himantopus* (Linnaeus)	J, K, U	O
Family: SCOLOPACIDAE			
Sub family: SCOLOPACINAE			
22.	Common Sandpiper *Actitis hypoleucos* Linnaeus	J, K, U	O
Family: CHARADRIIDAE			
Sub family: CHARADRIINAE			
23.	Red Wattled Lapwing *Vanellus indicus indicus* (Boddaert)	J, K, U, D, P, R	C
24.	Little Ringed Plover *Charadrius dubius*	J	R
Family: JACANIDAE			
25.	Pheasant Tailed Jacana *Hydrophasianus chirurgus*	J, U	R
Order: GRUIFORMES			
Family: RALLIDAE			
26.	Indian Moorhen *Gallinula chloropus indica* Blyth	J, K, U	O
27.	Purple Moorhen *Porphyrio porphyrio*	J, K	R
28.	Coot *Fulica atra*	J, K, U	R
29.	Indian White Breasted Waterhen *Amaurornis phoenicurus phoenicurus* (Pennant)	J, K, U	C
Family: GRUIDAE			
30.	Sarus Crane *Grus antigone antigone* (Linnaeus)	K	R
Order: GALLIFORMES			
Family: PHASIANIDAE			
31.	Grey partridge *Francolinus pondiecirianus* (Gmelin)	J, K, U	C
32.	Black partridge *Francolinus francolinus* (Linnaeus)	J, K, U, R, P	O

Contd...

Table 15.2–*Contd...*

Sl.No.	*Name*	*Station-wise Distribution*	*Abundance*
33.	Jungle Bush Quail *Perdicula asiatica* (Latham)	J, K	C
34.	Indian Red Jungle Fowl *Pavo cristatus* Linnaeus	J, K, U, R, P	O
36.	Kaleej Pheasant *Lophura leucomelanos hamiltonii* (J.E. Gray)	K, U, R, D, P	R
37.	Chir Pheasant *Catreus wallichi* (Hardwicke)	K, U, D, P, R	R
38.	Monal pheasant *Lophophorus impejanus* (Latham)	K, D	R
39.	Chukar Partridge *Alectoris chukar*	D, R, P	R
Order: COLUMBIFORMES			
Family: COLUMBIDAE			
40.	Little Brown Dove *S. senegalensis cambayensis* Gmelin	J, K, U, R, D, P	C
41.	Indian Ring Dove *Streptopelia decaocto decaocto* (Frivaldszky)	J, K, U, D, P, R	C
42.	Red Turtle Dove *S. tranquebarica tranquebarica* (Hermann)	J, K	R
43.	Rufous Turtle Dove *S. orientalis meena* (Sykes)	J, K, U, D, P, R	C
44.	Indian Spotted Dove *S. chinensis suratensis* (Gmelin)	J, K, U, D, P, R	C
45.	Indian Blue Rock Pigeon *Columba livia intermedia* Strickland	J, K, U, D, P, R	C
Order: PSITTACIFORMES			
Family: PSITTACIDAE			
46.	Large Indian Parakeet *Psittacula eupatria nipalensis* (Hodgson)	J, K, U	O
47.	Blossom Headed Parakeet *P. cyanocephala* (Forester)	J, K, U, D, P, R	O
48.	Rose Ringed Parakeet *P. krameri manillensis* (Bechstein)	J, K, U, D, P, R	C
49.	Slaty Headed Parakeet *P. himalayana* (Lesson)	K, U, D, P, R	O
Order: STRIGIFORMES			
Family: STRIGIDAE			
Subfamily: STRIGINAE			
50.	Northern Spotted Owlet *Athene brama indica* Franklin	J, K, U, D, P, R	C

Contd...

Table 15.2–*Contd...*

Sl.No.	*Name*	*Station-wise Distribution*	*Abundance*
51.	Great Horned Owl *Bubo bubo bengalensis* (Franklin)	J, K, U, D, P, R	R
52	Asian Barred Owlet *Glaucidium cuculoides*	J, K, U, D, P, R	R
Order: CORACIFORMES			
Family: CORACIIDAE			
53.	Blue Jay *Coracias bengalensis bengalensis* (Linnaeus)	J, K, U, D, P, R	C
Family: ALCEDINIDAE			
54.	Small Blue Kingfisher *Alcedo athis bengalensis* Gmelin	J, K, U	R
55.	White Breasted Kingfisher *Halcyon smyrnensis smyrnensis* (Linnaeus)	J, K, U, D, P, R	C
56.	Pied Kingfisher *Ceryle rudius leucomelanura* Reichenbach	J, K, U	O
57.	Himalayan Pied Kingfisher *Ceryle lugubris continentalis* Hartert	K	O
Family: MEROPIDAE			
58.	Indian Small Green Bee-eater *Merops orientails orientalis* Latham	J, K, U, D, P, R	C
59.	Blue tailed Bee-eater *M. phillipinus phillipinus* Linnaeus	J, K	O
Order: CUCULIFORMES			
Family: CUCULIDAE			
60.	Indian Koel *Eudynamys scolopacea scolopacea* (Linnaeus)	J, K, U, D, P, R	C
61	Pied Crested Cuckoo *Clamator jacobinus serratus* (Sparrman)	J, K, U, R	O
62.	Western Sirkeer Cuckoo *Phaenicophaeus leschenaultii sirkee* (J.E Gray)	J, K	R
Family: CENTROPODIDAE			
63.	Crow Pheasant *Centropus sinensis sinensis* (Stephens)	J, K	O
Order: UPUPIFORMES			
Family: UPUPIDAE			
64.	European Hoopoe *Pupa epops epops* Linnaeus	J, K, U, D, P, R	C
Order: PICIFORMES			
Family: CAPITONIDAE			
65.	Large Green Barbet *Megalaima zeylanica zeylanica* (Gmelin)	J, K, U	C

Contd...

Table 15.2–*Contd...*

Sl.No.	*Name*	*Station-wise Distribution*	*Abundance*
66.	Copper Smith *M. haemacephala indica* (Latham)	J, K, U, D, P, R	C
67.	Blue Throated Barbet *M. asiatica* (Latham)	K	R
68.	Himalayan Great Barbet *M. virens* (Boddaert)	J, K, U, D, P, R	C
Family: PICIDAE			
Subfamily: PICINAE			
69.	Lesser Golden- backed Woodpecker *Dinopium benghalense benghalense* (Linnaeus)	J, K, U	C
70.	Yellow Fronted Pied Woodpeacker *Picioides maharattensis maharattensis* (Latham)	J, K, U, D, P, R	O
71.	Northern brown pygmy woodpecker *Picoides nanus nanus* (Vigors)	J, K, U	R
Order: PASSERIFORMES			
Family: LANIDAE			
72.	Rufous-backed Shrike *Lanius schach erythronotus* (Vigors)	J, K, U, D, P, R	C
73.	Woodshrike *Lanius meridionalis*	J, K, U	O
Family: ORIOLIDAE			
74.	Indian Golden Oriole *Oriolus oriolus kundoo* (Sykes)	J, K, U	O
Family: DICRURIDAE			
75.	Black Drongo *Dicrurus adsimilis albirictus* (Hodgson)	J, K, U, D, P, R	C
76.	Ashy Drongo *Dicrurus leucophaeus*	U, D, R, P	R
Family: STURNIDAE			
77.	Bank Myna *A. ginginnianus* (Latham)	J, K, U, D	C
78.	Indian Myna *Acridotheres tristis tristis* (Linnaeus)	J, K, U, D, P, R	C
79.	Northern Jungle Myna *A. fuscus fuscus* (Wagler)	J, K, U	R
80.	Indian Pied Myna *Sturnus contra contra* Linnaeus	J, K	O
81.	Brahminy Myna *Sturnus pagodarum* (Gmelin)	J, K, U, D, P, R	C
82.	Starling *S. vulgaris indicus* Blyth	J, K, U	R

Contd...

Table 15.2–*Contd...*

Sl.No.	*Name*	*Station-wise Distribution*	*Abundance*
Family: CORVIDAE			
83.	House Crow *Corvus splendens splendens* Vieillot	J, K, U, R, P	C
84.	Himalayan Jungle Crow *Corvus macrorhynchos intermedius* Adams	J, K, U, D, P, R	C
85.	Indian Jungle Crow *C. macrorhynchos culminatus* Sykes	J, K, U, P, R	C
86.	Northwestern Tree Pie *Dendrocitta vagabunda* (Blyth)	J, K, U, D, P, R	C
87.	Yellow Billed Blue Magpie *Cissa flavirostris* (Blyth)	K, U, D, P, R	C
Family: CAMPEHAGIDAE			
88.	North Indian Scarlet Minivet *Pericrocotus flammeus speciosus* (Latham)	J, K, U, D, P, R	O
89.	Small Minivet *Pericrocotus cinnamomeus* (Linnaeus)	J, K, U	O
Family: PYCNONOTIDAE			
90.	Red Vented Bulbul *Pycnonotus cafer cafer* (Linnaeus)	J, K, U, D, P, R	C
91.	White-cheeked Bulbul *P. leucogenys leucogenys* (Grey)	J, K, U, D, P, R	C
92.	Black Bulbul *Hypsipetes madagascariensis* (P.L.S. Muller)	K, U, D, P, R	O
Family: MUSCICAPIDAE			
Subfamily: TIMALIINAE			
93.	Jungle Babbler *Turdoides striatus somervillei* (Sykes)	J, K, U, D	C
94.	Common Babbler *Turdoides caudatus caudatus* (Dumont)	J, K, U	C
Subfamily: MONARCHINAE			
95.	Paradise Flycatcher *Terpsiphone paradisii paradisii* (Linnaeus)	J, K, U, D, R.P	R
Subfamily: MUSCICAPINAE			
96.	Verditer Flycatcher *Muscicapa thalassina thalassina* Swainson	J, K, U, D, P, R	R
Subfamily: SYLVINAE			
97.	Indian Tailor Bird *Orthotomus sutorius guzuratus* (Latham)	J, K, U	O
Subfamily: TURDINAE			
98.	Blue Rock Thrush *Monticola solitarius* (Linnaeus)	K, D, R, P	R

Contd...

Table 15.2–*Contd...*

Sl.No.	*Name*	*Station-wise Distribution*	*Abundance*
99.	Himalayan whistling thrush *Myiophonus caeruleus* (Scopoli)	J, K, U, D, P, R	C
100.	Brown Rock Chat *Cercomela fusca* (Blyth)	J, K, U, D, P, R	C
101.	Collared Bush Chat *Saxicola torquata indica* (Blyth)	K, R, P	O
102.	Pied Bush Chat *Saxicola caprata bicolor* (Sykes)	J, K, U	C
103.	Indian Robin *Saxicoloides fulicata cambaiensis* (Latham)	J, K, U	C
104.	Indian Magpie Robin *Copsychus saularis saularis* (Linnaeus)	J, K, U, D, P, R	C
105.	White Capped Redstart *Chaimarrornis leucocephalus* (Vigors)	J, K, U, D, P, R	C
106.	Plumbeous Redstart *Rhyacornis fuliginosus* (Vigors)	J, K, U	O
107.	Kashmir Black Redstart *Phoenicurus ochruros*	J, K	C
108.	Western Spotted Forktail *Enicurus maculatus* (Vigors)	K, U, D, P, R	O
109.	Blue Throat *Erithacus* = (*Luscinia*) *svecicus svecicus* (Linnaeus)	J, K	R
Family:	**MOTACILLIDAE**		
110.	Indian White Wagtail *Motacilla alba dukhunensis* Sykes	J, K, U, D, P, R	C
111.	Indian Pied Wagtail *Motacilla madreraspatensis* Gmelin	J, K, U, D, P, R	C
112.	Blue Headed Yellow Wagtail *Motacilla flava beema* (Sykes)	K	R
113.	Grey Wagtail *Motacillla cinerea* Tunstall	J, K, U, D, P, R	C
Family:	**ZOSTEROPIDAE**		
114.	Indian White Eye *Zosterops palpebrosa palpebrosa* (Temminck)	J, K, U, D, P, R	O
Family:	**NECTARINIIDAE**		
115.	Purple Sunbird *Nectarinia asiatica asiatica* (Latham)	J, K, U, D, P, R	C
116.	Yellow-backed Sunbird *Aethopyga siparaja seheriae* (Tickell)	J, K, U, D, P, R	O
Family:	**PASSERIDAE**		
Subfamily:	**PASSERINAE**		
117.	Indian House Sparrow *Passer domesticus* (Linnaeus)	J, K, U, D, P, R	C

Contd...

Table 15.2–*Contd...*

Sl.No.	*Name*	*Station-wise Distribution*	*Abundance*
118.	Himalayan Cinnamon Tree Sparrow *P. rutilans* (Temminck)	K, U, D, P, R	O
Subfamily: PLOCEINAE			
119.	Indian Baya *Ploceus philippinus* (Linnaeus)	J, K, U, D, P, R	C
Subfamily: ESTRILDINAE			
120.	Spotted Munia *Lonchura punctulata* (Linnaeus)	J, K, U, D, P, R	C
121.	Red Munia *Estrilda amandava amandava* (Linnaeus)	J, K	R
Family: PARIDAE			
Subfamily: PARINAE			
122.	Grey Tit *Parus major* (Linnaeus)	J, K, U, D, P, R	O
123.	Green Backed Tit *P. monticolus* Vigors	K, U, D, P, R	R
Family: FRINGILLIDAE			
Subfamily: EMBERIZINAE			
124.	Himalayan Rock Bunting *Emberiza cia* Linnaeus	K, U, D, P, R	O
125.	Crested Bunting *Melophus lathami* (Gray)	J, K, U, R, P	R
Family: HIRUDINIDAE			
126.	Red Rumped Swallow *Hirundo daurica erythropygia* Sykes	J, K, U, D, P, R	C
127.	Wire Tailed Swallow *Hirundo smithii* Stephens	J, K	O
Family: CINCLIDAE			
128.	Himalayan Brown Dipper *Cinclus pallasii* Bonaparte	K, U, D, P, R	R
Family: ALAUDIDAE			
129.	Crested Lark *Galerida cristata* (Franklin)	J, K, U	C
Family: CERTHIDAE			
130.	Himalayan Tree creeper *Certhia himalayana* Vigors	J, K, U	R
Family: SITTIDAE			
131.	Wall creeper *Tichodroma muraria*	J, K, U	R

Contd...

Table 15.2–*Contd...*

Sl.No.	*Name*	*Station-wise Distribution*	*Abundance*
Order: BUCEROTIFORMES			
Family: BUCEROTIDAE			
132.	Common Grey Hornbill *Tockus* (= *Ocyceros*) *birostris* (Scopoli)	J, K, U	R
Order: APODIFORMES			
Family: APODIDAE			
133.	House Swift *Apus affinis*	J, K, U	O
Order: ANSERIFORMES			
Family: ANATIDAE			
134.	Mallard *Anas platyrhynchos* (Linnaeus)	J, U	R
135.	Shoveller *Anas clypeata*	J, U	R
136.	Pintail *Anas acuta* (Linnaeus)	J, U	R
137.	Common Pochard *Aythya ferina* (Linnaeus)	J, U	R
138.	Red Crested Pochard *Rhodonessa rufina*	J, U	R
139.	Tufted Duck *Aythya fuligula* (Linnaeus)	J, U	R
140.	Gadwall *Anas strepera* Linnaeus	J, U	R

K: Kathua; J: Jammu; U: Udhampur; D: Doda; P: Poonch; R: Rajouri.

Defining the Status:

For assigning status to the faunal elements observed during the survey of study area based on sight records, the terminology of Khan (2002) was followed:

C – Common: means it can invariably be seen in that habitat where it occurs with the provision that the season is also appropriate.

F – Frequent: means that even visiting appropriate habitat, it will not be seen or heard invariably, perhaps only in one visit out of three.

O – Occasional: means seen or heard only in one visit out of six.

R – Rare: means even less likelihood of occurrence.

It is hoped that the present work will serve as the field guide for various other researchers in different fields of animal sciences and will help to develop new strategies for conservation and management of these wild creatures.

Acknowledgements

We are thankful to the Dept. of Zoology, University of Jammu, Jammu for providing various facilities to carry out the present study.

References

Ahmed, A., 2004. Diversity and community structure of the birds of Tehsil Doda, Jammu. *M.Phil. Dissertation.* Univ. of Jammu, Jammu.

Alfred, J.R.B., Kumar, A., Tak, P.C. and Sati, J.P., 2001. *Water Birds of Northern India.* Zoological Survey of India.

Ali, S., 1996. *The Book of Indian Birds,* 12th enlarged centenary edition. J. Bomb. Nat. Hist. Soc. Oxford Univ. Press, New Delhi.

Ali, S. and Ripley, S.D., 1968–74. *The Handbook of Birds of India and Pakistan.* Ten Volumes. Oxford University Press, Bombay.

Ali, S. and Ripley, S.D., 1983. *Compact Handbook of Birds of India and Pakistan.* Oxford Univ. Press, Bombay.

Choudhary, V., 2002. Studies on the Avian Diversity of Jammu District of J&K State. *Ph.D. Thesis.* University of Jammu, Jammu.

Grewal, B., Harvey, V. and Pfister, O., 2002. *A Photographic guide to the Birds of India.* Periplus Editions (HK) Ltd., Singapore.

Grimmett, R., Inskipp, C. and Inskipp, T., 1998. *Birds of the Indian Subcontinent.* Oxford Univ. Press, Delhi.

Inskipp, T., Lindsey, N., and Duckworth, W., 1996. *An Annotated Checklist of the Birds of the Oriental Region.* Oriental Bird Club, Sandy.

Khan, M.A., 2002. Avifauna of Kaghan Valley Northwest Frontier Province, Pakistan. *Tiger Paper.* 29(3): 16–19.

Kumar, S., 2006. Diversity of avian and mammalian fauna of district Kathua, (J&K). *Thesis* submitted to the University of Jammu, Jammu.

Kumar, S. and Sahi, D.N., 2006. Diversity and status of avifauna of Jasrota Wildlife Sanctuary, Kathua (J&K state). *J. Himalayan. Ecol. Sustain. Dev.*, 1: 95–104.

Osmaston, B.B., 1927. Notes on birds of Kashmir, Part–I. *J. Bomb. Nat. Hist. Soc.*, 31: 975–999.

Sahi, D.N., 1985. Breeding ecology of Blue Rock Pigeon, Columba Livia Gmelin. *Jammu University Review*, 3: 64–75.

Sharma, B., 2003. Faunal Diversity of Ramnagar Wildlife Sanctuary, Jammu. *M.Phil.* Dissertation, University of Jammu, Jammu.

Wani, A.A. and Sahi, D.N., 2005. Diversity and status of birds of tehsil Doda, Jammu. *J. Natcon.*, 173(1): 135–143.

Ward, A.E., 1906a and 1906b. Birds of provinces of Kashmir and Jammu and adjacent districts, Part I and Part II. *J. Bomb. Nat. Hist. Soc.*, 17: 108–113, 479–485.

Ward, A.E., 1907a and 1907b. Birds of provinces of Kashmir and Jammu and adjacent districts, Part III and Part IV. *J. Bomb. Nat. Hist. Soc.*, 17(3): 723–729, 17(4): 943–949.

Whistler, H., 1928. *Popular Handbook of Indian Birds*. Gurney and Jackson, London.

2013, Animal Diversity, Natural History and Conservation, Vol. 3 *Pages* **253–281**
Editors: **V.K. Gupta and Anil K. Verma**
Published by: **DAYA PUBLISHING HOUSE, NEW DELHI**

Chapter 16

Ecology and Status of the Ganges Dolphin (*Platanista gangetica gangetica*): India's National Aquatic Animal, in Southern West Bengal

Jayanta Kumar Mallick
Wildlife Wing (Headquarters), Forest Department,
Government of West Bengal, Bikash Bhawan, North Block,
Third Floor, Salt Lake City, Kolkata – 70 0091, W.B., India

ABSTRACT

The Bhagirathi-Hooghly-Matla distributaries of the Ganges in southern West Bengal, flowing through nine districts over a stretch of about 540 km up to the Bay of Bengal, including few tributaries, are last strongholds of the Ganges dolphin (*Platanista gangetica gangetica*), commonly known as *susu(k)*, in the region. Previous reports highlighted their dwindling populations in the fragmented and degraded habitats during last two decades [Vulnerable (1988), Endangered (1996)]. A reconnaissance survey was conducted in the riverine tracts of southern Bengal in 2010 and 2011 with assistance of the local forest-staff to identify and evaluate the ecology of this national aquatic animal, their seasonal habitat preferences and status. This exercise was preceded by a sample questionnaire survey among the local communities, tourists and NGOs. Engineering constructions, reduced flow, sedimentation, salinity, water-traffic, domestic, agricultural and industrial pollution,

* *Corresponding author.* E-mail: jayantamallick2007@rediffmail.com

river-resources extractions, gillnet-entanglement, stranding and poaching have been threatening their survival. Declaration of a Dolphin Sanctuary is recommended for intensive and systematic conservation of this flagship species as well as the overall riverine ecosystem-health and biodiversity.

Keywords: *Degraded habitat, Low population, Anthropogenic threats, Conservation measures.*

Introduction

The Ganges (river) dolphin was being revered through ages [Rigveda (1700-1000 BC), verse I.116.18-19: vehicle of the Ganges; the Mahabharata (origin of Jahnavi), Srimad-Bhagavatam (Chapter 23): *sisumara*] and is still worshipped as the 'cow of the Ganges', especially during the annual Ganga *Puja* (worship) in the study area, as may be observed at Mayapur, Nadia.

For the genus *Platanista,* diversity is very low (only one living species) with no fossil record, but many stem platanistoid lineages include diverse species ranging back to 28–29 Ma (Perrin *et al.,* 2008). But, its scientific discovery was deferred due to an erroneous record in 72 AD by the Roman Naturalist-Historian Gaius Plinius Secundus or Pliny the Elder, who first described *Platanista* sp. in *The Natural History of Fishes* (9: 46), which was observed in the River Ganges and identified as a large fish (16 cubits long), with remarks that it had the muzzle and tail like the dolphin (Ranneft *et al.,* 2001).

Since the recognition of Ganges dolphin as a mammalian species, its taxonomy and authorship were controversial. Taxon *Delphinus* (the name for dolphin in Latin) *gangetica* (type locality: the River Hooghly, a distributary of the Ganga in the delta, lower Bengal) was recorded in 1801 by both William Roxburgh, father of the Indian Botany and the then Superintendent of the Calcutta Botanical Gardens as well as Julius Lebeck, the Dutch Scientist. Whereas Pilleri (1972, 1978) attributed the original description of this species to Roxburgh, following Home (1818), Kinze (2000), based on a deduced latest date of publication (24 August 1801), gave priority over the Lebeck's authority. However, the International Commission on Zoological Nomenclature (1989) accepted the genus *Platanista* Wagler, 1830 and species *gangeticus* Roxburgh, 1801 and suppressed gensus *Susu* Lesson, 1828 (Wilson and Wilson, 2005). The valid taxonomy of the Gangetic dolphin is *Platanista gangetica* (Mammalia: Cetartiodactyla: Odontoceti: Platanistidae) and two erstwhile separate taxa (Pilleri and Gihr, 1971) were reduced to subspecies of a single taxa- the Ganges dolphin *Platanista gangetica gangetica*, native to Bangladesh, India and Nepal (uncertain in Bhutan) (Smith *et al.,* 2004) and the Indus dolphin *Platanista gangetica minor* Owen, 1853, native to Pakistan (Braulik *et al.,* 2004).

The Ganges dolphin appeared to be protected for the first time in the 3rd century BC by the Great Mauryan Emperor Ashoka, who issued a decree (Pillar Edict II) not to kill the *'Gangā–puputaka'.* It is included in Schedule I of the Indian Wildlife (Protection) Act, 1972, Appendix I of the Convention of International Trade in Endangered Species of Wild Fauna and Flora (CITES), Appendix II of the Convention on Migratory Species and categorized as Endangered in 2004 (Vulnerable in 1988) on the Red List of

International Union for the Conservation of Nature and Natural Resources (IUCN), for the species population being declined in its entire distribution range. But its legal protection in India is described as "completely ineffective" (Sinha, 2002).

Since the Ganges dolphins live mostly within the human-dominated landscape beyond the protected areas (Wildlife Sanctuaries or National Parks), it has not been given due priority for conservation like the terrestrial mega-fauna- the tiger, elephant or rhinoceros. The proceedings of the Patna High Court (C.J.W.C. No. 5628 of 2001) directed the state and federal governments to allocate funds for supporting efforts to conserve and monitor the dolphins in the Ganges (Sinha *et al.,* 2010a). As a sequel to declaration of the Prime Minister in the first meeting of the National Ganga River Basin Authority on 5th October 2009, the Ministry of Environment and Forests, Government of India, notified the Ganges dolphin as the National Aquatic Animal on 10th May 2010 (http://moef.nic.in/downloads/public-information/Dolphins_press_n.pdf., downloaded on 14th January 2011). The Conservation Action Plan for the Ganges dolphin (2010-2020) has also been prepared (Sinha *et al.,* 2010a).

The grayish brown adult Ganges dolphin is distinguished by its long and slender rostrum, laterally compressed and widest at the tip with notable sexual dimorphism (proportionally longer in the females after attaining sexual maturity than in the adult males, making up to 20 per cent of the length of the animal), tiny eyes functioning as light detectors and lack of crystalline eye-lenses (rendering it effectively blind), unfused neck vertebra for head-rotation, longitudinally positioned blowhole, presence of a caecum at the junction of the small and large intestines, more dorsal position of the testis, subcutaneous muscle between two layers of blubber, echolocation to navigate-cum-hunt and side-swimming characteristics (Herald *et al.,* 1969; Sinha *et al.,* 2010a].

They bend slightly upward and can reach a length of about 2 m– 2.2 m (males) and 2.4 m–2.6 m (females), weighing between 70 kg and 90 kg. At the time of birth, the neonate measures 70 cm– 90 cm and weighs between 4 kg and 7.5 kg. They have long flippers (up to 18 per cent of total body length), quite large tail fluke (up to 46 cm or roughly a quarter of total body length) and the dorsal fin resembling a fleshy hump (a few centimeters in height) on the back. They characteristically have several folds of skin forming a wattle. The highly asymmetrical skull has a distinctly steep forehead and a longitudinal ridge. The upper and lower parts of the jaw have long, sharp teeth (between 26 and 39 on the upper and 26 to 35 on the lower, the latter being comparatively longer), which are visible even when the mouth is closed, but are worn down and become flat in old age.

The Ganges River dolphins are known to feed on several species of fishes and invertebrates (Sinha *et al.,* 1993). In the first detailed account of Platanista in life, John Anderson (1878) wrote of finding plant and insect fragments in the stomach. They grab fishes, shrimps, turtles and crabs either from the water column or from the bottom. They also have aquatic bird-eating proclivities. They are often observed in the vicinity of cremation grounds, an indication of its scavenging habit (Gupta, 1986). Generally they are solitary animals, but they occasionally congregate in loose aggregations of up to a dozen individuals, especially at the tributary junctions, where the prey fishes congregate. They exhibit active foraging behaviour in the morning (07:00 h–10:00 h) and after noon (15:00 h–17:00 h).

The breeding season of the Ganges dolphin extends from January to June. While mating usually takes place between March and June, it has been observed even in July. Only a single baby is born after a gestation period of about 9 months. The mother and the lactating calf remain together for about one year. The male attains sexual maturity at an age of about 10 years when they reach a length of 1.7 meters while the females are known to attain sexual maturity at 10 or less years (Kasuya, 1972), when they are around 2 m long (Harison, 1972). The maximum longevity may be close to 30 (Smith *et al.*, 2004).

Study Area (Figure 16.1)

The Ganges is the most important river system in India. Out of a total length of 2,525 km, 1,425 km is in Uttaranchal and Uttar Pradesh, 475 km in Bihar and 625 km in West Bengal. It enters West Bengal near Rajmahal and is divided into two distributaries. Until the 15th century AD, its major flow passed through the south-western channel, the Bhagirathi-Hooghly (right arm). A great volume of water passed through the Bhairab-Jalangi channel in Nadia district. Afterwards, more and more water started flowing into its south-eastern course, *i.e.*, the Padma (left arm).

The Ganga basin drains some 46 per cent of surface water of West Bengal, of which the left bank tributaries contribute 11.18 per cent, south Bengal distributaries 16.01 per cent and right bank tributaries 19.11 per cent (Ghosh, 1998), as shown in Table 16.1. In the extreme south, the rivers are tidal. From the Deccan shield zone many flood-prone rivers like the Damodar flow into this basin as tributaries to the Bhagirathi-Hooghly.

The Rivers Bhagirathi-Hooghly flow through nine civil districts- Murshidabad, Nadia, North 24-Parganas, Kolkata and South 24-Parganas on the left bank (east) and Burdwan, Hooghly, Howrah and Purba Medinipur on the right bank (west). south of Beharampur and north of Palashi it used to form the border between Bardhaman District and Nadia District at south of Baharampur and north of Palashi it used to form the border between Bardhaman district and Nadia district. The river then flows south past Katwa, Nabadwip and Kalna. At Kalna, it originally formed the border between Nadia district and Hooghly district, and then further south between Hooghly district and North 24-Parganas district. After it flows past Kamarhati, it enters the twin cities of Kolkata and Howrah and turns to the southwest. At Nurpur, it enters an old channel of the Ganges and turns south to empty into the Bay of Bengal.

The hydrological regime in the study area (known habitat of the Ganges dolphin) is outlined below.

Distributaries of the Ganges

Bhagirathi

The Bhagirathi (non-tidal), which is reduced to a spill channel of the Padma, runs parallel to it for about 2 km and then branches off near Ramakantapur. On the

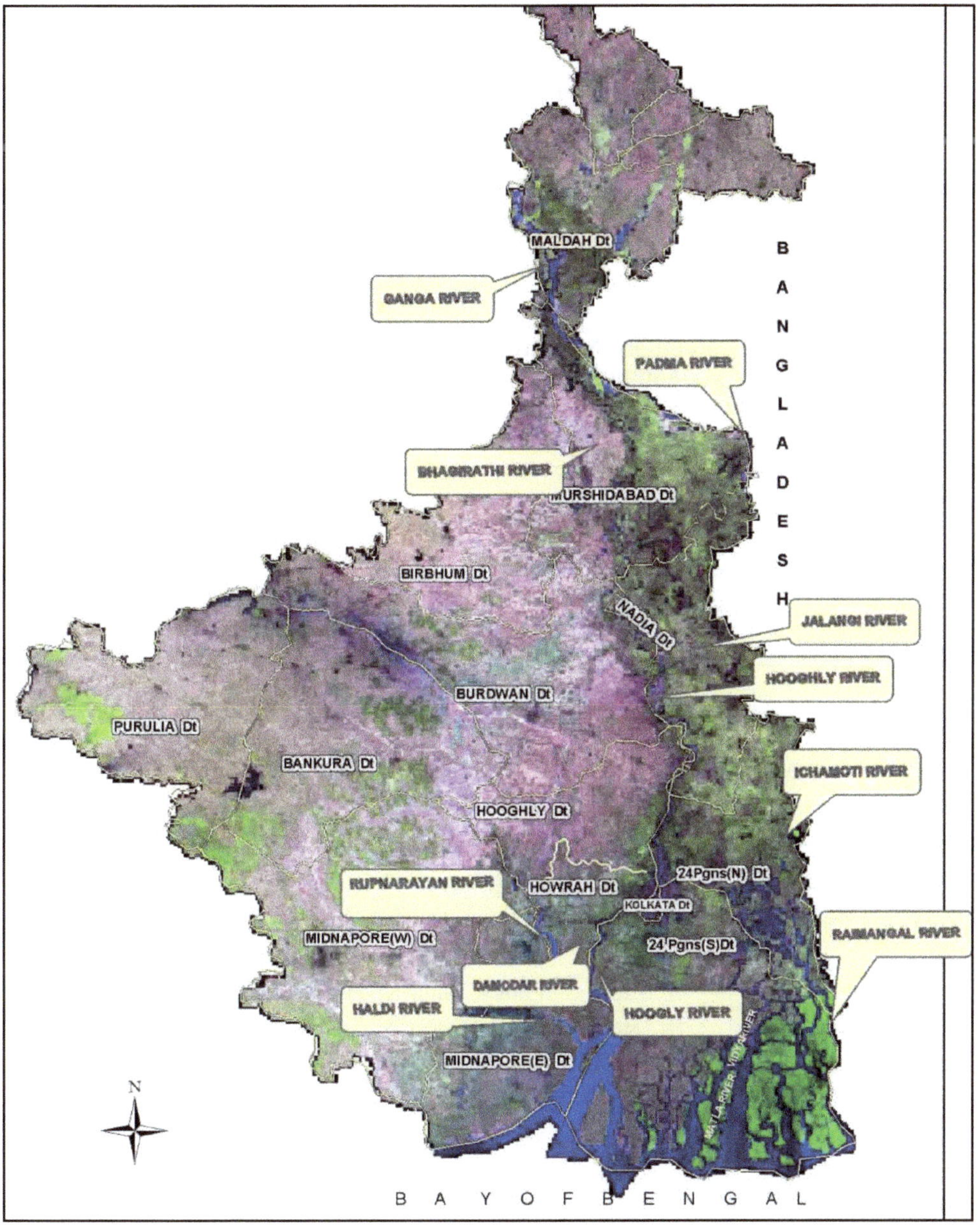

Figure 16.1: Ecological Map of the Study Area

north-west of the study area, a 2,245 m long barrage with rail-cum-road bridges was constructed at Farakka and inaugurated in 1975. The water flows through a feeder canal (38.38 km long), originating off the upstream of the Farakka Barrage. The river originates at Jangipur (24.30°N, 88.4°E), where another 213 m long barrage was constructed, and flows through the following three districts (approximate coordinates given in parenthesis):

1. Murshidabad district on both banks [Farakka (24.49°N, 87.55°E) to north of Plassey].
2. Nadia district on the left bank [Plassey (23.47°N, 87.55°E) through Mayapur (22.27°N, 88.8°E) up to the confluence of the river Jalangi at Swarupganj (23.24°N, 88.23°E)].
3. Burdwan district on the right bank [Ketugram (23.42°N, 88.03°E), confluence of the river Mor or Mayurakshi, through Katwa (23.39°N, 88.11°E), confluence of the river Ajay, up to Purbasthali (23.27°N, 88.20°E), confluence of the old Ganga.

Table 16.1: Catchment Area and Quantum of Surface Water of the Ganges River Basin in South Bengal

Name of the Sub-Basin	*Catchment Area (km²)*	*District(s)*	*Surface Water (MCM)*
South Bengal distributaries			
Bhagirathi-Hooghly	8,61,000	Murshidabad, Nadia, Burdwan, Hooghly,Howrah, Kolkata, North and South 24-Parganas, Purba Medinipur	13,643
Tidal rivers	15,650	Midnapore, 24-Parganas	3,929
Left bank tributaries			
Jalangi	5,640	Murshidabad, Nadia	3,707
Churni	NA	Nadia	NA
Right bank tributaries			
Haldi	8,920	Midnapore	327
Bagmari-Pagla-Bansoli	1,250	Birbhum, Murshidabad	591
Brahmani-Dwaraka	NA	Murshidabad	1,957
Mayuarakshi-Babla	5,470	Birbhum, Murshidabad, Burdwan	2,590
Ajoy	2,490	Birbhum, Murshidabad, Burdwan	2,509
Damodar	5,250	Purulia, Burdwan, Hooghly, Howrah	8,924
Dwarakeshwar	NA	Purba Medinipur	3,330
Shilabati	NA	Purba Medinipur	2,068
Kangsabati	NA	Paschim Medinipur	3,233
Keleghai	NA	Purba Medinipur	818
Rupnarayan	10,930	Purulia, Bankura, Midnapore, Howrah	1,188
Rasulpur	2,620	Midnapore	401
Pichabhanga	NA	South 24-Parganas	462

NA.: Not Available.

*Hooghly

Below the Jalangi confluence, the river flows in the southern direction under the name of Hooghly (tidal) through the following eight districts-

1. Nadia district on the left bank from Swarupganj to Kalyani (22.98°N, 88.48°E).
2. Burdwan district on the right bank [through Samudragarh (23.21°N, 88.20°E) up to Kalna (23.13°N, 88.22°E), confluence of the Khari on the north].
3. Hooghly district on the right bank [Guptipara (23.12°N, 88.24°E) through Balagarh (23.6°N, 88.24°E), Tribeni (22.59°N, 88.24°E) to Uttarpara (22.40°N, 88.20°E), confluence of the river Kunti at Kuntighat (23.1°N, 88.25°E)].
4. North 24-Parganas district on the left bank [Kanchrapara (22.90°N, 88.41°E) to Kamarhati (22.40°N, 88.22°E)].
5. Howrah district on the right bank [Bally (22.64°N, 88.34°E) to Uluberia (22.50°N, 87.95°E), confluences of the Kana Damodar at Phuleswar (22.28°N, 88.07°E), the Damodar at Garhchumuk (22.35°N, 88.08°E) and the Rupnarayan at Gadiyara (22.23°N, 88.04°E)].
6. Kolkata district on the left bank [Dakshineswar (22.39°N, 88.22°E) to Metiaburuz (22.55°N, 88.3°E)].
7. Purba Medinipur district on the right bank [Sutahata (22.12°N, 88.11°E) up to Bay of Bengal near Mandarmani (Kanthi) (21.65°N, 87.65°E)], confluence of the Haldi at Haldia (22.02°N, 88.05°E) and the Rasulpur at Hijli (21.79°N, 87.91°E).
8. South 24-Parganas district on the left bank [Alipur (22.55°N, 88.11°E) to Sagar Island (21.75°N, 88.5°E)], confluence of the Rupnarayan at Gadiyara-Geonkhali-Nurpur (22.12°N, 88.03°E). The marine zone of the estuary starts below Kakdwip (21.53°N, 88.11°E). Namkhana and Fraserganj islands are located in between the freshwater (right bank) and estuarine (left bank) river systems.

Depth of the Hooghly-Bhagirathi from Tribeni to Farakka, the prime dolphin habitat in the study area, is shown in Table 16.2. The tidal effect stretches from the Bay of Bengal up to Nabadwip (23.24°N, 88.23°E), where the rivers Bhagirathi and Jalangi, both taking off at different sites from the upstream of the Ganges, meet and discharge their rain water (freshets) into the Hooghly. This tidal regime has its cyclical and seasonal variations. The coast-mean is 2.0 m (spring 2.7 m) at the mouth of the river. But the tidal range increases inland and as the tidal discharge moves into a much smaller cross-section associated with decreasing channel widths, the mean increases to 3.7 m (spring 4.9 m).

Other important rivers in North and South 24-Parganas districts, draining into the Hooghly-Matla river systems, are:

***Icchamati**

It originates from the Padma and joined by the tidal Icchamati in Nadia district, flows in south-eastern direction through North 24-Parganas district bordering Bangladesh (Bongaon to Basantapur) and ultimately meets the Raimangal estuary in the Sundarbans.

Table 16.2: Mean Depths (in mtrs) in the Hooghly-Bhagirathi (Tribeni-Farakka) during April 2007 and March 2008

River Stretch (Chainage in kms)	*Apr*	*May*	*June*	*July*	*Aug*	*Sept*	*Oct*	*Nov*	*Dec*	*Jan*	*Feb*	*Mar*
Tribeni-Balagarh (193-221)	2.7	2.8	2.9	3.8	5.3	4.1	3.5	3.2	3.0	3.0	3.0	3.1
Balagarh-Kalna (221-245)	3.2	3.0	2.9	2.8	3.8	4.5	4.7	3.2	2.6	3.0	2.8	3.0
Kalna-Samudragarh (245-263)	3.0	3.0	3.2	3.8	4.7	4.5	4.4	2.7	2.8	2.8	2.8	2.8
Samudragarh-Nabadwip (263-280)	2.4	2.5	2.7	3.5	4.7	4.4	4.2	2.7	2.5	2.6	2.7	2.8
Nabadwip-Patuli (280-322)	2.8	2.8	3.0	3.1	5.3	3.7	3.9	2.5	2.5	2.6	2.7	2.7
Patuli-Katwa (322-345)	2.5	2.8	3.1	3.7	3.7	4.2	2.7	3.0	3.0	3.0	3.0	3.0
Katwa-Plassey (345-371)	2.3	2.5	2.7	2.9	3.4	2.8	3.1	2.5	2.5	2.6	2.5	2.7
Plassey-Chaurigacha (371-400)	2.9	3.0	3.2	3.3	3.2	3.5	4.1	2.9	3.0	2.6	2.8	2.8
Chaurigacha-Berhampur (400-421)	3.1	3.1	3.3	3.5	3.7	3.3	4.3	3.4	3.7	3.6	3.4	3.4
Berhampur-Mohammadpur (421-449)	3.2	3.0	3.4	3.5	3.6	4.2	3.4	3.2	3.1	3.1	3.0	2.8
Mohammadpur-Nasirpur (449-479)	3.5	3.4	3.3	3.5	3.3	3.5	3.4	3.1	3.6	3.1	3.4	3.3
Nasirpur-Jangipur Lock (479-505)	4.5	3.4	4.1	4.8	3.6	4.1	3.8	3.1	2.9	3.9	3.5	3.5
Jangipur Lock-Farakka Lock (505-544)	2.5	2.5	2.6	2.8	2.5	2.0	2.7	2.6	2.4	2.5	2.7	2.5

***Raimangal**

It originates from the Icchamati and forms a tidal estuarine river in the Sundarbans bordering Bangladesh.

***Bidyadhari**

It is a tidal river with circuitous course. It begins near Kalinagar and passes through Malancha-Minakhan-Ghusighata, Ranigachi and Haroa, where it is known as Haroa Gang and then joins Nona Khal. Previously it was connected with by Tolly's Nullah and Beliaghata cannal (Kolkata), but later silted up.

***Matla**

It was previously connected to the Piyali and Bidyadhari at Canning, which were gradually silted up. It forms a wide estuary in the Sundarbans and ultimately discharges in the Bay of Bengal.

Tributaries of the Bhagirathi-Hooghly

During its journey from Farakka up to sea face, the Bhagirathi-Hooghly receives waters from many tributaries and begins to spread into many small distributaries, forming the great Gangetic delta, extending to the Matla river system.

River Bansoli

It flows through Murshidabad district and falls into the Bhagirathi, north of Jangipur.

***River Jalangi**

It is a branch of the Padma along with the Mathabhanga/Bhairab, flowing through Murshidabad and Nadia districts. It joins the Bhagirathi near Swarupganj, south of Mayapur.

River Churni

It is a distributary of the Mathabhanga flowing through Nadia district and meets the Hooghly at Chakdaha (23.13°N, 88.50°E) in Nadia district, opposite Balagarh in Hooghly district.

***River Mayurakshi (Mor)**

Originating in Jharkhand, it is fed by the Brahmani, Dwarka, Bakreshwar and Kopai and joins the river Hooghly near Ketugram in Burdwan district. Dams were built on this river for irrigation.

River Ajay

It rises in the Santhal Parganas hills. At Mangalkot, it is joined by the Kunur and falls into the Bhagirathi near Katwa-Sankhai in Burdwan district.

River Behula

It falls into the Hooghly near Sabujdwip at Behula, Hooghly district.

River Kunti

It falls into the Hooghly at Kuntighat above Tribeni.

***River Damodar**

It rises in Palamou and joined by the Barakar, flows in West Bengal through Bankura, Burdwan, Hooghly and Howrah districts and then joins the river Hooghly at Beonkhali near Falta point. Four dams and a barrage were constructed at Durgapur, from which a canal takes off for irrigation.

River Dwaraka (Babla)

It rises in the Birbhum hills, flows through Murshidabad district along eastern boundary of Kandi and drains into the Bhagirathi after leaving the district.

River Dwarakeshwar

Originating from the Chota Nagpur plateau, it is joined by the Silai near Ghatal in Purba Medinipore district and together known as the Rupnarayan.

***River Haldi**

It is an important tributary of the Kansai on which the Kangsabati project was constructed to irrigate nearly 0.4 million ha. The Keleghai joins this river, then flows through Purba Medinipur district and falls into the river Hooghly at Haldia below the confluence of the Rupnarayan.

***River Kangsabati (Kasai)**

It flows through Paschim Medinipur district before draining into the Haldi.

River Keleghai

It flows through Purba Medinipur district and joins the Kasai to form the river Haldi.

River Mundeswari

It flows through Purba Medinipur, Hooghly, and Howrah districts and flows into the river Rupnarayan.

River Rasulpur

It flows through Purba Medinipur district and drains into the Hooghly near Hijli.

***River Rupnarayan**

It rises in the Tilabi hills and drains into the Hooghly near Gadiyara/Nurpur.

***River Shilabati**

It originates in the Chota Nagpur plateau, meets the Dwarakeshwar near Ghatal and joins the Rupnarayan.

A few tidal canals are also linked to the river Hooghly on its both banks.

Habitats and Value Assessment

The Ganges dolphin is a major freshwater river species endemic to the Indian sub-continent (South Asia) and has a fairly extensive distribution range. In India, it is confined to the Ganges-Brahmaputra river systems. During 19th century, it occurred in the entire length of the Ganga and Brahmaputra, and their tributaries from the delta at the Bay of Bengal till the Himalayan foothills, where only rocky barriers, high velocities or shallow water limited its distribution.

Dolphins in the main channel of the Ganga were split into two subpopulations in 1975 when the Farakka Barrage, located approximately at the center of the subspecies' overall range, was commissioned in West Bengal for irrigation and electricity generation. The Ganges dolphin is found in three disjunct habitats in West Bengal (Mallick, 2010):

(a) Freshwater distributaries of the Ganges, *i.e.*, the Bhagirathi-Hooghly and their tributaries;

(b) Tidal distributaries in the Hooghly-Matla river systems (downstream limit depending on low salinity); and

(c) Tributaries of the river Brahmaputra in northern Bengal (*e.g.*, Raidak, Sankosh, Dima, Checko, Kaljani and Torsa; extinct in the Teesta).

The Bhagirathi-Hooghly is replete with hundreds of other aquatic fauna including the critically endangered Gharial (*Gavialis gangeticus*), the vulnerable semi-aquatic smooth-coated otter (*Lutrogale perspicillata*), more than hundred species of resident and migratory wetland birds, twelve freshwater turtles including the soft-shell turtles (Trionychidae), hundreds of fish species including the migrants, crustaceans, shrimps and zooplanktons. The Ganges dolphin is the top predator in this tract (Perrin *et al.*, 2008).

Table 16.3: Wildlife Values (WLV) with Conservation Significance Factor (CSF) for the Bhagirathi-Hooghly and Three Tributaries in South Bengal

Faunal Group	*CSF Values*	*Rupnarayan*		*Damodar*		*Ajay*		*Bhagirathi-Hooghly*	
		WLV Score	*WLV with CSF*	*WLV Score*	*WLV with CSF*	*WLV Score*	*WLV with CSF*	*WLV Score*	*WLV with CSF*
Fishes	0	2	2	1	1	2	2	2	2
Turtles	5	1	5	0	0	1	5	1	5
Crocodiles	5	1	5	0	0	0	0	1	5
Migratory waterfowl	5	0	0	1	5	1	5	1	5
Aquatic mammal (Dolphin)	10	1	10	1	10	0	0	1	10
Total		**5**	**22**	**3**	**16**	**4**	**12**	**6**	**27**

The Wildlife Institute of India, Dehra Dun, assessed the Wildlife values (WLV) with Conservation Significance Factor (CSF) for the Ganges (Bhagirathi-Hooghly distributaries) and three tributaries in South Bengal (Table 16.3). Here, the CSF value assigned for the Ganges dolphin was 10 against a rating of 5 for all other species excluding fishes (WII 1994).

Material and Methods

A study on the ecology of the Ganges dolphin was conducted along the Bhagirathi-Hooghly-Matla river systems in 2010 and 2011 with the help of the field staff of seven forest divisions [Nadia-Murshidabad, Burdwan, Howrah, North and South 24-Parganas Forest Divisions, Sundarban Tiger Reserve and the Wildlife Wing (Headquarters)]. Based on the literature review and the sample questionnaire survey on a set *pro forma* among the concerned forest officials and staff, the local communities including river bank dwellers, fishermen, tourists and NGOs (97 respondents), the habitats of the Ganges dolphin were primarily identified. The information was then verified through direct field survey during the daylight hours by using a motor boat for locating dolphins in the identified confluences, meanders, sandbars and ox-bow lakes as well as evaluating these habitats and highlighting the anthropogenic threats in order to recommend suitable measures for conserving this river cetacean in the study area.

The study area was divided into four water channel transects for sampling: 1. Farakka up to the mouth of the Bhagirathi at Ahiron- 45 km; 2. Ahiron to Nabadwip-Mayapur (confluence of the Jalangi and Bhagirathi)- 185 km; 3. Nabadwip to Tribeni and Bansberia to Kolkata (Rabindra Setu or Howrah Bridge): 210 km; 4. Kolkata to outfall to sea near Sagar Island- 100 km.

During the study, approximately 500 hours (excluding inclement weather conditions [stormy, rainy and foggy], unproductive traversing and holidays), equivalent to 62.5 person-days, were used for sample questionnaire survey, habitat identification and evaluation.

Results

In all, dolphin encounters were recorded at forty locations (Table 16.4) in six distributaries of the Ganga (Bhagirathi: 9, Hooghly: 20, Icchamati: 1, Matla: 1, Bidyadhari: 1, Raimangal: 1) and five tributaries of the Bhagirathi-Hooghly (Jalangi: 1, Silabati: 1, Damodar: 2, Rupnarayan: 2, Haldi: 1) in the study area. Of them, the most important habitat appears to be the Bhagirathi-Hooghly, especially from Farakka up to Tribeni. This stretch is chiefly non-industrial and less polluted. Moreover, the salinity is comparatively low here. The depth of the rivers in this tract varies seasonally from 2.3 m during the dry season to 5.3 m during the monsoon. The dolphins were recorded mostly in the upstream (partly tidal) and rarely in the downstream (industrial, polluted and high saline zone).

The Bhagirathi-Hooghly is full of bends, turns, heavy siltation, long sand-bars (*chars* or river islands), oxbow lakes, meander scars, undercurrents and the confluences of seventeen tributaries. The meandering stretches are separated by many braids. Four prominent braid reaches (21 km) were identified in the prime dolphin habitat between Jangipur and Tribeni-Bansberia. They are-

(i) The outfall of Raghunathpur (24.46°N and 88.05°E) in Murshidabad district: 3 km;

(ii) Chowrigachha (23.93°N and 88.18°E) to Suti (24.37°N and 88.01°E) in Murshidabad district: 3 km;

(iii) Katwa (23.39°N and 88.08°E) in Burdwan district (right bank) to Beldanga (23.51°N and 88.11°E) in Murshidabad district (left bank): 12 km; and

(iv) Jirat (23.5°N and 88.27°E) to Bansberia (22.56°N and 88.24°E) (right bank) in Hooghly district (opposite Nadia district): 3 km.

Table 16.4: Site-Specific Encounters with the Ganges Dolphin in the Study Area (with approximate coordinates)

Location	*Latitude*	*Longitude*	*Periodicity*	*Source*
River Bhagirathi (Murshidabad, Nadia)				
Jangipur (20 km south)	24.47° N	88.07° E	November	Tourist party
Azimganj (10 km north)	24.14° N	88.16° E	December	Tourist party
Bazarpara, Berhampur	24.01° N	88.25° E	March	Forest staff
South Berhampur	24.06° N	88.19° E	July	Forest staff
Goaljan, Budurpara, Berhampur	24.06° N	88.19° E	June	Forest staff
Arpara beel, Nakashipara	23.39° N	88.26° E	April	Local villagers
Kulepota, Nakashipara	23.35° N	88.24° E	October	Local villagers
Mithipur, Raghunathganj II	23.06° N	88.29° E	August	Local villagers
Hulorghat, Mayapur (confluence of Jalangi)	22.27° N	88.08° E	Frequent	Zoological Survey of India, Gopakdasi (ISKON), Rimli Sengupta, V. Solaiyappan, K.M.A. Haldar, T. Bhattacharya, Author
River Jalangi (Nadia)				
Haldarghat near Baro Andulia	22.25° N	88.22° E	August	Local villagers
River Hooghly (Burdwan, Hooghly, Nadia, North 24-Parganas, Kolkata, South 24-Parganas)				
Samudragarh	23.20° N	88.19° E	October	Forest staff
Kalna	23.13° N	88.25° E	December	Forest staff
Sabujdwip, Somrabazar	23.08° N	88.26° E	November	Author
Khayaramarirchar, Khamargachhi	23.03° N	88.26° E	Frequent	IREBEL
Kalyanpurghat	22.24° N	88.24° E	Frequent	IREBEL
Bansberia	22.58° N	88.24° E	November	Debashis Sen
Hooghlyghat (Jubilee bridge)	22.54° N	88.24° E	January	Author
Chinsurah	22.53° N	88.24° E	January, April	Forest staff, Author
Chandannagar (Collegeghat, south of ferryghat)	22.50° N	88.24° E	May	Author

Contd...

Table 16.4–*Contd...*

Location	*Latitude*	*Longitude*	*Periodicity*	*Source*
Debitalaghat, Icchapore	22.48° N	88.18° E	May	Forest staff
Shrirampur			March	Author
Lalbabar Ashram ghat, Belur	22.38° N	88.18° E	Summer (c 3)	Bhaswaran
Ahiritolaghat	22.35° N	88.19° E	Summer	Forest staff
Near Howrah ferryghat	22.34° N	88.20° E	February	N.K. Chandra
Chandpalghat	22.32° N	88.22° E	November	Forest staff
Babughat	22.31° N	88.22° E	November	Forest staff
Budge Budge	22.28° N	88.10° E	October	Forest staff
Botanical Gardens, near ferryghat	22.65° N	88.26° E	September	Forest staff, Panu
Diamond harbour	22.19° N	88.11° E	September	Forest staff
Kakdwip	22.52° N	88.11° E	April	Forest staff
River Haldi (Purba Medinipur)				
Haldia	22.5° N	88.7° E	July	Forest staff
River Rupnarayan (Howrah)				
Kolaghat	22.26° N	88.53° E	Winter	Forest staff, villagers
Irrigation canal, Tilkhoja, 42km away from the river, Purba Medinipore	–	–	January, February	Forest staff, villagers
River Damodar (Burdwan, Howrah)				
Irrigation channel at Rondiha	23.21° N	87.28° E	December	Subrata Pal Chowdhury
Gargchumuk near Uluberia	23.28° N	88.6° E	August	Biswajit Roy Chowdhury
River Icchamati (North 24-Parganas)				
Faridkati, Basirhat	22.39° N	88.52° E	January	Forest staff
River Raimangal (South 24-Parganas)				
Exact location not available	–	–	December	Forest staff
River Matla (South 24-Parganas)				
Canning	22.19° N	88.40° E	January	Forest staff
River Bidyadhari (South 24-Parganas)				
Akandaberia, Haroa	22.36° N	88.40° E	October	Forest staff
River Shilabati (Purba Medinipur)				
Sitakunda, Daspur, Ghatal	22.40° N	88.43° E	May	Forest staff

In three of these braided reaches, the course is divided by the pointed bars in the mid-stream. There are many channels in Katwa-Beldanga reach, where point-bars are also intercepted by cross-channels. Sixteen important sandbars or *chars,* some of which also create oxbow lakes, in the Bhagirathi-Hooghly from Jangipur up to Chandan Nagar, are:

1. Satui-Chaurigachha, south of Beldanga (23.51°N and 88.11°E),
2. Nayachar, 4 km south of Katwa (23.37°N and 88.10°E),
3. Patuli (23.36°N and 88.15°E),
4. Chupichar, Purbasthali (23.46°N and 88.36°E),
5. Balagarh/Madanpur (23.27°N and 88.20°E),
6. Fakirdanga below Nabadwip (23.20°N and 88.20°E),
7. West of Shantipur (23.15°N and 88.23°E),
8. Sabujdwip (Green Island with plantations),
9. Somrabazar, confluence of the Behula (23.14°N and 88.44°E),
10. Shantipur (23.13°N and 88.26°E),
11. Kalna (23.12°N and 88.21°E),
12. Rajdwip and Mangaldwip at Chakdaha/Balagarh (23.7°N and 88.30°E),
13. Khayaramarirchar, between Khamargachhi and Dumurdaha (23.3°N and 88.26°E),
14. Bansberia (22.56°N and 88.24°E),
15. Shyambaburghat, Chinsurah (22.51°N and 88.25°E), and
16. College Ghat, Strand Road, Chandannagar (22.49°N and 88.24°E).

Most of the dolphins were encountered in the above mentioned micro-habitats, which seem to be their preferred foraging tracts. Pods consisting of three to nine dolphins were also found at six (Satui-Chaurigachha, Nayachar, Chupichar, Fakirdanga, Sabujdwip and Khayaramarirchar) out of above sixteen locations.

Out of 59 meander loops found in the study area, the tortuosity (twisted curves with many turns) ratio exceeded two in seven cases- six in the Bhagirathi and one in the Hooghly, all forming acute bends, where the water is deeper. In the Bhagirathi, the loops near Diara-Balagachhi and Char-Chakundi in Murshidabad district and near Purbasthali in Burdwan district have very high tortuosity ratio from 4 to 5. At two ends of these loops, they often try to join by forming cut-offs, two of which were found in the study area- one at Baidyanathpur in Murshidabad district, at about 95 km downstream, which was formed in 1984 and the other at Purbasthali, about 210 km downstream of the Bhagirathi off-take, which was formed in 1990. These meander loops with high tortuosity ratio are mostly preferred by the Ganges dolphins.

The straight reaches in this tract extends over about 140 km, which unite either a braided or a meander reach at the ends. Three major and lengthy reaches were found- (i) from Murshidabad to Sonagai in the Bhagirathi, (ii) from Shrirampore to Garden Reach and (iii) Hooghly Point to Diamond Harbour in the river Hooghly. Dolphin encounters along these comparatively polluted tracts are infrequent and these tracts are used by the dolphins for seasonal migration.

To sum up, the following five segments were found to be the preferred habitats of the Ganges dolphin in the study area:

(i) In the Bhagirathi, between Plassey on the left bank (23.47°N and 88.15°E) in Nadia district and Ballavpara opposite confluence at Katwa on the right bank (23.38°N and 88.7°E) in Burdwan district.

(ii) Mayapur on the left bank (23.26°N and 88.23°E) and Nabadwip on the right bank (23.24°N and 88.23°E) in Nadia district, *i.e.*, around the confluence of the Jalangi and the Bhagirathi. This habitat is the best one not only in Nadia district but also in the whole study area.

(iii) Ketugram-II on the right bank (23.70°N and 88.05°E) downstream in Burdwan district: Best congregation in the district and second best in the whole study area, where highest encounters took place at four locations- (a) Purbasthali or Chupichar with oxbow lakes (23.28°N and 88.17°E): three individuals, (b) Nayachar (23.37°N and 88.10°E): six individuals, Shankai: four individuals and Kalyanpur Ghat: seven individuals.

(iv) Pyradanga (left bank) in Nadia district and Jirat (right bank) in Hooghly district; best encounters near the sandbanks at Sabujdwip (23.8°N and 88.26°E) opposite Somrabazar ferry ghat and a tourist spot in Hooghly district: three individuals, Mongaldwip (23.7°N and 88.30°E) near Pyradanga, Rajdwip (23.6°N and 88.29°E) near Balagarh.

(v) Jirat to Bansberia: Best encounter including a breeding pod at Khayaranarur *char* near Khamargachi in Hooghly district: four individuals including two calves.

At the lower stretch of the Hooghly, dispersed single individuals were sighted from Kalyani to Kakdwip. No pods were sighted in any of the tributaries in the study area.

Eighteen cases of dolphin mortality (two poaching and sixteen accidental) were recorded in the study area during the last seven years and seven dolphins were rescued and rehabilitated by the forest department. In India, West Bengal is the pioneer in successful application of an innovative technique for such rehabilitation.

During the survey, it was also observed that the stretch of river Hooghly for about 125 km from Kalyani (Nadia district) to Birlapur (Garden Reach, Kolkata district) on the eastern bank and Tribeni (Hooghly district) to Uluberia (Howrah district) on the western bank used to receive maximum untreated sewage from approximately 270 outfalls. There are more than 150 large industrial plants on both banks of this river belt, wherein more than 70 per cent of Biochemical Oxygen Demand (BOD) load of average 8187.80 kg/hr (domestic and industrial sources 75 per cent and others 25 per cent) is reportedly contributed by 26 major outfalls, half of which comes from the pulp and paper industries (WBPCB, 2003).

The uppermost stretch of the Bhagirathi-Hooghly exhibited better quality of water in comparison to the downstream stretch. The water quality was found to be poorest during the monsoon when the dolphins used to migrate from the main river to the tributaries.

Discussion

The Bhagirathi-Hooghly-Matla distributaries of the Ganges, flowing through nine districts over about 540 km up to Bay of Bengal, including tributaries in West Bengal, is the last stronghold of the only Indian freshwater cetacean, the Ganges dolphin, which inhabits the human-dominated landscape. Spread over an area of 1785 km^2, the Kolkata metropolitan area has a population of over 14.6 million people by the 2001 count, which has multiplied during the last one decade.

The habitat preferences of the Ganges dolphin in the study area appear to have changed during the last two decades. Whereas during early 1980s the lower stretch of the Hooghly river from Palta (North 24-Parganas) downstream was their known preferred habitat, now the habitats in the upper stream are being preferred by them for foraging as well as breeding. The size and nature of their habitats, temporary or permanent, depending on presence of adequate food resources vis-à-vis their migratory pattern, are the most important criteria related to their ecology. Two typical dolphin habitats were identified during the present study:

(i) Primary or Dry Season Habitat

This habitat is determined by an eddy counter-current system in the main river flow caused by a fine sand/silt point bar formed from sediment and deposits, a convergent stream branch or an upstream meander. The dolphins concentrate in the locations of high prey (fishes and prawns) availability and reduced flow. Relatively high density of dolphins was found at the confluences in the main river and its tributaries or just downstream of shallow stretches where the current is relatively weak, off the mouths of the irrigation canals, and low density was observed near the disturbed human settlements and ferry routes. Dolphins were also found below the sand bars and the bridge-pilings where eddies are formed.

(ii) Marginal or Rainy Season Habitat

The Ganges dolphins migrate to the tributaries of the Bhagirathi-Hooghly during the high flood situation, particularly during the rainy season. These marginal habitats are determined by a smaller eddy counter-current system caused by an upstream meander. The marked seasonal changes in *susu* distribution and density over much of its range are due, at least in large part, to fluctuations in water levels. During the dry season from October to April, when the floodwaters recede, the dolphins leave the tributaries to congregate in the main channels, only to return to the tributaries the following rainy season or often during the high tides. They are often stranded in the canals, pools and river branches during this period.

Seasonal Migration

The Ganges dolphins mostly inhabit the non-protected areas and also to meet their habitat requirements, they often prefer the areas where commercial fishing is most intense. Though presence of the dolphin is reported throughout the year in the Bhagirathi-Hooghly, it is found more frequently in the cold weather than during the height of summer and the rains. It is exceedingly rare from May to the end of June. Appearance of this species in lower stretches of the river Hooghly has been explained as seasonal migration (Gupta, 1986). However, the fishermen also stated that traditionally with the approach of the hot weather the dolphins used to ascend the river, returning with the rains. In July and up to the end of September, its existence was demonstrated by its being frequently caught in the fine-mesh gillnets of the fishermen. In the tidal zone, it usually descends during the monsoon season, when salinity is low, but appears to decrease in number with the advent of summer. This migration also seemed to be associated with the migration and dispersal of fishes, which are their main prey. In course of their upstream migration, they often enter the tributaries or irrigation canals during high tide and get trapped during the low tide. Earlier, the dolphins used to migrate during the breeding season. But since the construction of the Farakka barrage, their upstream journeys crossing the boundary of West Bengal have been halted.

Impact of High and Low Tides

The dolphins were seen moving upstream towards the non-tidal zone during the high tides and downstream during the low tide. They were observed swimming through the midstream to avail advantage of the deeper stretch.

Population Estimates

Population of this obligate river dolphin has been imperiled by extensive anthropogenic threats and declined sharply from tens of thousands to estimated 2000 individuals in its entire range (Sinha *et al.,* 2010b), of which the population in the study area is estimated to be less than 200. But during the early 19th century, they were found in great numbers in the navigable Ganges, mostly in the labyrinth of the rivers and creeks, which intersect the delta to the south, south-east and east of Calcutta (Roxburgh, 1801). Though initially skulls and skeletons of twelve specimens of both sexes were collected from the lower stretch of the river Hooghly [Anderson, 1878], no instance of its occurrence below the Mud Point in Sagar (Saugor) Island located at the mouth of the river Hooghly was ever recorded.

Jones (1982) observed that the Hooghly is the typical habitat of the *susu* from Palta, about 30 km north of Calcutta, to Diamond Harbour at the main land about 60 km down to the south. Gupta (1986) first surveyed the lower Ganges region (Farakka Barrage to Kakdwip) and recorded only nine encounters of the Ganges dolphin- a single individual each at Jagannath Ghat and Metiaburuz of Kolkata, two at the confluence of Bhagirathi and Feeder Canal and five at Farakka Barrage. Sinha (1997, 2000) estimated a maximum population of about 150 *susus* in the entire Bhagirathi-Hooghly river system below Farakka Barrage [86 groups composed of 104-132 dolphins (best estimate= 119 dolphins) from the Jangipur Barrage to Tribenighat; 10 groups for a total of 10-16 dolphins (overall best estimate= 12 dolphins) from Tribenighat to the Botanical Gardens, Shibpur, Howrah, excluding 18 km segment between Barrackpur and the Howrah bridge covered during the night, resulting in no sighting records, and only a single dolphin near the mouth of the Hooghly delta at Kakdwip (30 km past Diamond Harbour)].

Bahuguna and Mallick (2010) reported 112 Ganges dolphins in four zones of South Bengal on the basis of survey done during February-April 2002 by the Centre for Environment and Development, Kolkata. The zonewise population is shown below.

Zone I: Ganges-Bhagirathi (Murshidabad district): (i) Farakka Barrage: 2; (ii) Feeder Canal (24.5° to 25°N) in between Rajnagar and Nimtala: 12; Total: 14;

Zone II (23° to 24°N): Bhagirathi-Hooghly: (i) in between Plassy and Ballabhpara (Nadia and Burdwan districts): 21, (ii) Mayapur (Nadia district): 10, (iii) in between Pyradanga (Nadia district) and Jirat (Hooghly district): 8 and (iv) in between Jirat and Tribeni (Hooghly district): 6; Total: 45;

Zone III (22° to 22.5°N): (1) Damodar: in between Garhchumuk and Gadiyara: 8; (2) Hooghly: Gadiyara to Diamond Harbour: 26, Kulpi: 1; and (3) Rupnarayan: (i) in between Mankur and Kolaghat: 17 and (ii) in between Swatighat and Gadiyara: 1; Total: 53;

Zone IV: Sundarbans: No data are available.

However, Sharma (2010) gave a higher estimate of 170 dolphins (at best individual count 139) in the river Hooghly [from Mayapur to Howrah bridge: 98, Howrah bridge to Kakdwip (left bank): 41 as against Kakdwip to downstream Kolkata (right bank): 31] on the basis of survey during 2005 and 2007. But no detailed site-specific sightings were recorded in his report.

In 2008, 36 dolphins were observed at six sample locations over 170 km stretch of the river Hooghly, with congregations between four and eleven dolphins at a few places (Ghosh *et al.*, 2009).

Choudhury *et al.* (2010) estimated the population figure of the Bhagirathi-Hooghly as 152 (119 in the stretch of 320 km from Jangipur to Tribeni, 32 from Tribeni to Kolkata and 1 between Kolkata and Sagar Island) plus 21 in the Farakka feeder.

Status of the Ganges Dolphin in different Habitat Zones

In terms of sightings of the Ganges dolphin in the lower Ganga Basin in West Bengal during the current study, the area may be stratified into six distinct zones:

1. **Padma-Deltaic non-tidal plains:** Farakka to Jalangi Bazar (India-Bangladesh border)– 105 km: Sighting is very rare.
2. **Bhagirathi-Deltaic non-tidal plains:** (a) Farakka to Ahiron (mouth of Bhagirathi)– 45 km: Common; (b) Ahiron to Nabadwip (confluence of Jalangi river)– 185 km: Frequent sighting including pods.
3. **Hooghly-Outfall deltaic tidal plains-** 210 km: (a) Nabadwip to Bansberia: Frequent sighting along with pods; (b) Bansberia to Kolkata (Rabindra Setu or Howrah bridge) and (c) Kolkata to Kakdwip– 80 km: Sighting mostly seasonal. Even during the late 20th century, 5-7 dolphins were sighted at Babughat (Kolkata), but now it has become rare at this location, where an overall elevated level of BHC, DDT and PCB was recorded in 2010 (Sarkar *et al.*, 2010).
4. **Kakdwip to outfall to sea near Sagar Island** - 20 km: No sighting record is available.

5. **Tidal channel network in the Hooghly-Matla river systems (Sundarbans):** Only a few dolphins were recorded in the river system of northern Sundarbans, where salinity is comparatively low and evidences of occurrence of a stable and permanent population in this estuarine tract were not found. Out of seven main rivers and innumerable creeks, the dolphins were recorded only in the upper stretches of four rivers, namely the Haldi, Icchamati, Bidyadhari and Matla, particularly during the winter. Here, its distribution is subject to low salinity, high turbidity and moderate depth during the high freshwater flow.
6. **Damodar basin**: No surveys have been conducted in the Damodar River system, but the Ganges dolphins were reported in the Rupnarayan and Damodar (once in Shilabati). During monsoon, when the water-level of these rivers rises to 3-6 m, they are safe, but prone to stranding during the dry season when the water-level becomes low (WII 1994).

Threats to Survival of Ganges Dolphins

The Ganges dolphins and the ecological integrity of their riverine environment are vulnerable mainly to various biotic and abiotic threats. The identified threats are:

1. *Poaching for meat, to reduce competition to capture fish resources and lack of awareness:* Though killing the dolphins is not a general trend in the study area, they often fall prey to the inquisitive rural people because they are thought to be 'matsyakanya' or mermaid. Dolphin meat is often sold illegally in some local markets in Murshidabad. Two cases of such poaching were detected in the Bhagirathi-Jalangi region. To reduce competition for fish catches by the dolphins, the fishermen often deliberately kill them. With very low population of this species, even a few catches will have devastating effects on the breeding potential and long-term survival of this endemic species.
2. *Increasing intensity of capture fishery, fatal entanglements in fishing gear such as nylon gillnets*: Accidental killing is a severe problem for the Ganges dolphin because the primary fishing grounds, where gillnets are extensively deployed, are also the preferred habitat of the Ganges dolphins, *i.e.*, the counter-current pools below the confluences and meanders. The fishermen often block the tributary mouths by erecting gillnets during the return period, resulting in dolphins getting entangled in the mono-filament gillnets. It is wrapped around the long crocodile-like rostrum of the dolphin. As a result, its lower jaw is broken and the animal dies due to profuse bleeding, as was the incident in the river Icchamati near Basirhat in North 24-Parganas District on 5th June 2009. Ghosh *et al.* (2009) also recorded several such deaths in the lower stretch of the river Hooghly.
3. *Depletion of prey-base:* River course modifications including construction of the Farakka Barrage bring about a change in hydrobiology and fish resources, especially migrants, both upstream and downstream. For example, the River Churni has become so poisonous that plenty of fishes

died. Das and Chakrabarty (2007) also recorded depletion of the fish species in the Churni (63.6 per cent) due to higher degree of pollution, compared to the less polluted Jalangi. As a result, no record of the dolphin was available in the Churni. Further, in the estuarine Sunderbans large-scale destruction of biomass during prawn seed collection and resultant reduction of the dolphin food species affects their migration.

4. *Stranding*: The dolphins are mostly stranded in the rivers like the Kangsabati, Damodar and Matla. On 30th May 2007, a female dolphin entered the river Shilabati from the Rupnarayan during high tide and during the low tide got stuck in the mud and died due to respiratory failure. The dolphins also occasionally enter the irrigation canals during the high tide, but rarely could successfully return to the main channel of the river during the low tide. Most of them get stranded or stuck in the shallow, muddy water and are prone to death with fractured snout. A number of stranded Ganges dolphins, mostly females, were rescued and rehabilitated from the shallow waters of the irrigation canals and the rivers Kangsabati and Damodar. In January 2000, a single dolphin was rescued after becoming stranded in a deep pool at Rondiha, when the flow was diverted during the dry season by the upstream barrage and it was translocated to Ambika Kalna in Burdwan district, located at a distance of 150 km by road and released in the river Hooghly) (Ganguli, 2007).

5. *Discharge of domestic, agricultural, industrial and other pollutants into the rivers*: The lower stretch of the Hooghly, flowing through the industrial metropolis of the Greater Kolkata, is the worst affected river, where waste-water is disposed, pollutants are loaded, chemical and biological parameters are changed, leading to mortality of the aquatic fauna. The study area was reportedly contributing 868 million litres of waste water everyday in addition to 119.53 tonnes of Biochemical Oxygen Demand (BOD) load, 330.96 tonnes of Chemical Oxygen Demand (COD) load and 405.99 tonnes of Suspended Solid (SS) load. The domestic waste water discharge was estimated to be 338.8 million litres per day. Daily the BOD stood at 22.24, the COD at 101.21 and the SS at 81.49 tonnes, according to the records maintained by the State Environment Department. Kar *et al.* (2008) also assessed the heavy metal pollution (low in the upstream and increasingly high in the downstream). Annual mean BOD concentration at three sampling stations in the study area (data collected from West Bengal Pollution Control Board) is shown in Table 16.5. In the downstream, impact of the heavy metals on the shrimps, one of the main food items found in the stomach of the dead Ganges dolphin in the study area, is also discussed by Mitra *et al.* (2010). Moreover, oil spill often threatens the aquatic lives. However, effects of such pollution are considered deleterious to dolphin populations because of low metabolism capacity of their bodies. Untreated sewage, rotting carcasses and industrial effluents find their way into the dolphins because of their preferred habitat in the counter-current pools downstream of confluences and sharp meanders exposing them in close proximity to the point sources in major metropolitan areas.

Table 16.5: Annual Mean BOD Concentration at Three Sampling Stations on the Lower Hooghly

Sampling Stations	*1991*	*2001*	*2002*
Palta (North 24-Parganas)	0.91 mg/l	2.41 mg/l	2.25 mg/l
Dakshineswar (North 24-Parganas)	0.87 mg/l	2.95 mg/l	2.96 mg/l
Uluberia (Howrah)	0.96 mg/l	2.21 mg/l	2.41 mg/l

6. *Immersion of idols:* The Bhagirathi-Hooghly is being poisoned by the practice of immersing Hindu idols, made of toxic materials, in the sacred river waters. Lakhs of brightly painted idols of the goddess Durga and other deities are dunked in the rivers every year. Whereas the old statues were made of clay and decorated with vegetable colours and other biodegradable materials, the modern idols are made of synthetic paints containing lead, cadmium and strontium and decorated with non-biodegradable materials. The use of modern paints and glues to colour the 10-armed effigies of Durga is causing serious river pollution. It is estimated that each year up to 9.8 tons of paint goes into the holy river at the end of only Dusshera. Other occasions would definitely multiply this figure. Among the most obviously affected species is the Ganges river dolphin.
7. *High sedimentation, low freshwater-flow and velocity and reduced water-holding capacity, shifting of river course resulting in new human settlements*: Due to gradual shift in the river course, reduction in both quantum and duration of the upland freshwater discharge, imbalance created between the tides and the upland discharge, loss of vegetation cover in the catchment area and floodplains, the Bhagirathi-Hooghly-Matla river systems are silting up fast, leading to rise in the river bed, reduction of the water-retaining capacity, depth and prey base of the dolphin. Accelerated sedimentation is observed mostly in the zone of saline and freshwater interface. The estimated quantity of sediment moving in the Ganga at Farakka was 790×10^6 t/yr, of which 328×10^6 t/yr was transported down the river Hooghly (Wasson, 2003). Dolphins usually prefer deep waters but due to high siltation their nose is often got stuck in the silt mounds resulting in accidental mortality.
8. *Construction of dams and barrages*: Due to heavy siltation just below its point of origin, the river Bhagirathi used to receive low freshwater discharge. The Farakka Barrage [2.58 km in length, with 109 bays (approximately 237 m each)] was constructed at the border of Malda and Murshidabad districts and commissioned in April 1975 to feed the Bhagirathi through a man-made feeder canal (41 km long and 300 m wide, with about hundred lock gates). But this has dramatically changed the hydrological regime downstream in areas that were important, at least historically, to the dolphin population (Reeves and Leatherwood, 1994; Sinha, 2004; Parua, 2009; Haque, 1976). This has created an impenetrable barrier to the dolphin movement and segregated the lower Ganges population in West Bengal from those of the upper stream in Bihar and Uttar Pradesh and reduced

their gene pool, leading to loss of genetic diversity and increasing the probability of extinction. Moreover, construction of embankments for flood control has led to shrinkage of riparian eco-tonal zones for natural feeding and breeding of the Ganges dolphin and other species.

9. *Other engineering constructions and thermal plants*: Six rail / road bridges have been constructed on the Bhagirathi-Hooghly, namely Jubilee Bridge (1500 ft) between Hooghly Ghat (Hooghly district) and Garifa (North 24-Parganas district) [1887], Vivekananda Setu between Bally (Howrah district) and Dakshineswar (North 24-Parganas district [1932], Rabindra Setu between Howrah and Kolkata [1943], Gouranga Setu between Nabadwip and Krishnagar in Nadia district, Iswar Gupta Setu between Bansberia (Hooghly district) and Kalyani (Nadia district) [1989], Vidyasagar Setu between Alipore (South 24-Parganas district) and Shibpur (Howrah district) [1992] and Nivedita Setu between Bally and Dakshineswar [2007]. Two ports (Kolkata and Haldia) and two thermal power plants (Bandel and Kolaghat) were also built up on this river. Free passage and breeding of the migratory dolphins and other aquatic fauna is also threatened by such extensive construction works.
10. *Irrigation*: Excess extraction of the river water for irrigation has lowered the water level, particularly during the dry season, which is also catastrophic for the migration of the Ganges dolphins.
11. *Increasing salinity*: The freshwater Ganges dolphin is not generally known to occur in salinities greater than 10 ppt, although they have been recorded in waters as saline as 23 ppt (Smith *et al.*, 2004). Salinity at three stations in the river Hooghly is given in Table 16.6. The study area may be divided into the following salinity zones:
    (*a*) Bhagirathi-Hooghly: low saline water for the freshwater discharge;
    (*b*) Hooghly above Diamond Harbour: low to medium saline water;
    (*c*) Hooghly below Diamond Harbour: higher saline water; and
    (*d*) Estuarine Sundarbans: highest saline water.

    The Ganges dolphins are mostly found in the upstream because of low salinity.

    Water extraction for irrigation and domestic purposes as well as the sea-level rise has caused the salinity of the dolphin habitat to increase. Moreover, since late 20th century, there was slow increase in salinity at the freshwater zone between Nabadwip and Barrackpore. It is alarming that the saline mangrove species like *Sonneratia caseolaris* were found invading on both banks of the river Hooghly from Diamond Harbour in South 24-Parganas district up to Uttarpara in Hooghly district. Reduced discharge of freshwater from Farakka also helps increase of salinity in the lower course.

    (*e*) *Riverine resource extraction including dredging, indiscriminate sand mining on the river bed and destruction of sandbanks for use in brick factories* (*dozens of such factories operating on both banks*): Large-scale sand-mining continues to deepen the riverbed. This eliminates the counter currents, one of the most preferred foraging sites of the dolphin and contributes to decline in their population in the affected areas.

(*f*) *Increasing heavy river traffic, mainly the motorized rivercrafts* and *cargo steamers*: Some of the important ferry service jetties constructed on the River Hooghly are located from Tribeni to Kolkata. In addition, innumerable motorized ferry boats used to travel at a speed of 5–6 mph, crisscrossing the dolphin movement routes from early morning to midnight and cause bow-waves. Their propeller blades often inflict severe injuries to the slow-moving dolphins and ultimately cause their mortality. During the survey, the dolphins were found dead on a number of occasions due to collision with the mechanized boat/launch/ship at Chinsurah (Hooghly), Babughat (Kolkata), the Botanical Garden (Shibpur, Howrah), near Kidderpore docks (Kolkata) in the Hooghly and also the Bidyadhari near Akandaberia village (Haroa, North 24-Parganas). This heavy river traffic has a great impact on the dolphins leading to their habitat restriction, depletion of prey base and changes in feeding behaviour. The Ganges dolphins were seen to avoid the disturbed river stretch with heavy river traffic in the river Hooghly in and around Kolkata, preferring relatively undisturbed areas, where a lower number of mechanized vessels ply (Sinha, 1997). Since the dolphins depend on echolocation, sonic pollution may adversely affect their well-being.

Table 16.6: Trend of Salinity (ppt) in the River Hooghly

Location	*1985*	*1995*	*1997*	*2000*	*2005*	*2009*
Sagar Island(South 24-Parganas)	–	17	22	22.24	–	26
Uluberia (Howrah)	0.028	0.042	–	–	0.094	–
Barrackpore (North 24-Parganas)	0.019	0.054	–	–	0.055	–

Sources: Central Pollution Control Board, Central Inland Captured Fisheries Research Institute, School of Oceanographic Studies, Jadavpur University.

Conservation Measures

The dolphins and humans equally need the healthy rivers for their sustenance. Intensive and systematic conservation of the Ganges dolphin is, therefore, crucial to overall riverine ecosystem health and biodiversity (Sinha, 1999). Their presence in the rivers indicates the availability of fish and pollution-level of that river.

The Bhagirathi-Hooghly river system is the only viable habitat of the Ganges dolphin in West Bengal. But human disturbances are directly influencing its habitat and population. As a result, there is high possibility of disappearance of this single species of freshwater dolphin in this riverine tract in future. Since the Ganges river dolphin population inhabits mostly outside Wildlife Protected Areas, the freshwater riverine network in the study area could not provide adequate protection to the dolphin habitat and population. Survival of the dolphins depends on improvement of the degraded habitat and special protection of their habitats in the Bhagirathi-Hooghly and their tributaries.

The most important conservation measure recommended is to declare the prime dolphin habitat in between Farakka and Tribeni as a dolphin sanctuary and develop and implement a management plan, as was done in case of Vikramshila declared in 1991, a 50 km stretch in Patna, Bihar. Other measures recommended are:

1. Launching of a "Project Dolphin" by the Government of India.
2. Habitat improvement and implementation of the existing laws, rules and regulations in collaboration with the civil, fishery, forest/wildlife, water development and barrage authorities.
3. Monitoring on the dolphin movements with fluctuations in the water level between dry and wet season.
4. Awareness campaign among the local communities on the critical status and importance of conserving the Ganges dolphins.
5. Prohibition on sand mining in the identified micro habitats of the dolphins.
6. Strict enforcement of the fishing regulations in the dolphin habitats including prohibition of the round-the-clock fishing during dolphin birthing season (October-March) especially at the time of peak birthing (December-January) months.
7. Prohibition on discharge of untreated pollutants into the critical dolphin habitats.
8. Prohibition of use of chemical fertilizers and organochlorine pesticides on farms located on the hinterland of these habitats.
9. Restrictions on the motorized ferry services in the sensitive ecological areas of the upper stretch of the river Hooghly.
10. Initiation of community-based management practices for conservation of the Ganges river dolphin and the ecosystem.
11. Arrangement for alternative employment for the fishermen evicted for the sake of dolphin conservation, such as eco-tourism because the dolphin hotspots like those at Nayachar, Chupichar (20 km^2), Burdwan district, Mayapur/Nabadwip, Nadia district and Sabujdwip (2000 m long and 12 m wide), developed in 1993, Hooghly district, are visited by thousands of tourists including the foreigners.
12. Capacity building of the local communities through meetings, training and workshops.
13. Involving the NGOs and the concerned government departments in such conservation efforts.

Acknowledgements

I am specifically grateful to all the forest staff involved in conducting the reconnaissance survey and the respondents to the questionnaire survey. I also acknowledge contributions of I. Mitra (GIS mapping) and S. Chakrabarti (data retrieval) of the Wildlife Wing (Headquarters). Thanks are also due to the NGOs like WWF-India, Eastern Region and IREBEL, who are working for conservation of the Ganges dolphins in the study area.

References

Anderson, J., 1878. *Anatomical and zoological researches: comprising an account of zoological results of the two expeditions to western Yunnan in 1868 and 1875, and a monograph of the two Cetacean Genera, Platanista and Orcella*. Volumes I and II. Bernard Quaritch, London, UK.

Bahuguna, N.C. and Mallick, J.K., 2010. *Handbook of the Mammals of South Asia*. Natraj Publishers, Dehradun.

Braulik, G.T., Smith, B.D. and Chaudhry, A.A., 2004. *Platanista gangetica ssp. minor*. In: IUCN 2010, IUCN Red List of Threatened Species, version 2010.4, www.iucnredlist.org, downloaded on 9 January 2011.

Choudhury, B.C., Behera, S. and Wakid, A., 2010. A review of the conservation status and protected areas for the susu *Platanista gangetica gangetica*, Bhulan *Platanista gangetica minor* and Irrawaddy dolphin *Orcaella brevirostris* in the Ganges, Brahmaputra and Beas rivers, and Chilika lagoon, India. In: *Establishing protected areas for Asian freshwater cetaceans: Freshwater cetaceans as flagship species for integrated river conservation management, Samarinda*, 19–24 October 2009, Kreb, D., Reeves, R.R., Thomas, P.O., Braulik, G.T., and Smith, B.D. (Eds.). Final Workshop Report. Yayasan Konservasi RASI, Samarinda, Indonesia, pp. 107–119.

Das, S.K. and Chakrabarty, D., 2007. The use of fish community structure as a measure of ecological degradation: a case study in two tropical rivers of India. *Biosystems*, 90(1): 188–196.

Ganguli, U., 2007. Rescue of a Ganges river dolphin (*Platanista gangetica*) in West Bengal. In: *IFAW–WTI Emergency Relief Network Digest 2006–2007*, Ashraf, N.V.K. and Pillay, R. (Eds.). Wildlife Trust of India, New Delhi, pp. 45–49.

Ghosh, A.K., 1998. *Report on the State of Environment in West Bengal*. Department of Environment, Government of West Bengal.

Ghosh, T., Mukherjee, S., Chowdhury, G., Paul, S., Bhattacharya, P., Ghosh, M., Ghosh, S., Mukherjee, S. and Sil, D., 2009. *Expedition– in search of Gharials and Gangetic dolphins in the river Hooghly*, unpublished.

Gupta, D., 1986. The Gangetic dolphin *Platanista gangetica* (Lebeck, 1801). In: *Wildlife Wealth of India* (*Resources and Management*), Majupuria, T.C. (Ed.), Teepress Service, L.P., Bangkok, pp. 553–562.

Haque, A.K.M. Aminul., 1976. *Comments on the abundance and distribution of the Ganges susu Platanista gangeticaus, and the effects of the Farakka barrage on its population*. FAO ACMRRR, Scientific consultation on Marine mammals, AGMRR/MM/SC 132.

Harison, R.J., 1972. Reproduction and reproductive organs in *Platanista indi* and *Platanista gangetica*. *Invest. Cetacea*, 4: 71–82.

Herald, E.S., Brownell, R.L., Frye, F.L., Morris, E.J., Evans, W. and Scott, A., 1969. Blind river dolphin: first side–swimming cetacean. *Science*, New Series, 166(3911): 1407–1410.

Home E. 1818. A description of the teeth of the *Delphinus gangeticus. Phil. Trans. R. Soc.*, 108: 417–419.

Jones, S., 1982. *The present status of the gangetic susu, Platanista gangetica* (*Roxburgh*), *with comments on the Indus susu, P. minor Owen*. FAO Advisory Committee on Marine Resources Research, Working Party on Marine Mammals, FAO Fisheries Series 5(4): 97–115.

Kar, D., Sur, P., Mandal, S.K., Saha, T. and Kole, R.K., 2008. Assessment of heavy metal pollution in surface water. *Int. J. Environ. Sci. Tech.*, 5(1): 119–124.

Kasuya, T., 1972. Some information on the growth of the Gangetic dolphin with a comment on the Indus dolphin. *The Scientific Reports of the Whales Research Institute*, 24: 87–108.

Kinze, C.C., 2000. Rehabilitation of *Platanista gangetica* (Lebeck, 1801) as the valid scientific name of the Ganges dolphin. *Zool. Med. Leiden*, 74(11): 193–203, figs. 1–3.

Lebeck, H.J. 1801. *Delphinus gangeticus* beschrieben von Heinrich zu Trankenber. *Der Gesellschaft naturforschender Freunde zu Berlin Neue Schriften*, 3: 280–282, Tab. II.

Mallick, J.K., 2010. New national aquatic animal, Ganges dolphin, 2010. *West Bengal*, 42(9&10): 27–31.

Mitra, A., Banerjee, K., Ghosh, R. and Ray, S.K., 2010. Bioaccumulation pattern of heavy metals in the shrimps of the lower stretch of the river Ganga, Mesopot. *J. Mar. Sci.*, 25(2): 1–14.

Parua, P.K., 2009. Decline of Bhagirathi–Hooghly channel. *The Ganga*, 64: 61–77.

Perrin, W.F., Wursig, B. and Thewissen, J.G.M., 2008. *Encyclopedia of marine mammals*. Second edition, Academic Press, Burlington, USA.

Pilleri, G., 1972. Original description of the Gangetic dolphin, *Platanista gangetica*, attributed to William Roxburgh, Bull. *Br. Mus. nat. Hist.* (*Zool.*), 21(8): 345–348.

Pilleri, G., 1978. William Roxburgh (1751–1815), Heinrich Julius Lebeck (†1801) and the discovery of the Ganges dolphin (*Platanista gangetica* Roxburgh, 1801). *Investigations on Cetacea*, 9: 11–21, pl. 1.

Pilleri, G. and Gihr, M., 1971. Differences observed in the skulls of *Platanista gangetica* (Roxburgh, 1801) and *indi* (Blyth, 1859. *Investigations on Cetacea*, 3: 13–21.

Ranneft, D.M., Eaker, H. and Davis, R.W., 2001. A guide to pronunciation and meaning of cetacean taxonomic names. *Aquatic Mammals*, 27.2: 183–195.

Reeves, R.R. and Leatherwood, S., 1994. *Doplphins, porpoises, and whales: 1994–1998 action plan for the concervation of cetaceans*. IUCN, Gland, Switzerland.

Roxburgh, W. 1801. An account of a new species of *Delphinus*, an inhabitant of the Ganges. *Asiat. Res. Trans. Soc.* (Calcutta Ed), 7: 170–174, pl.5.

Sarkar, S.K., Alam, M.A. and Bhattacharya, B., 2010. Trace organics in sediments in the lower stretch of Ganges estuary and their ecotoxicological significance. In: *Abstracts of the proceedings of National Seminar on applied chemistry in marine sciences: current and future trends and annual general body meeting: 2010 of Indian Society of Applied Geochemists (ISAG), Hyderabad, February 24–25, 2010,* pp. 31–33.

Sharma, G., 2010. Current status of susu (*Platanista gangetica gangetica,* Roxburgh, 1801) in river Hooghly in West Bengal, India. *Rec. Zool. Surv. India,* 110 (Part I): 61–69.

Sinha, M., 2004. Farakka barrage and its impact on the hydrology and fishery of Hooghly estuary. In: *The Ganges Water Diversion: Environmental Effects and Implications,* (Ed.) M.M.Q. Mirza. Kluwar Academic Publishers, Netherlands, pp. 103–121.

Sinha, R.K., 1997. Status and conservation of Ganges river dolphin in Bhagirathi–Hooghly river systems in India. *International Journal of Ecology and Environmental Sciences,* 23(4): 343–355.

Sinha, R.K., 1999. The Ganges river dolphin– a tool for baseline assessment of biological diversity in River Ganges, India. *Final technical report,* Patna.

Sinha, R.K., 2000. Status of the Ganges river dolphin (*Platanista gangetica*) in the vicinity of Farakka Barrage, India. In: *Biology and conservation of freshwater cetaceans in Asia*, Reeves, R.R., Smith, B.D.and Kasuya, T. (Eds.). IUCN Species Survival Commission, Occasional Paper 23, Gland, Switzerland and Cambridge, UK, pp. 42–48.

Sinha, R.K., 2002. An alternative to dolphin oil as a fish attractant in the Ganges River system: conservation of the Ganges River Dolphin. *Biological Conservation,* 107: 253–257.

Sinha, R.K., Behera, S.K. and Choudhary B.C., 2010a. *The conservation action plan for the Ganges river dolphin 2010–2020*. Ministry of Environment and Forests, Government of India, unpublished.

Sinha, R.K., Das, N.K., Singh, N.K., Sharma, G. and Ahsan, S.N., 1993. Gut content of the Gangetic dolphin *Platanista gangetica. Investigations on Cetacea,* 24: 317–321.

Sinha, R.K., Verma, S.K. and Singh, L., 2010b. Population status and conservation of the Ganges river dolphin (*Platanista gangetica gangetica*) in the Indian Subcontinent. In: *Biology, Evolution, and Conservation of River dolphins within South America and Asia*, Ruiz–Garcia, M. and Shostell (Eds.). Nova Science Publishers, Inc., New York, USA, (2010).

Smith, B.D., Braulik, G. and Sinha, R.K., 2004. *Platanista gangetica ssp. gangetica*. In IUCN 2010, IUCN Red List of Threatened Species, version 2010.4, www.iucnredlist.org, downloaded on 9 January 2011.

Wasson, R.J., 2003. A sediment budget for the Ganga–Brahmaputra catchment. *Cur. Sci.*, 84(8): 1041–1047.

West Bengal Pollution Control Board (WBPCB)., 2003. *Rapid estimation of major wastewater discharges to river Hooghly between the stretch of Palta to Dhankheti khal*. Salt Lake, Kolkata.

Wildlife Institute of India (WII)., 1994. *Impact assessment of Haldia–Barauni pipeline project on wildlife values: a report to Engineers India Limited, New Delhi*. WII–EIA Technical Report no. 4, unpublished.

Wilson, D.E. and Wilson, D.E. (Eds.)., 2005. *Mammal species of the world: a taxonomic and geographic reference*. Volume I, 3rd ed. Johns Hopkins University Press, Baltimore, Maryland.

2013, Animal Diversity, Natural History and Conservation, Vol. 3 *Pages* **283–322**
Editors: **V.K. Gupta and Anil K. Verma**
Published by: **DAYA PUBLISHING HOUSE, NEW DELHI**

Chapter 17

Ecology of Spotted Owlet (*Athene brama* Temminck, 1821) in Madurai District, Tamil Nadu, India

R. Santhanakrishnan[1] and A. Mohamed Samsoor Ali[2]*
[1]Department of Zoology,
Saraswathi Narayanan College Perungudi, Madurai – 625 022, Tamil Nadu, India
[2]Division of Environmental Impact Assessment,
Sálim Ali Centre for Ornithology and Natural History (SACON)
Anaikatty, Coimbatore – 641 108, Tamil Nadu, India

ABSTRACT

Various aspects of the ecology of Spotted Owlet (*Athene brama*) was studied in Madurai District, Tamil Nadu between 2007 and 2009. Regardless of habitats and years, Spotted Owlet densities varied from 10.4 to 35.6/km^2 and among the habitats rural and agricultural lands supported comparatively higher number of owlets. Moreover, within habitats, density of owlet varied among seasons and years. A total of 292 roosting-sites were located: 229 in/on trees, 43 in human-made structures, 12 in artificial wooden nest boxes and 8 in unused wells. Spotted Owlets selected 15 species of trees for roosting of which majority of them were in *Ficus bengalensis* (42.9 per cent). The surrounding ecological variables *viz.*, distance to nearest tree, building, footpath, electric line, light source, groves and water sources were very closer, *i.e.*, within the vicinity of 100m radius to all types of

---

* *Corresponding author.* E-mail: amsamoor@yahoo.co.in

roost-sites. Spotted Owlet used a variety of perch-sites during foraging in which electric power lines were the most favoured perch type (51.3 per cent). The perch and perching height used by Spotted Owlet was classified into 2 meter categories, up to 16m. A total of 4,181 regurgitated pellets were analyzed and we recorded 40,278 individual prey remains of 16 prey species. The number of prey items per pellet varied from 1 to 31 with an overall mean prey of 9.6 ± 3.03. The analysis revealed that invertebrates were predominant prey (insecta 83 per cent; arachnida 1.1 per cent; myriapoda 0.2 per cent) followed by vertebrates (reptiles 3.3 per cent; mammals 2.9 per cent). Mammalian prey species in the diet included *Mus* spp., *Rattus rattus, Suncus murinus, Tatera indica, Bandicota bengalensis* and micro chiropteran bats. Totally 32 nests-sites were located and largely found in holes/ cavities in trees, highest numbers of nests were located in *Ficus bengalensis,* rather than other sites for nesting. Clutch size varied between 2 and 5 with a mean of 2.6 ± 0.72 eggs. The mean incubation period of the Spotted Owlet was 26.2 ± 1.31 days and both sexes took part in incubation. The hatching and fledging success of the Spotted Owlet were 83 per cent and 71 per cent, respectively. Of the 26 nest boxes erected in various sites, 13 nest boxes were used by Spotted Owlet for roosting and nesting. Mean occupation of nest boxes during the study was 50 per cent.

Keywords: *Spotted Owlet, Population, Roost-site, Food, Breeding, Nest box, Madurai.*

Introduction

Owls, a unique group of birds, are better documented in folklore, legend and historical accounts than any other group of bird species. They are nocturnal in habit and characterized by their large forward facing eyes surrounded by a facial disc of short stiff feathers and an upright position. They occupy almost an equivalent niche to the diurnal birds of prey such as hawks, falcons, eagles and buzzards but they are not closely related. The owls are successful birds and dispersed all over the world except Antarctica and some remote oceanic islands. Owls are an essential component of the ecosystem. They feed on small mammals, birds, frogs, lizards and insects and are at the apex of the food chain. Owls are included in two families namely Tytonidae and Strigidae. The family Strigidae is cosmopolitan in distribution and composed of 22-27 genera, while Tytonidae has only 2 genera that include species of Barn Owl (*Tyto alba*), Grass Owl (*Tyto capensis*) and Bay Owl (*Phodilus badius*).

Though considerable account of research on various owl species are available throughout the world, there are wide gaps in our knowledge on the biology and ecology of owls in India. Most of the available literature on owls in India generally pertains to their natural history, occasional sightings, distribution and status, prey preferences, nesting-sites, vocalization, behaviour, nest box utilization and predation (Neelakantan, 1971, 1979; Balasubramanian, 1990; Sugathan and Jacob, 1995; Iyer, 1997; Rasmussen and Collar, 1998; King and Rasmussen, 1998; Chandrasekhara and Nameer, 2003; Kanakasabai *et al.,* 1996; Nagarajan *et al.,* 1995, 2002; Ramanujam, 2000a, b; Agoramoorthy and Minna, 2001; Ramanujam, 2003a,b; Verzhutskii and Ramanujam, 2002; Ramanujam, 2007; Srinivasulu and Srinivasulu, 2007; Vanitha and Kanakasabai, 2009). However, detailed studies on their status, distribution, population, roosting-sites, food and feeding, nesting biology and their role in biological control are very few and sporadic.

India has 33 species of owls (Manakadan and Pittie, 2002) out of a total of about 225 species known from the world (http:\\www.globalowlproject.com). Among these Spotted Owlet (*Athene brama*) is a most common small-sized owl distributed throughout south-east Asia including India (Ali and Ripley, 1983). It is common in open habitats including farmland and human habitations and has adapted to live in crowded cities also. There are five widely recognized sub-species namely *albidia* (Koelz, 1950) of western Asia in Iran and Pakistan; *indica* (Franklin, 1831) of northern India; *brama* (Temminck, 1821) of southern India; *ultra* (Ripley, 1948) of north-eastern India and *pulchra* (Hume, 1873) of south-east Asia from Myanmar and Thailand extending into Cambodia and Vietnam. The Spotted Owlet is characterized by squat, white-spotted, greyish-brown plumage with a large round head and forward-staring yellow eyes. The Spotted Owlets are mainly nocturnal and roost in the day under thick shady vegetation, cavities in trees, human-made structures including functional or non-functional irrigational wells and crevices found in rock cliffs. It is a generalist predator that feeds on diverse prey such as rodents, small birds, reptiles, amphibians and invertebrates such as insects and annelids (Kumar, 1985). Except for an ecological study carried out in Hyderabad (Kumar, 1985), little is known about their roosting-sites, feeding (Jain and Advani, 1984; Jadhav and Para-sharya, 2003; Pande *et al.*, 2004) and breeding (Jadhav and Parasharya, 2003; Pande *et al.*, 2006, 2007) and most studies have been reported from northern and western India. In this chapter we discuss the population, roost-site, perch-site, food habits, breeding biology and nest box utilization of Spotted Owlet in Madurai District, Tamil Nadu, Southern India.

Study Area

Madurai, the second largest urban centre in Tamil Nadu located on the banks of River Vaigai which lies betwen 9°552 24.122 N and 78° 07217.022 E of Tamil Nadu, South India (Figure 17.1). It is bounded on the North by the districts of Dindigul, Thiruchirapalli and on the East by Sivagangai and on the West by Theni and South by Virudhunagar. The total geographical area of the district is about 10,88,622 km^2 with an overall population of 25.6 lakhs (2001). Madurai remains a centre for economic, educational, cultural and industrial activities in Southern part of Tamil Nadu state.

In general, Madurai has a flat topography with mild slopes here and there. The hillocks such as Pasumalai, Yanamalai, Nagamalai, Alagarmalai and Thiruparankundram present a panoramic landscape of Madurai. The texture of the soil varies from silty clay to loamy sand on the surface and the colour varies from red, brown to grey in all shades. The prime source of water to the study area is the rainfall which is confined to only monsoon period and is unevenly distributed both in space and time. Apart from this, the Vaigai river basin acts as an important ground water source to cater to the needs of Madurai people for domestic, agricultural and other industrial purposes.

With regard to agricultural practice, it could be stated that the rice is the predominant crop. However, cash crops namely sugarcane, banana, jasmine and

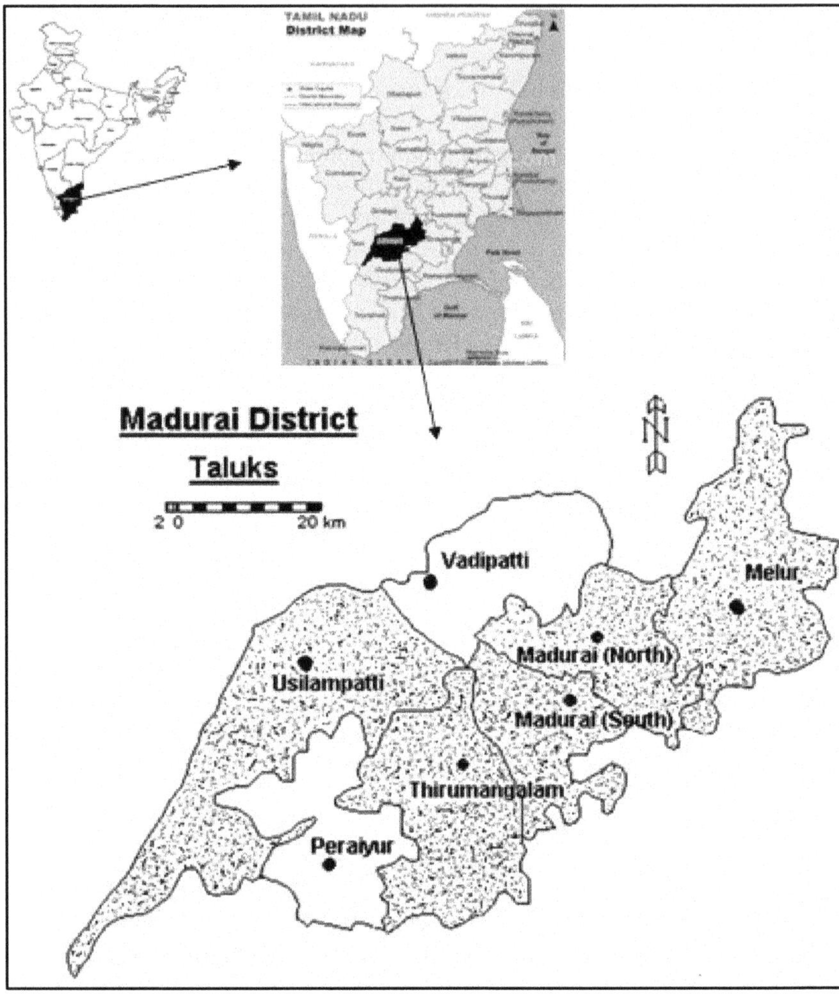

Figure 17.1: Study Area of Madurai District, Tamil Nadu, India. Dot fill indicated in Madurai District is intensive study area.

betlevine are localized in few areas, where the ground water facility is adequate. The crops such as groundnut, vegetables, sorghum and finger millet are cultivated in single crop system in small scales with the help of ground water.

The mean monthly minimum and maximum temperatures in the study area during the study period was in the range of 23.0°C (range: 19.7°C–26.9°C) to 33.4°C (range: 29.1°C–38.8°C). The study area receives most of its precipitation during the north-east monsoon which sets usually between October and December. The annual mean monthly rainfall during the study period ranged between 2.3 mm and 271.4 mm with an overall mean precipitation of 79.8 mm (Figure 17.2).

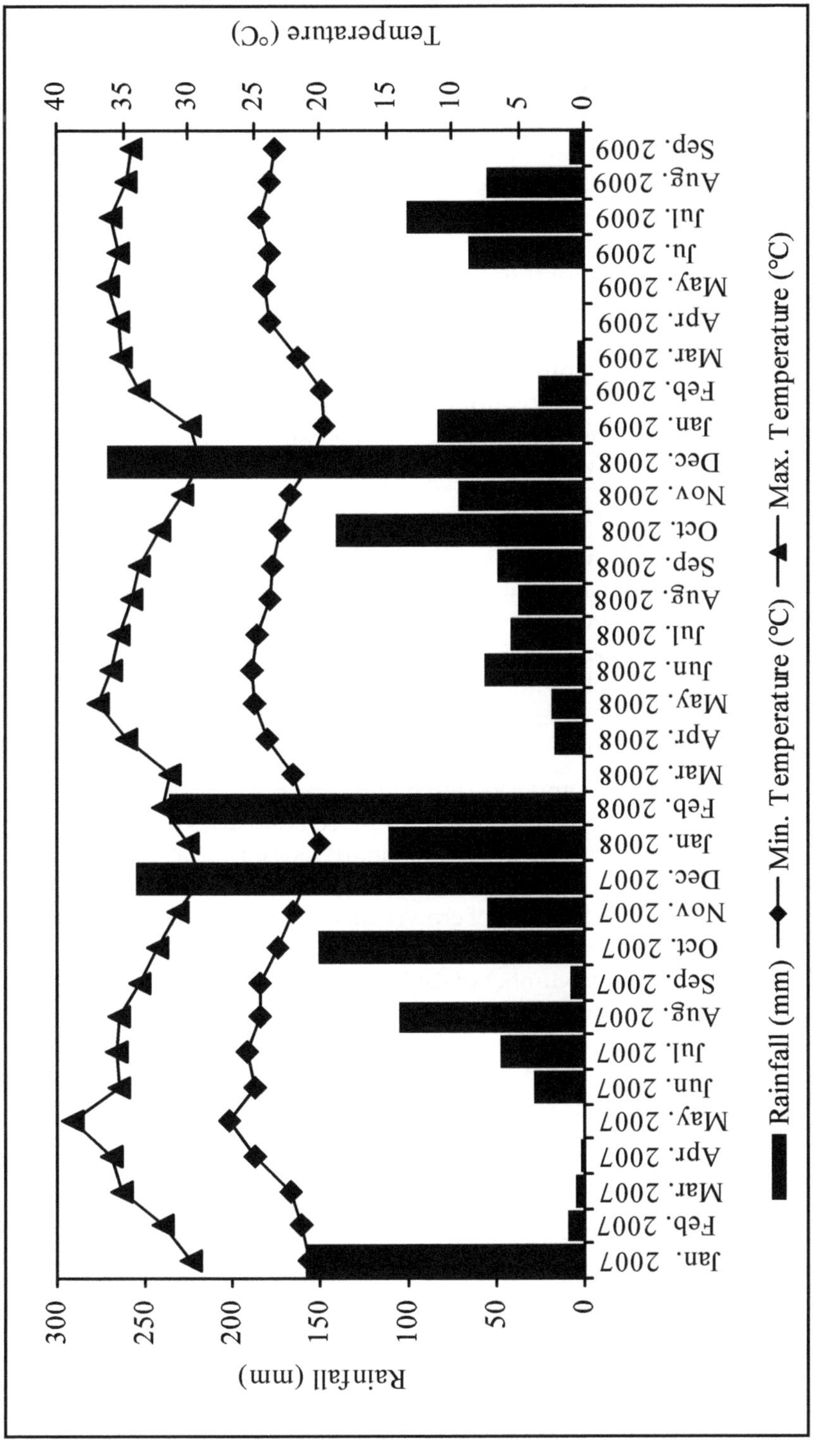

Figure 17.2: Minimum and Maximum Temperature (°C) and Total Rainfall (mm) in the Study Area during 2007-2009

Material and Methods

Population Density

The population of Spotted Owlet was estimated by line transect method (Buckland *et al.,* 1993; Lloyd 2003). Census was carried out in four different habitats namely rural, urban, agricultural lands and riverine. In each habitat, one 4500 m long transects were selected following a preliminary survey undertaken earlier. The Spotted Owlets were censused within a 50 m belt on either side of transects (100 m wide strip). Each nocturnal census was conducted between 1800 and 2200 hrs. Censuses were carried out monthly once in all the habitats. However, owl counts were not carried out on rainy or strong windy days, but were deferred to the following day. Transect was walked at a normal and regular speed. At each transect, we recorded the time of each bird contact, number of individuals and angular distance. Spotted Owlet calls were not taken into account for the calculation of density estimation. To estimate Spotted Owlet densities as number per sq km the following formula: D = Number of birds/(2×L×W), where L= Length of transect and W= ½ width of transect.

Roost-Site Preference

Day roosts of Spotted Owlets were identified systematically while surveying the study area by make use of the available roads. Surveys were conducted either by foot or by motorized vehicle. Information provided by the local people was considered during each visit and all those sites were all checked. All the roost-sites such as trees, ruined buildings, wells and other human-made structures were inspected periodically. Fresh deposition of regurgitated pellets, white droppings and vocalization were also taken as a clue for the presence of Spotted Owlets.

Total height of roost-site and roosting height was measured in meters from the ground by using an altimeter. Basal circumference and diameter at breast height (DBH) of roost trees were measured using a standard measuring tape. Canopy volume of roosting tree was calculated using the following formula: $V = 4/3\pi (D/2)^3$, where, V = Volume of canopy; D = Diameter of crown. Mean crown diameter was estimated by averaging the measurements of four canopy cover estimates (north, east, west and south) around the roost trees. Number of cavities and number of primary branches of each roost-site was counted visually. In order to quantify the factors influencing the roost-site preference the following ecological variables *viz.*, the distance from each roost-site to the nearest tree, agricultural lands, fallow lands, water sources, groves, human habitations, building, road, foot path, electric line and light source was measured around 1000m radius centered at each roost-site (Pande *et al.,* 2007).

Perch-Site Preference

The perch types of Spotted Owlet was categorized as electric power lines, trees, buildings, shrubs, compound wall, fence posts, and ground. Percentage of use of the above perches was calculated by following the method of Bell and Ford (1990). In the present study, we used two heights: perch-site height (overall height of a perch-site) and perching height (the height at which a Spotted Owlet was seen perched while feeding). The total height of perch-site used by Spotted Owlet was grouped into various class intervals *viz.*, 0-2 m, 2-4 m, 4-6 m, 6-8 m, 8-10 m, 10-12 m, 12-14 m, 14-16 m and >16 m. The perching height were also grouped into the 2-m intervals (0-2 m, 2-4 m, 4-6 m, 6-8 m, 8-10 m, >10 m) and their percent was calculated.

Food Habits and Prey Spectrum

Food habits of Spotted Owlets were determined by analyzing their regurgitated pellets collected from their roosting and nesting-sites. Spotted Owlet pellets were collected fortnightly in six different roosting/nesting-sites. Fresh pellets were collected separately during each visit and packed in polythene bags along with tags indicating the site name and date of collection before being brought to the laboratory. Pellets were kept at 70°C in a hot air oven for 24 hours to kill the associated invertebrates. The length and width of the pellets were measured by using a Vernier Caliper (to nearest mm) and the weight was measured with a digital balance with an accuracy of 1 mg. Pellets were analyzed following the method of Schueler (1972) and Yalden (2003).

Mammalian prey items were identified up to species level by comparing their bony remains (*e.g.*, skulls, mandibles) with museum specimens collected at the same study area. Reptiles were identified by their skulls and mandibles (Naranthiran, 1989). Invertebrate prey items were identified up to order level on the basis of undigested anatomical pieces such as heads, mandibles, wings, legs, stings and body rings (Naranthiran, 1989; Yalden, 2003, Asokan *et al.*, 2009a).

The biomass of mammalian prey species recovered from the pellets was estimated by using a standard log-log regression of the right mandible length as a function of body weight following Hamilton (1980) and Santhanakrishnan (1987, 1995). For this study, the locally available mammalian species such as House Rat (*Rattus rattus*), Lesser Bandicoot Rat (*Bandicota bengalensis*), Soft-furred Filed Rat (*Millardia meltada*), Indian Gerbil (*Tatera indica*), Indian Mice (*Mus* spp.) and the Grey Musk Shrew (*Suncus murinus*) and their corresponding log-log regression values for the calculation of biomass were used as described by Santhanakrishnan (1995). Mandible measurements recovered from Spotted Owlet pellets were then applied to the following regression equation to estimate the biomass of each prey species represented in the pellets. $\text{Log } y = \text{Log } a + b (\log x)$, where Log y is unknown body weight; log a known mandible length of museum specimen; log b known body weight of museum specimen; and log x mandible length of rats obtained from pellets (Hamilton, 1980).

Breeding Biology

To study the breeding biology of Spotted Owlet, the study area was surveyed intensively for active nests. The onset of nesting was determined by the presence of egg or young in the nest and nests containing one or more eggs or nestlings were defined as active nests. Once an active nest was located, it was periodically monitored (at least once in a day) to record the reproductive chronology.

Nest-site and nest location height above the ground were measured by using an altimeter. Diameter at breast height (DBH) and the basal circumference of nesting tree was measured with a standard measuring tape. Number of primary branches in each nesting tree was counted visually. In order to quantify the factors determining the nest-site preference the following ecological variables, *viz.*, the distance from each nest to the nearest tree, agricultural lands, fallow lands, water sources, groves, human habitations, building, road, foot path, electric line and light source was measured around 500m radius centered at each nest-site. The radius of 500m was chosen because Spotted Owlets usually forage near the nest-sites within this range (Pande *et al.*, 2007).

The hole or cavity height and width was measured by a centimeter scale or standard measuring tape. Cavity depth was measured from the cavity entrance to the bottom of the nest by a centimeter scale or tape. The thickness of tree wall at hole was measured by using a Vernier calliper (Nagarajan, 1998).

The length and width of egg was measured with a Vernier caliper and weighed to the nearest 500mg with a spring balance. The egg shape index (ESI) was calculated by using the formula breadth x 100/length (Prasanth *et al.*, 1994). The number of eggs in each nest was enumerated for clutch size. The incubation period was determined from the date in which the first egg was laid till the first egg hatched. The chick rearing period was interval between first day of chick and date in which the first chick fledged from the nest. Number of nestlings present in each nest was considered as brood size.

The hatching success (Balant, 1993) and fledging success (Dabert, 1987) was calculated by using the following formulas

Hatching success (per cent) = No. of eggs hatched/Total no. of eggs laid × 100

Fledging success (per cent) = No. of nestlings fledged/Total no. of nestlings hatched × 100

Nest Box Utilization

The Spotted Owlet nest-boxes were constructed by using ¾" thickness country wood as suggested by Neelanarayanan *et al.* (1995). The nest-box size and entrance hole were identical in all nest-boxes with dimensions of 40 cm height × 23 cm width with an opening of 13 cm at a height of 12 cm on the bottom of the box. Limited funding and some other difficulties restricted our efforts to 26 artificial wooden nest-boxes which were erected in the study area. The nest-boxes were placed in trees and human-made sites at various heights above ground level and were regularly checked to see if the Spotted Owlet were roosting or nesting in them.

General Public Owl Knowledge

In Madurai district, 183 persons were interviewed between 2008 and 2009. The questionnaires were typed in English and Tamil language and circulated among different target groups like farmers (n=73), temple workers (n=30), land owners (n=36), students (n=24) and public peoples (n=20). The interviewed persons were 37.7 per cent females and 62.3 per cent males and ranged in age from 15 to 69 years. The interviewed persons were asked their education level and they are classified as illiterate (28.4 per cent), semi-literate (32.8 per cent) and literate (38.8 per cent). The feedbacks of the above have been analyzed and interpreted.

Statistical Analysis

Arithmetic mean and standard deviation were calculated for all the replicate variables and are given as mean ± SD. One-way Analysis of Variance (ANOVA) was used to find out the mean of different groups (habitats, seasons and years) of data. Student 't' test was used to compare the means of two different groups of data. Principal Component Analysis (PCA) was performed on the roost tree characters to determine the most important factors in delimiting the habitat preference of Spotted Owlet. The multiple regression equation was use to test the any relationship between nest dimensions (length, width and depth) and clutch size. The statistical software MINITAB (version 13.1) was used for data analysis. Results are reported as significant if they are associated with a value $p<0.05$. Statistical inferences were made by following Sokal and Rohlf (1981).

Results

Population Density

The Spotted Owlet density varied between 10.4/km^2 (in summer 2007 at riverine habitat) and 35.6/km^2 (in post-monsoon 2009 at agricultural lands). In general, the rural and agricultural lands supported relatively higher number Spotted Owlet than other habitats (Table 17.1). Within habitats densities were higher during monsoon of 2007 (20.7/km^2), pre-monsoon of 2008 (33.3/km^2) and post-monsoon of 2009 (28.1/km^2) in rural and during monsoon of 2007 (20.0/km^2), summer of 2008 (27.4/km^2) and pre-monsoon of 2009 (27.4/km^2) in urban habitats. In agricultural lands the highest densities were recorded during monsoon of 2007 (27.4/km^2), summer of 2008 (3.4/km^2) and post-monsoon of 2009 (35.6/km^2) and in riverine the densities were highest during monsoon of 2007 (20.7/km^2) and summer of 2008 (25.3/km^2) and 2009 (30.4/km^2) (Table 17.1).

Table 17.1: Habitat-wise Variations in the Density (owls/km^2) of Spotted Owlet in different Habitats

Year	*Season*	*Habitats*			
		Rural	*Urban*	*Agricultural Lands*	*Riverine*
	Summer	17.8 ± 6.65	13.3 ± 2.25	20.0 ± 2.20	10.4 ± 3.36
2007	Pre-monsoon	18.5 ± 5.58	16.3 ± 2.60	23.7 ± 6.78	14.8 ± 6.78
	Monsoon	20.7 ± 2.54	20.0 ± 3.81	27.4 ± 3.40	20.7 ± 3.36
	Post-monsoon	21.5 ± 5.58	17.8 ± 2.20	23.7 ± 2.60	17.0 ± 1.27
2008	Summer	30.4 ± 2.54	27.4 ± 2.60	33.3 ± 3.87	25.3 ± 3.42
	Pre-monsoon	33.3 ± 2.25	25.9 ± 3.42	28.1 ± 3.42	25.2 ± 3.42
	Monsoon	31.9 ± 5.59	25.2 ± 3.42	30.4 ± 3.36	22.2 ± 4.45
	Post-monsoon	28.1 ± 5.59	26.7 ± 4.45	35.6 ± 5.88	26.7 ± 5.14
2009	Summer	26.7 ± 4.45	26.7 ± 2.25	27.4 ± 3.40	30.4 ± 2.20
	Pre-monsoon	27.4 ± 3.40	27.4 ± 3.40	34.1 ± 4.64	23.0 ± 2.54
	Monsoon	20.7 ± 2.83	20.7 ± 2.20	29.6 ± 3.21	22.2 ± 1.88

Seasonal variations in Spotted Owlet density inferred that rural (26.4/km^2) and urban (23.2/km^2) habitats has comparatively higher during pre-monsoon and the agricultural lands (29.6/km^2) and riverine (22.6 km^2) habitats had higher during post-monsoon. However, there were no significant differences (One-way ANOVA; P>0.05) in the mean density of Spotted Owlet among seasons (Table 17.2).

Table 17.2: Seasonal Variations in the Density (owls/km^2) of Spotted Owlet in different Habitats

Habitats	*Season*				*ANOVA*	
	Post-monsoon	*Summer*	*Pre-monsoon*	*Monsoon*	*F*	*P*
Rural	24.8 ± 6.17	24.9 ± 7.00	26.4 ± 7.33	24.4 ± 5.16	2.21	0.106
Urban	22.2 ± 5.78	22.4 ± 7.15	23.2 ± 5.89	22.0 ± 4.51	2.23	0.103
Agricultural lands	29.6 ± 7.66	26.9 ± 6.42	28.6 ± 6.34	29.1 ± 3.61	1.96	0.138
Riverine	22.6 ± 6.92	21.4 ± 8.89	20.9 ± 6.19	21.7 ± 4.22	1.61	0.205

Yearly variations of population density showed that the urban (25.4/km^2), agricultural lands (31.7/km^2) and riverine (25.6/km^2) had higher mean densities during 2009. In rural the density was relatively higher (29.2/km^2) in 2008. Significant yearly variations in Spotted Owlet densities was existed in all the habitats (One-way ANOVA; P<0.05) (Table 17.3).

Table 17.3: Yearly Variations in the Density (Owls/km^2) of Spotted Owlet in different Habitats

Habitats	*Year*			*ANOVA*	
	2007	*2008*	*2009*	*F*	*P*
Rural	19.0 ± 4.71	29.2 ± 6.04	25.7 ± 3.63	14.21	0.000*
Urban	16.5 ± 3.86	24.0 ± 4.62	25.4 ± 4.27	12.5	0.000*
Agricultural lands	23.7 ± 5.08	28.8 ± 4.64	31.7 ± 3.40	6.93	0.003*
Riverine	15.3 ± 6.11	22.4 ± 4.48	25.6 ± 3.08	11.57	0.000*

* Differences among years (One-way ANOVA; P<0.05).

Roost-Site Preference

A total of 292 roost-sites of Spotted Owlet were recorded during 2007-2009. For the convenience, these sites were grouped and summarized into four broad categories *viz.*, roost-sites in trees (n=229), human-made structures (n=43), artificial nest boxes (n=12) and unused wells (n=8) (Figure 17.3).

Roost-Sites in Trees

In the study area Spotted Owlet used 15 different species of woody trees for roosting in which *Ficus bengalensis* was the most preferred tree (42.9 per cent). Other species of trees used were *Tamarindus indicus* (15.7 per cent), *Enterelobium saman* (9.2 per cent), *Azadirachta indica* (8.7 per cent), *Acacia nilotica* (4.8 per cent), *Ficus religiosa* (3.9 per cent), *Cocos nucifera* (3.5 per cent), *Borassus flabellifer* (3.5 per cent), *Polyalthia longifolia* (2.6 per cent), *Moringa oleifera* (1.3 per cent), *Lannea coromandelica* (1.3 per

Figure 17.3: Different Roost-Sites of Spotted Owlet in the Study Area

cent), *Pongamia pinnata* (1.3 per cent), *Mangifera indica* (0.4 per cent), *Morinda tinctoria* (0.4 per cent) and *Terminalia catapa* (0.4 per cent).

Height of trees used for roosting ranged between 4.5 m to 30.0 m (16.4 ± 5.29 m) from the ground, of which Spotted Owlet preferred to roost at a mean height of 9.0 ± 2.69 m (range: 3.0-16.5 m). Mean DBH and basal circumference of the trees were 141.3 ± 36.1 cm (range: 12.0-415.0 cm) and 306.1 ± 16.42 cm (range: 18.0-985.0 cm) respectively. The Spotted Owlet used well branched trees for roosting and the number of primary branches varied from 0 to 23 with an overall mean of 9.9 ± 4.67 (Table 17.4).

Roost-Sites in Human-Made Structures

On ranking next to trees Spotted Owlet used a variety of human-made structures including 24 in buildings, 18 in holes on compound walls and one in temple tower. Mean total height of all the above structures was 7.1 ± 2.14 m (range: 4.0-16.0 m) in which roosting was observed at a mean height of 6.4 ± 1.92 m (range: 3.5-14.5 m) (Table 17.4).

Table 17.4: Roost-Site and Habitat Characteristics of Spotted Owlet in the Study Area

Variables	*Trees*	*human-Made Structures*	*Nest Boxes*	*Wells*
Total height of roosting-site (m)	16.4 ± 5.29	7.1 ± 2.14	10.8 ± 2.87	12.0 ± 2.55
Roost height (m)	9.0 ± 2.69	6.4 ± 1.92	5.9 ± 1.38	4.8 ± 1.60
DBH (cm)	141.3 ± 36.10	—	—	—
Basal circumference (cm)	306.1 ± 16.42	—	—	—
Canopy volume (m^3)	431.0 ± 48.53	—	—	—
No. of branches	9.9 ± 4.67	—	—	—
No. of holes	2.0 ± 1.04	2.0 ± 1.98	—	—
No. of birds / roosting-site	1.9 ± 0.69	1.6 ± 0.62	1.2 ± 0.45	1.5 ± 0.53
Distance to tree (m)	4.6 ± 2.61	10.0 ± 9.55	7.5 ± 4.64	7.4 ± 3.30
Distance to agricultural lands (m)	249.2 ± 32.51	285.9 ± 31.72	213.9 ± 60.38	241.9 ± 49.35
Distance to fallow lands (m)	275.4 ± 36.81	310.3 ± 32.76	264.3 ± 72.56	275.0 ± 47.95
Distance to foot path (m)	3.5 ± 2.83	2.5 ± 0.27	2.2 ± 1.20	3.5 ± 6.68
Distance to water sources (m)	90.8 ± 42.31	114.1 ± 21.85	55.6 ± 27.06	15.1 ± 4.91
Distance to groves (m)	131.4 ± 78.12	158.2 ± 15.54	212.8 ± 34.75	107.9 ± 37.60
Distance to human habitations (m)	78.9 ± 68.60	42.8 ± 51.73	70.8 ± 28.33	31.6 ± 21.7
Distance to building (m)	55.8 ± 28.99	32.7 ± 9.05	17.2 ± 4.70	49.8 ± 15.52
Distance to road (m)	84.9 ± 27.26	47.4 ± 10.73	128.7 ± 33.95	37.4 ± 11.69
Distance to light source (m)	32.7 ± 13.55	21.4 ± 5.67	25.9 ± 6.57	25.9 ± 17.09

Roost-Sites in Nest Boxes

Out of 26 wooden nest boxes installed, 12 nest boxes were used for roosting. Nest boxes were kept in various trees and buildings at a height between 6.0 and 15.0 m. The mean height of nest boxes preferred by Spotted Owlet was 5.9 ± 1.38 cm (range: 3.0-8.0 m) (Table 17.4).

Roost-Sites in Wells

The total depth of the well varied from 6.5 to 15.5 m with a mean depth of 12.0 ± 2.55 m. The Spotted Owlet roosted in the holes found in the inner wall of the well at a mean depth of 4.8 ± 1.60 m (range: 2.5-8.0 m). Number of birds per roost-site varied between one and two with a mean of 1.5 ± 0.53 (Table 17.4).

Ecological Variables

Among ten ecological variables studied around each Spotted Owlet roost-site, at least seven variables such as distance to nearest tree, building, footpath, electric line, light source, groves and water sources were very closer (within 100 m) to all types of roost-sites (Table 17.4).

Principal Component Analysis (PCA)

For this analysis only roost tree characteristics were alone taken. PCA resulted in the extraction of three main components. First three components, which accounted for 77.8 per cent of total variance, were considered. First component (PC-I) accounted for 43.5 per cent of total variance, correlated highly with roost height, canopy volume, basal circumference and distance to groves. Second component (PC-II) accounted for 21.2 per cent of total variance correlated highly with roost tree height, DBH and distance to human habitations. Third component (PC-III) accounted for 13.1 per cent of total variance and correlated highly with number of branches, distance to agricultural lands and distance to light source (Table 17.5).

Table 17.5: Principal Components Derived from the Roost-Site Variables of Spotted Owlet in the Study Area. Underlined values represent high correlations with the respective principal components.

Variables	*Principal Components*		
	PC-I	*PC-II*	*PC-III*
Roost tree height	0.415	<u>0.83</u>	-0.403
Roost height	<u>0.804</u>	0.001	-0.17
DBH	0.22	<u>0.633</u>	-0.545
Basal circumference	<u>0.763</u>	0.484	0.435
Canopy volume	<u>0.794</u>	-0.421	-0.006
No. of branches	0.01	0	<u>-0.602</u>
No. of holes	0.003	0	-0.001
Distance to tree	0.001	-0.002	0.431
Distance to agricultural lands	0.039	0.203	<u>0.881</u>
Distance to fallow lands	0.017	0.118	0.083
Distance to foot path	0.031	-0.369	0.104
Distance to water sources	-0.034	-0.221	-0.478
Distance to groves	<u>0.615</u>	-0.191	0.214
Distance to human habitations	0.013	-0.568	-0.415
Distance to building	-0.001	0.002	-0.17
Distance to road	-0.241	0.104	-0.133
Distance to light source	0.321	0.271	<u>-0.562</u>
Explained variance (%)	43.5	21.2	13.1
Cumulative percent variation explained	43.5	64.7	77.8

Perch-Site Preference

Perch-Site

The Spotted Owlet used a variety of perch-sites for foraging (hunting), of these electric power lines were mostly used sites (51.3 per cent) followed by trees (28.6 per cent). The other perch-sites included buildings, shrub, compound wall, fence posts and ground; all were used by less than 10 per cent (Figure 17.4).

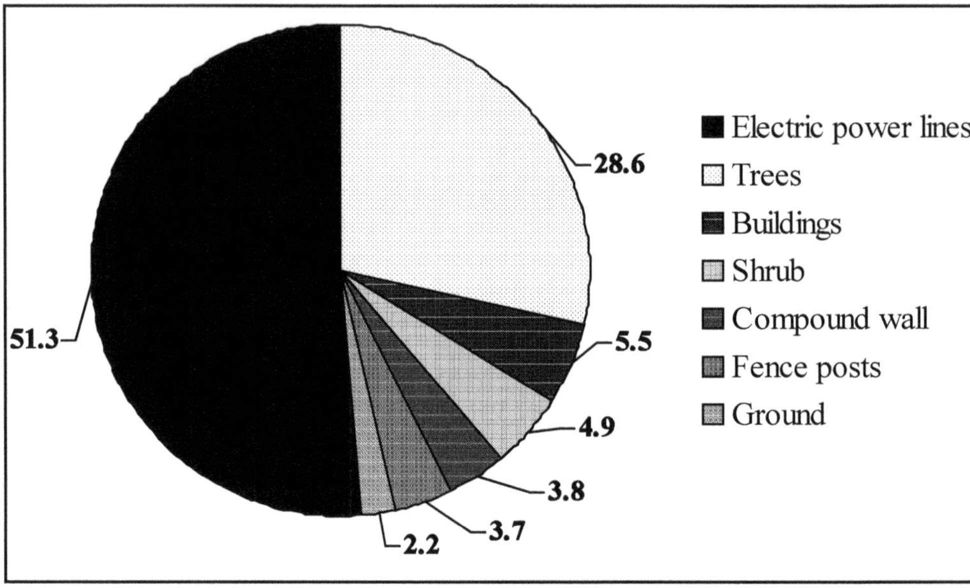

Figure 17.4: Per cent Use of Various Perch Types by Spotted Owlet in the Study (n=1660)

Perch-Site Height

The height of perches varied from 0 to 25m however, 6-8 m height category (54.4 per cent) was most preferred by Spotted Owlet. The second dominant perch height preferred by Spotted Owlet was 8-10 m (11.0 per cent) while all other height categories used by less than 10 per cent (Figure 17.5).

Perching Height

The perching height used by Spotted Owlet ranged from 0 to 16 m and most preferred perching height category was 6-8 m (60.0 per cent). The other perching height categories used by Spotted Owlet were 4-6 m (14.8 per cent) and 2-4 m (13.0 per cent) (Figure 17.6).

Food Habits and Prey Spectrum

Pellets

The fresh pellets of Spotted Owlet were compact and dark in colour because they contained masses of undigested materials mostly insects; on the contrary, the old pellets were pale in colour and held loosely (Figure 17.7). Intact pellets were measured 26.3 ± 0.29mm in length (n=4181; range: 18.0-46.0mm), 13.8 ± 0.16 mm in width (n=4181; range: 10.0-26.0mm) and weighed 0.50 ± 0.16g (n=275; range: 0.10-1.74 g).

Prey Composition

In all pellets, insects representing the order Coleoptera were predominant and represented in 91.3 per cent of the pellets followed by Orthoptera (80.8 per cent), Hemiptera (27.0 per cent), Diptera (14.8 per cent) and Lepidoptera (13.3 per cent)

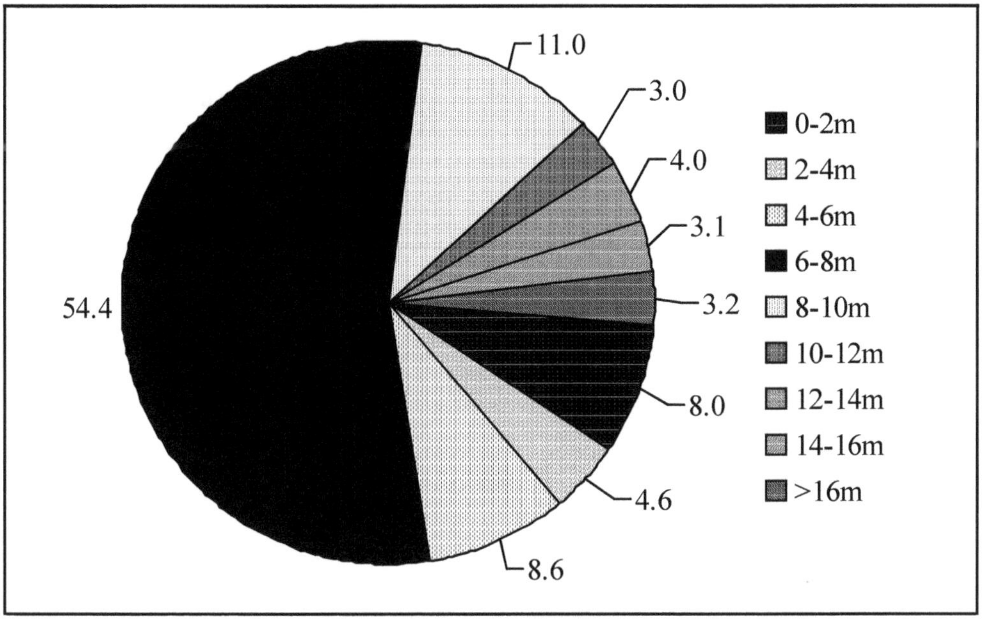

Figure 17.5: Per cent Use of Various Perch Site Height Categories by Spotted Owlet in the Study Area (n=1160)

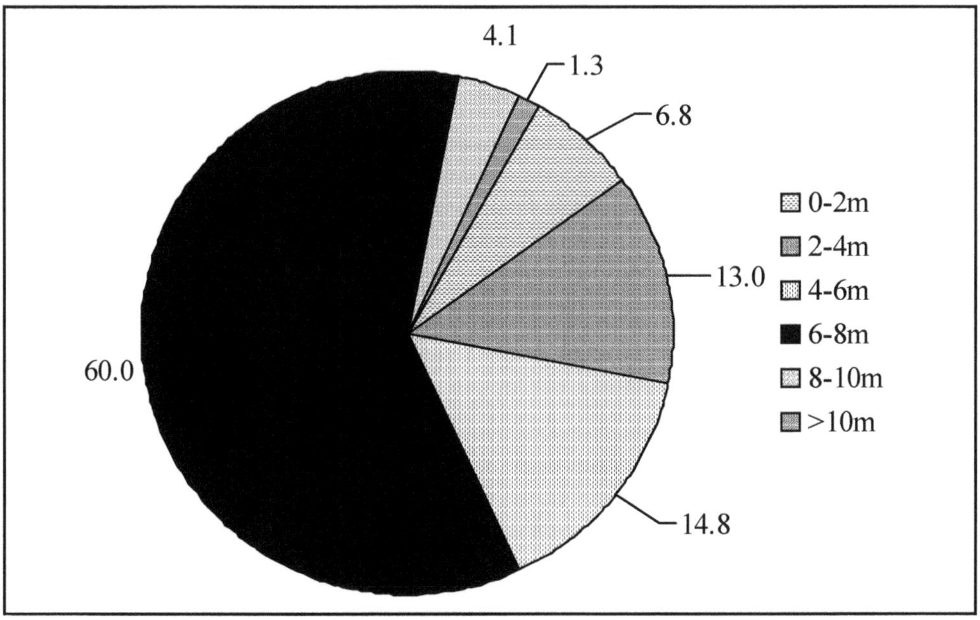

Figure 17.6: Per cent Use of Various Perching Height Categories by Spotted Owlet in the Study Area (n=1160)

(Figure 17.8). Among vertebrates, the reptilian remains were found only in 27.2 per cent followed by *Mus* sp. (15.7 per cent) (Table 17.6).

Table 17.6: Comparative Statistics for different Prey Items Found in Spotted Owlet Pellets (n=4181) in the Study Area

Prey Items	*Frequency of Occurrence of Prey Items*	*Percent Frequency of Occurrence of Prey Items*	*Range*	*Sum*
INVERTEBRATES				
Coleoptera	3821	91.3	1-14	16858
Orthoptera	3377	80.8	1-16	13356
Hemiptera	1125	27	1-6	1396
Diptera	618	14.8	1-4	704
Hymenoptera	607	14.5	1-5	700
Lepidoptera	560	13.3	1-3	618
Arachnida (Scorpions)	438	10.4	1-2	443
Myriapoda (Millipedes)	25	0.6	1-5	67
VERTEBRATES				
Reptiles	1139	27.2	1-2	1345
Bats	14	0.3	1-2	15
Mus spp.	656	15.7	1-2	678
Rattus rattus	193	4.7	1-2	201
Tatera indica	53	1.2	1-2	54
Bandicota bengalensis	47	1.2	1-2	50
Suncus murinus	154	3.7	1-2	157

A total of 40,278 individual prey items were obtained from 4,181 pellets in the study area between 2007 and 2009. The number of prey items per pellet ranged from 1 to 31 with an overall mean of 9.6 ± 3.03. Invertebrates were the most common prey represented 84.8 per cent of the total diet followed by vertebrates (6.2 per cent). Among invertebrates, insects accounted for 83.5 per cent of all individual prey (Table 17.7) with Coleopterans and Orthoptera being the most frequent items (41.9 per cent and 33.2 per cent, respectively).

Spotted owlets consumed a narrow spectrum of vertebrate prey including only reptiles and mammals in its diet (3.3 per cent and 2.8 per cent, respectively); although mammalian prey was constituted by four rodent species (*Rattus rattus, Bandicota bengalensis, Mus* spp., and *Tatera indica*), an insectivore (*Suncus murinus*) and an unidentified bat (Table 17.7 and Figure 17.9).

Biomass

Estimated biomass of all five identified small mammals prey species was 8875.6 g (Table 17.7) with *Mus* spp. constituting the greatest proportion (66 per cent) followed by *Rattus rattus* (22 per cent). Biomass of bat, reptiles and insects prey were not calculated because of the unavailability of museum specimens for identification.

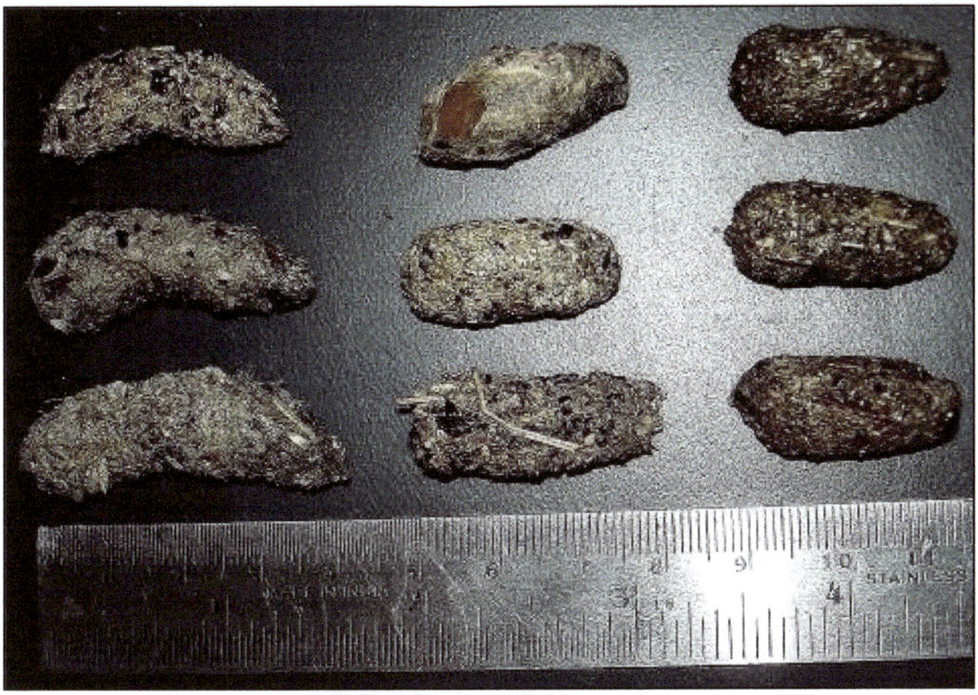

Figure 17.7: Various Shape and Sizes of Spotted Owlet Pellets Recorded in the Study Area

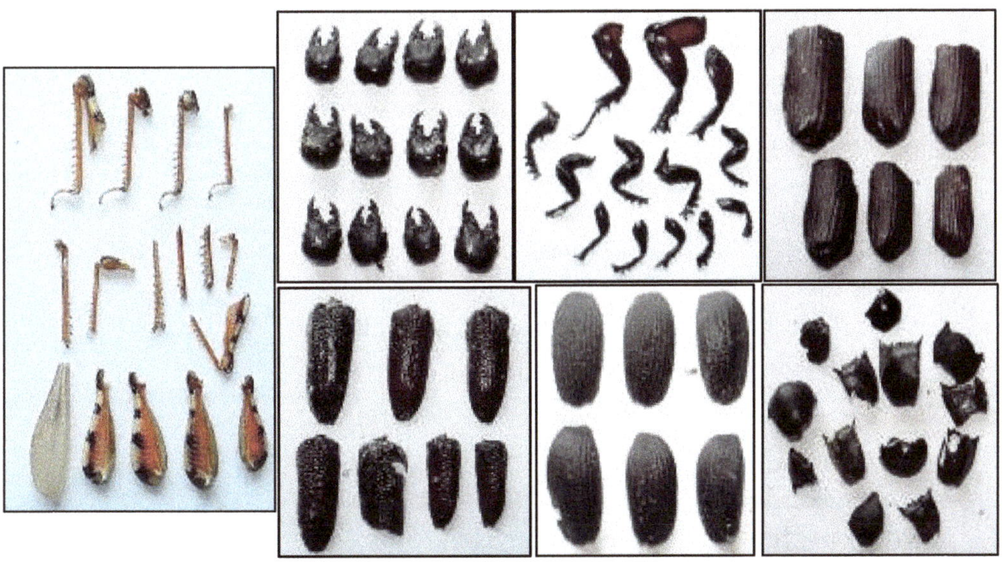

Figure 17.8: Orthopteran and Coleopteran Prey Remains Recovered in the Regurgitated Pellets of Spotted Owlet

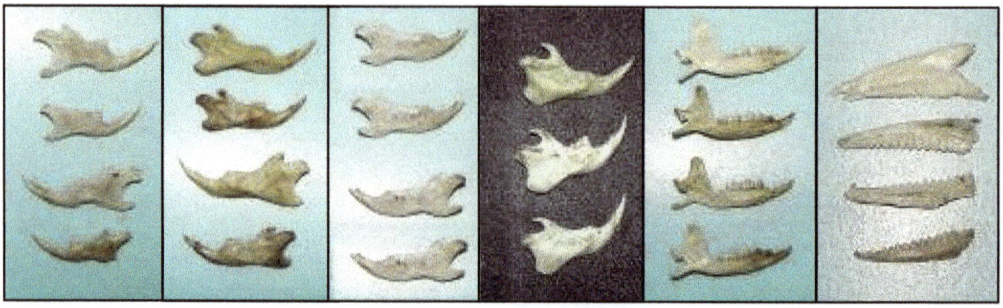

Mus spp. | *Rattus rattus* | *Tatera indica* | *Bandicota bengalensis* | *Suncus murinus* | Unidentified bat

Figure 17.9: Mammalian Prey Remains Identified in the Regurgitated Pellets of Spotted Owlet

Table 17.7: Overall Frequency and Proportions of Various Prey Items in the Pellets of Spotted Owlet (n=4181)

Prey Items	*Frequency of Occurrence*	*Percentage of Occurrence*	*Biomass (Per cent)*
INVERTEBRATES			
Class: Insecta			
Order: Coleoptera	16858	41.9	NC
Order: Orthoptera	13356	33.2	NC
Order: Hemiptera	1396	3.5	NC
Order: Diptera	704	1.8	NC
Order: Hymenoptera	700	1.7	NC
Class: Arachnida			
Order: Scorpionida	443	1.1	NC
Class: Myriapoda			
Order: Diplopoda	67	0.2	NC
VERTEBRATES			
Class: Reptilia	1345	3.3	NC
Class: Mammalia			
Order: Micro Chiroptera	15	0.04	NC
Order: Rodentia			
Mus spp.	678	1.7	66.3
Rattus rattus	201	0.5	21.9
Tatera indica	54	0.1	1.2
Bandicota bengalensis	50	0.1	0.7
Order: Insectivora			
Suncus murinus	157	0.4	9.9
Unidentified prey items	3636	9	NC
Total 40278 100 100			

NC: Not calculated.

Seasonal Variations of Prey

During all the three years we studied diet of Spotted Owlets, a total of 9,018 prey items were identified from post-monsoon, 10,957 from summer, 13,092 from pre-monsoon and 7,211 in monsoon. The invertebrates were the most common prey in all the seasons which were at its maximum in post-monsoon. Among invertebrates, the insects were mostly consumed in all the seasons which mean accounted for 83.4 per cent. Among insects, coleopterans were the most consumed prey and in a similar proportion in all seasons. Reptiles and mammals constituted over 4 per cent of all individual prey in all seasons (Table 17.8).

Table 17.8: Seasonal Variations in the Frequency and Proportion of Various Prey Items in the Pellets of Spotted Owlet

Prey Items	*Post-monsoon (Jan–Mar)*		*Summer (Apr–Jun)*		*Pre-monsoon (Jul–Sep)*		*Monsoon (Oct–Nov)*	
	N	*%*	*N*	*%*	*N*	*%*	*N*	*%*
INVERTEBRATES								
Class: Insecta								
Order: Coleoptera	3881	43	4559	41.6	5386	41.1	3032	42
Order: Orthoptera	2960	32.8	571	32.6	4362	33.3	2463	34.1
Order: Hemiptera	287	3.2	408	3.7	466	3.6	235	3.3
Order: Diptera	161	1.8	216	2.0	252	1.9	75	1.0
Order: Hymenoptera	132	1.5	205	1.9	266	2	97	1.4
Order: Lepidoptera	135	1.5	189	1.7	203	1.5	91	1.3
Class: Arachnida								
Order: Scorpionida	111	1.2	124	1.1	151	1.2	57	0.8
Class: Myriapoda								
Order: Diplopoda	67	0.7	—	—	—	—	—	—
VERTEBRATES								
Class: Reptilia	306	3.4	377	3.4	442	3.4	220	3.1
Class: Mammalia								
Order: Micro Chiroptera	6	0.1	6	0.1	3	0.02	—	—
Order: Rodentia								
Mus spp.	157	1.7	205	1.9	212	1.6	104	1.4
Rattus rattus	53	0.6	60	0.6	61	0.5	27	0.4
Tatera indica	5	0.1	29	0.3	18	0.1	2	0.03
Bandicota bengalensis	9	0.1	16	0.2	20	0.2	5	0.05
Order: Insectivora								
Suncus murinus	26	0.3	45	0.4	59	0.5	27	0.4
Unidentified prey items	7223	8.0	947	8.5	1191	9.1	776	10.8
Total prey items	9018		10957		13092		7211	
No. of pellets	940		1147		1319		775	
Mean prey/pellet	9.6 ± 2.86 (Range: 2-22)		9.6 ± 3.02 (Range: 1-22)		9.9 ± 3.07 (Range: 4-31)		9.3 ± 3.16 (Range: 4-21)	

Annual Variations of Prey

Total number of prey items identified during 2007, 2008 and 2009 was 4,487, 18,842 and 16,949 individuals respectively. During all year's invertebrates were the dominant prey items in the diet of Spotted Owlet among which insects were the most numerous. Among insects, Coleopterans constituted a virtually stable proportion of prey in all years (Table 17.9).

Table 17.9: Yearly Variations in the Frequency and Proportion of Various Prey Items in the Pellets of Spotted Owlet

Prey Items	*2007*		*2008*		*2009*	
	N	*%*	*N*	*%*	*N*	*%*
INVERTEBRATES						
Class: Insecta						
Order: Coleoptera	1868	41.6	7991	42.4	6999	41.3
Order: Orthoptera	1397	31.1	6412	34	5547	32.7
Order: Hemiptera	143	3.2	605	3.2	648	3.8
Order: Diptera	54	1.2	286	1.5	364	2.2
Order: Hymenoptera	35	0.8	283	1.5	382	2.3
Order: Lepidoptera	45	1	259	1.4	314	1.9
Class: Arachnida						
Order: Scorpionida	23	0.5	190	1	230	1.4
Class: Myriapoda						
Order: Diplopoda	—	—	—	—	67	0.4
VERTEBRATES						
Class: Reptilia	149	3.3	640	3.4	556	3.3
Class: Mammalia						
Order: Micro Chiroptera	—	—	—	—	15	0.1
Order: Rodentia						
Mus spp.	54	1.2	357	1.9	267	1.6
Rattus rattus	16	0.4	89	0.5	96	0.6
Tatera indica	—	—	24	0.1	30	0.2
Bandicota bengalensis	—	—	17	0.1	33	0.2
Order: Insectivora						
Suncus murinus	38	0.9	51	0.3	68	0.4
Unidentified prey items	665	14.8	1638	8.7	1333	7.9
Total prey items	4487		18842		16949	
No. of pellets	590		1967		1624	
Mean prey / pellet	7.6 ± 2.03 (Range: 3-20)		9.6 ± 3.19 (Range: 3-31)		10.4 ± 2.78 (Range: 1-22)	

Mammalian Prey Size

Spotted Owlets consumed prey species with a wide range of weight classes ranging from 1 g (insects) to >80 g (mammals). The weight of the mammalian prey items captured by Spotted Owlet ranged from 5 g (*Mus* spp.) to 84 g (*Rattus rattus*). Among the various prey weight classes, Spotted Owlet took mammalian prey 01-20 g (68.1 per cent) weight classes followed by 20-40 g (19.8 per cent). The other prey weight classes of, >40 g is represented by 12.1 per cent (Table 17.10).

Table 17.10: Percentage Occurrence of Various Mammalian Weight Classes in the Diet of Spotted Owlet

Weight Range (g)	*Mus spp.*		*R. rattus*		*T. indica*		*B. bengalensis*		*S. murinus*		*Total*	
	N	*%*	*N*	*%*	*N*	*%*	*N*	*%*	*N*	*%*	*N*	*%*
01–20	678	100	43	21.3	20	37.0	4	8	32	20.3	777	68.1
20–40	—	—	67	33.3	29	53.7	8	16	121	77.1	225	19.8
40–60	—	—	74	36.9	4	7.4	33	66	4	2.6	115	10.1
60–80	—	—	13	6.5	1	1.9	5	10	—	—	19	1.7
80–100	—	—	4	2.0	—	—	—	—	—	—	4	0.3
Total	678		201		54		50		157		1140	

Breeding Biology

A total of 32 nests were located and monitored during the entire study period: 12 in 2008 and 20 in 2009. The breeding activities of Spotted Owlet were noticed between 6th January and 26th April throughout the study period. It was observed that Spotted Owl performed only one brood per year. Most of the egg laying took place from the early January (29 nests) through February. The young ones were recorded mostly in mid March and fledglings were recorded in late April.

Nest-Site Characteristics

Of all the nests, 27 were found in trees and the remaining was in human-made structures (holes and roof of buildings) including one in wooden nest box (Figure 17.10). The trees used by Spotted Owlet for nesting were *Ficus benghalensis* (48 per cent), *Enterolobium saman* (30 per cent), *Cocos nucifera* (15 per cent) and *Tamarindus indicus* (7 per cent). The mean nest-site height and nest location height above the ground were 13.3 ± 5.44 m and 6.9 ± 1.92 m respectively. The DBH of the nest tree ranged between 30 and 412 cm with a mean of 194.3 ± 22.49 cm. Among the 11 microhabitat variables estimated around each nest-site, at least seven variables such as distance to nearest tree, building, footpath, electric line, light source, groves and water sources were very closer (<100 m) to the nest-site (Table 17.11).

Figure 17.10: Different Nest-Sites of Spotted Owlet in the Study Area

Table 17.11: Nest-Site and Habitat Characteristics of Spotted Owlet in Madurai District, Tamil Nadu, India

Variables	*Range*	*Mean ± SD*
Total height of nest-site (m)	4.5–23.0	13.3 ± 5.44
Height of nest location (m)	3.0–10.5	6.9 ± 1.92
Diameter at breast height (cm)	30.0–412.0	194.3 ± 22.49
Basal circumference (cm)	41.0–985.0	401.9 ± 55.25
Canopy volume (m3)	0.0–758.3	526.7 ± 48.14
Total number of primary branches	0.0–18.0	9.5 ± 5.35
Height of first branch	0.0–5.0	2.1 ± 1.30
Distance to tree (m)	1.0–10.0	3.8 ± 2.39
Distance to agricultural lands (m)	8.0–500.0	105.1 ± 34.52
Distance to fallow lands (m)	16.0–500.0	190.6 ± 32.29
Distance to water sources (m)	6.5–150.0	61.9 ± 26.36
Distance to groves (m)	15.0–200.0	96.5 ± 21.50
Distance to human habitations (m)	5.0–300.0	117.5 ± 19.96
Distance to building (m)	2.0–85.0	30.0 ± 13.10
Distance to road (m)	1.5–315.0	100.4 ± 21.25
Distance to footpath (m)	0.5–8.0	1.4 ± 1.28
Distance to electric line (m)	4.0–20.0	9.7 ± 4.32
Distance to light source (m)	3.5–40.0	18.0 ± 4.69

Nest Dimensions

The Spotted Owlets are secondary hole or cavity nesters (does not construct the nest), laid their eggs in readily available natural cavities or holes, nests of other hole nest birds and artificial nest boxes. Nest dimensions were measured for 29 nests which had the following mean dimensions: cavity/hole height: 34.5 ± 10.42 cm (range: 8-59 cm), cavity/hole width: 24.9 ± 9.24 cm (range: 6-45 cm), cavity/hole depth: 80.9 ± 15.74 cm and thickness of tree wall at hole: 3.7 ± 1.85 (range: 1.9-8 cm). In the present study 37.5 per cent of the nests do not contains any nesting materials and remaining 62.5 per cent nests consists of variety nesting materials chiefly fibres, grasses, leaves, polythene papers and feathers. The nest materials were found in the nests was collected by other hole nesting birds.

Egg Morphology

Spotted Owlet eggs were roundish oval and appeared milky white colour without any markings. Clutch size varied from 2 to 5 (Figure 17.11), with an overall mean of 2.6 ± 0.72 (Table 17.12). More nests had two egg clutches (53.1 per cent) than others (Figure 17.12). The multiple regression analysis inferred that no correlation between nest dimensions and clutch size (r=0.7801; p=0.859; df=31). Eggs averaged 26.8 ± 1.89 mm in length (range: 23.0-31.0 mm), 22.9 ± 1.61 mm in width (range: 20.0-26.0 mm) and 13.6 ± 1.47 g in weight (range: 10.0-17.0 g). The egg shape index value ranged between 75.0 and 91.6 with an overall mean of 85.3 ± 3.21 (Table 17.13).

Figure 17.11: Clutch Sizes of the Spotted Owlet Eggs in the Study Area

Table 17.12: Clutch and Brood Size and Nesting Success of Spotted Owlet in the Study Area

Year	*N*[1]	*Clutch size*	*Brood size*	*Nesting Success*[2]
		Mean ± SD	*Mean ± SD*	
2008	12	2.7 ± 0.89	2.3 ± 1.14	83 per cent
2009	20	2.5 ± 0.61	2.1 ± 0.76	95 per cent
Total	32	2.6 ± 0.72	2.1 ± 0.91	91 per cent

1. A nest was defined as an active nest with at least one egg.
2. A successful nest was one that fledged at least one nestling.

Incubation and Brood Size

Incubation started after laying the first egg and the mean incubation period for all clutches (n=18) was 26.2 ± 1.31 days (range: 25-30 days). Both partners involved in incubation. During incubation, while one bird incubates, the other partners either male or female perches nearby the nest within the vicinity and watching over the nest. The mean brood size for the two years together was 2.1 ± 0.91 (Table 17.2). Brood size of two was common (43.8 per cent) (Figure 17.13) and no significant differences was noticed between years ($t_{17} = 0.540$, $p = 0.297$). More than 90 per cent of the nests were successful (Table 17.12) of which 100 per cent success were found in 20 nests.

Hatching and Fledging Success

Of the 82 eggs laid in 32 clutches, 68 eggs hatched (2.1 ± 0.90 eggs/nest) and thus gave a mean hatching success of 83 per cent (Table 17.14). There was no significant difference ($t_{23} = 0.444$; $p = 0.482$) in hatching success between years. A

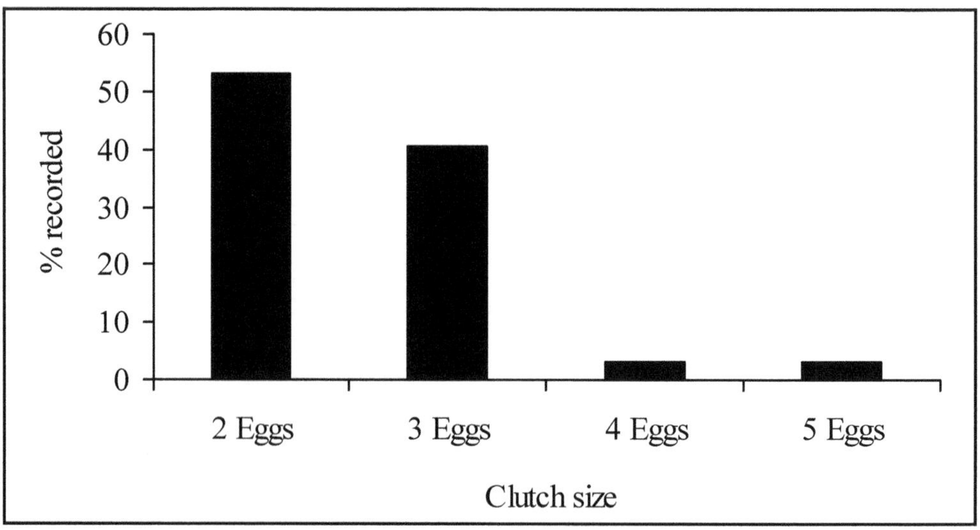

Figure 17.12: Clutch Size (n=32) Variations of Spotted Owlet in the Study Area

Table 17.13: Egg Morphology of Spotted Owlet in the Study Area

Year	N	Length (mm)		Width (mm)		Weight (g)		Egg Shape Index	
		Range	Mean ± SD	Range	Mean ± SD	Range	Mean ± SD	Range	Mean ± SD
2008	32	26.0-31.0	28.2 ± 1.31	21.0-26.0	23.9 ± 1.34	10.0-16.01	13.1 ± 1.30	77.8-89.7	84.7 ± 3.44
2009	50	23.0-30.0	25.9 ± 1.68	20.0-26.0	22.3 ± 1.49	1.0-17.0	14.0 ± 1.48	75.0-91.7	85.9 ± 3.35
Total	82		26.8 ± 1.89		22.9 ± 1.61		13.6 ± 1.47		85.3 ± 3.21

Table 17.14: Hatching and Fledging Success of Spotted Owlet in the Study Area

Year	No. Eggs Laid	No. Eggs Hatched	Hatching Success (%)	No. Young Fledged	Fledging Success (%)
2008	32	27	84	15	56
2009	50	41	82	33	80
Total	82	68	83	48	71 \

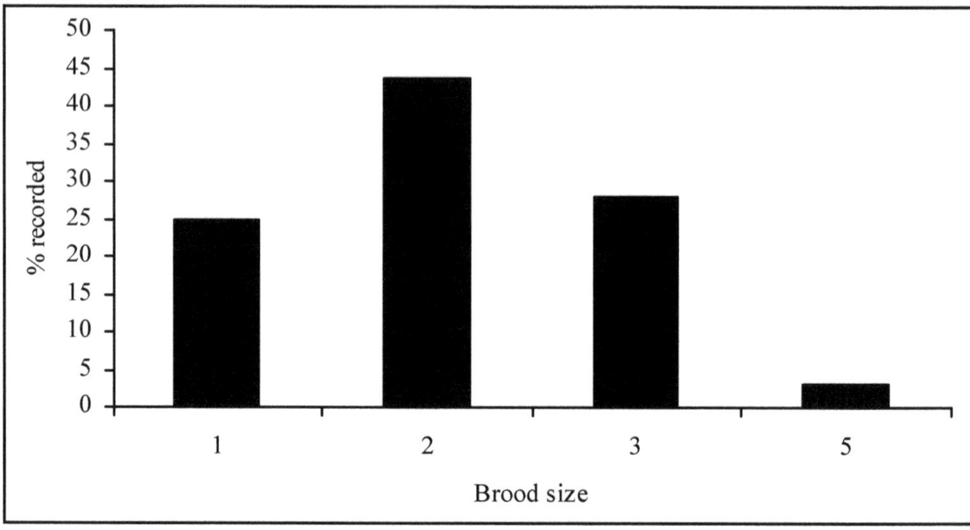

Figure 17.13: Variations in the Brood Size of Spotted Owlet in the Study Area, 2008-2009

total of 48 nestlings fledged (1.5 ± 0.72 nestlings/nest) which yielded a mean fledging success of 71 per cent (Table 17.14). The fledging success significantly differed (t_{21} = 2.513; p = 0.042) between years. The weaning period ranged between 30 and 33 days and 90 per cent of the nestlings fledged after 30 days.

Nest Box Utilization

A total of 26 Spotted Owlet nest boxes were installed in various sites like trees (n=16), buildings (n=9) and electric post (n=1) (Figure 17.14). Among these nest boxes, 12 nest boxes used for roosting and one for nesting. The overall occupancy rate nest boxes during the study period were 50.0 per cent (Table 17.15). All the nest boxes were kept at a height of 3-16 m from the ground level, in which Spotted Owlet shoed a preference of using nearly 70 per cent of nest boxes at a height of 6-8 m.

Table 17.15: Occupancy Preferences of Nest Boxes by Spotted Owlet in the Study Area

Site	*No. of Boxes Installed*	*No. of Nest Boxes Occupied*		*Total Occupied*	*% Occupied*
		Roosting	*Nesting*		
Tree	16	6	1	7	53.8
Building	9	6	0	6	46.2
Electric post	1	0	0	0	0
Total	26	12	1	13	

Figure 17.14: A View of Spotted Owlet Nest Boxes in different Sites

General Public Knowledge About Owl

Around 43 per cent of respondents liked owls because of their inestimable service as destroyers of insect and rodent pests. However, remaining respondents (57 per cent) dislike owls because of their threatening face with large eyes and scary screeching calls during the night hours. In Madurai, 54.2 per cent knew two species of owls (Spotted Owlet *Athene brama* and Barn Owl *Tyto alba*), 33.3 per cent knew only one species (Spotted Owlet) and 12.4 per cent knew three species (Spotted Owlet, Barn Owl and Indian Eagle Owl *Bubo bengalensis*). Majority of the respondents (77.7 per cent) disagree that owls attack human beings.

Owl Habitats and Food

The peoples listed different owl habitats in the study area and 40.6 per cent listed trees as major owl habitats (Table 17.16). Insects (30.1 per cent) and rodents (28.3 per cent) were listed as main owl prey in the study area. Interestingly, peoples also answered that owls eat lizards, snakes, birds and frogs (Table 17.17).

Owl Hunting

More than 60 per cent of the respondents answered owls are being hunted for food (30.3 per cent) and medicinal purposes (23.7 per cent) by a nomadic tribe "*kuravas*". Peoples also stated owl meat particularly liver (3.9 per cent), eyes (2.6 per cent) and flesh could cure the diseases associated with lungs (9.2 per cent) and eye associated problems (25.5 per cent).

Table 17.16: General Knowledge about Owl Habitats in Madurai District

Habitats	*% Respondents*
Trees	40.6
Temple towers	24.6
Holes in buildings and wells	12.3
Forests	10.1
Rocky mountains	6.2
No response	6.2

Table 17.17: General Knowledge about Owl Food in Madurai District

Food	*% Respondents*
Insects	30.1
Rodents	28.3
Lizards and insects	17.6
Snakes	6.5
Birds	5.8
Frogs	5.2
Rodents, birds and lizards	4.6
Fruits and vegetables	1.9

Owl Beliefs

Over 69 per cent of the respondents had strong beliefs and superstitions about owls. 80.6 per cent peoples replied that screech of owls at night hours in roof of house cause sudden death and 19.4 per cent peoples mentioned owls a sign of evil and misfortune. 72.3 per cent respondents believed that have not seen directly any such incident (death) but heard about such news from older peoples.

Owl Perches

Nearly 56 per cent of the respondents (mainly farmers and land owners) provided artificial perch sites in their agricultural field to attract owls. All such perching sites were made out of coconut stick (23.8 per cent), bamboo poles (17.4 per cent), casuarina poles (11.5 per cent) and available wooden sticks. 65 per cent of the farmers confirmed that such artificial perching sites were used by owls during night hours while searching the prey to swoop down the kill.

Owl Classification

Finally, the respondents were asked to classify owls based on their knowledge and beliefs. A majority of peoples classified owls as bad omen (53.2 per cent). Some famers found that owls as beneficiary birds and young students replied owls as just a bird (Table 17.18).

Table 17.18: Classification of Owl in Madurai District

Classification	*% Respondents*
Bad omen	53.3
Beneficial	20.5
Just a bird	14.5
Harmful	6.2
Fearing	3.5
Others	2.1

Discussion

Mean density of Spotted Owlet in the study area varied from 10.4 to 35.6/km^2 and this estimated density suggests that this species can be considered to be common in the study area. There is no previous report on the densities of Spotted Owlet that can be directly compared with the present findings. Spotted Owlet density among habitats and in general the density was comparatively higher in agricultural lands and rural habitat. Density fluctuations in different area reflect the difference at the habitat level (Redpath, 1995; Ranazzi *et al.,* 2000, 2002). For example, more Spotted Owlet in agricultural lands and rural habitat might be provided relatively rich supply of insects and rodents and low density of owlets in urban and riverine habitats might be due to lesser food availability coupled with greater human disturbances. Food availability is one of the main factors determining the habitat use by bird species (Block and Brennan, 1993), but other factors also determine habitat use. For example, occurrence of owls may be influenced by vegetation type rather than by food availability, because habitat structure is essential for protection, nesting and perching sites (Sparks *et al.,* 1994). Besides, the rural habitats provided adequate roosting and nesting sites, more foraging perches and camouflaged dense vegetation to Spotted Owlet which is highly essential during rainy season and these facilities were lesser in other habitats.

The seasonal variations of Spotted Owlet population inferred that density was more or less stable in all the habitats and seasons. But a slight increased pattern was noticed during post-monsoon (January-March), summer (April-June) and Pre-monsoon (July-September). This increased response could be related to the reproductive period of this species. The Spotted Owlet breeding season in the study area sets in January and ends in April. The high density estimates reported in summer and pre-monsoon may be reflected by the addition of recently emerged independent owls. During this time, food availability is high because insects and small vertebrates increase in abundance (Asokan *et al.,* 2003, 2010a; Ali *et al.,* 2010a,b). Smith *et al.* (1987) reported that density estimates of Eastern Screech Owl (*Megascops asio*) increase at the end of breeding season as juveniles are added to the population as independent individuals. However, other studies have reported that delectability of owls increase during the courtship period (Lynch and Smith, 1984). Relatively lower density estimates during monsoon (October-December) may be associated not only with the post-reproductive status and dispersal of owls and also with fewer food resources during the colder months.

Yearly variations of population densities differed among habitats. Various factors may be cited for yearly variations in avian density *viz.*, predation, intra and inter-specific competition, parasites and diseases, habitat availability, weather, food habits and migratory status (Asokan 2003; Chamberlain *et al.*, 1999).

The Spotted Owlet preferred structurally complex roost trees with well branched, high canopy densities with thick green foliage, large diameter with various height classes for roosting. They mostly preferred to roost in open limbs of trees, besides cavities or holes. Young *et al.* (1998) observed that Spotted Owl (*Strix occidentalis*) roost-sites in South-western Chihuahua, Mexico had more canopy layers, greater canopy closure and greater live tree basal areas. Dense tree canopies were provided well protection from predators, rain, wind, sun light exposure and concealment for the birds. Day time temperature is a one of the important factors in selecting roost-sites in Spotted Owl has been reported by Barrows and Barrows (1978). According to them Spotted Owls behaviourally facilitated heat loss at ambient temperature above 27°C. During the study period the mean minimum and maximum temperature in the study area recorded between 23°C and 33.4°C with a hike of 35°C and more in summer (April-June). To compensate this extreme temperature Spotted Owlet consistently chose well branched trees than that of open canopy trees. Many authors have been reported that canopy density is related to roost-site preference by Spotted Owl (*Strix occidentalis*) (Barrows, 1981), Mexican Spotted Owl (*Strix occidentalis lucida*) (Young *et al.*, 1998) and Eastern Screech Owl (*Otus asio*) (Trango *et al.*, 2001).

Next to trees the Spotted Owlet preferred to roost in a variety of human-made structures such as holes in compound walls, buildings and temple towers. Human-made structures provided some advantages over natural sites as they provided better protection against the environmental elements, *viz.*, temperature and rainfall. During colder months and rainy seasons such sites provided better insulation to Spotted Owlet. Human-made structures also offer safety from enemies and predators. Crows (*Corvus* spp.), Common Myna (*Acridotheres tristis*), Common Babbler (*Turdoides caudatus*) were noticed in many occasions mobbing Spotted Owlets whenever they come out from their roost-sites in day time. But these birds never tried to disturb the owlets inside the human-made structures. Using various kinds of human-made structures by Spotted Owlet in the study area might to be avoiding inter-specific competition that exists in the natural trees.

In addition to trees and human-made structures, inner wall and earth cuttings of wells were also used as roost-sites in the study area. Pande *et al.* (2006) reported similar roost-sites for Spotted Owlets in Maharashtra, India. Utilization of holes in wells, which are often away from human activities, appears to be an adaptive response to as well as to compensate the non-availability of suitable roost-sites.

The proximity of roost-sites to foraging ground with food availability was identified as one of the most important factors in roost-site preference of Spotted Owlet. The presence of agricultural lands, groves, human habitations, buildings and light sources directly governs and supports various kinds of prey species particularly insects and small mammals. These habitats were mostly present around the vicinity of roost-sites in the study area. Kumar (1985) and Naranthiran (1989) also reported that Spotted Owlet preferred roost-sites closer to agricultural lands. Carey *et al.* (1992) and Ward *et al.* (1998) suggested that Northern Spotted Owls (*Strix occidentalis caurina*) selected roost-sites based on the distribution of their prey items.

The Spotted Owlet used various perching objects both natural and artificial available in the study area during hunting prey. In the present study, two types of hunting strategies, *viz.*, sit-and-wait hunting strategy and sallying or hunting by wings. Of the various perch-sites used by Spotted Owlet, the electric power lines were mostly preferred. The electric power lines are a common feature in the study area especially around the human habitations and agricultural fields. These structures provided more suitable site to Spotted Owlet for searching and hunting the prey. In the present study, Spotted Owlet used electric power lines with close to light source (street light), because they attract variety of insects during night hours. Lammers and Collopy (2007) stated that avian predators are attracted to over-head utility power lines because they provide perches for various activities, including hunting prey. With the absence of electric lines in such areas, Spotted Owlet preferred and utilized other perches such as trees, buildings, shrubs, compound wall, fence post and ground. However, the preference of perch-sites changed over the habitat structure and type and behaviour of prey during foraging. In many occasions, Spotted Owlets were observed sitting on the ground while hunting the Scorpions and large Coleopteran beetles.

The Spotted Owlets mostly perched at 6-8 m height levels which is highly optimum for efficient foraging its prey species. Earlier, some researchers reported that similar perching height preference of insectivorous birds (Asokan, 1995; Asokan and Ali, 2010) and Great Horned Owl (Rohner and Doyle, 1992). Higher perching height may provide a larger field of view and increase chances of detecting prey. Butler *et al.* (2005) stated that distance to prey is a key factor influencing the perching height selection in sit-and-wait predators. However, the nature and height of perches used by Spotted Owlet is related to the type of prey captured, vegetation structure and availability of perches.

In Pakistan, Beg *et al.* (1990) classified Spotted Owlet regurgitated pellets based on the size: large, medium and small. Similar observations of such variations in the pellet sizes were noted in the present study. The size and shape of the pellets usually depend upon the size and the quantity of undigested materials found in the prey. The average measurements of the pellet sizes reported in other studies (Pande *et al.*, 2004; Tariq *et al.*, 2003) is different for our findings and the differences may be due to more insects prey per pellet.

The mean number of prey items per pellet in our study was 9.6 ± 3.03. Earlier, Naranthiran (1989) reported a mean number of 11.6 prey items in Tamil Nadu and Pande *et al.* (2007) registered a mean number of 6.9 prey items in Maharashtra. In contrast, Mahmood-ul-Hassan *et al.* (2007a) reported a mean number of 1.6 prey items at Pakistan. Variations in the number of prey items per pellet is usually influenced by a number factors such as size of the prey available, surface activity of prey, status of owls hunger, nutritional value, biomass of the prey, prey selection or prey diversity in an area (Santhanakrishnan, 1995). If the prey size is large (>20 g) the number of prey items per pellet may decrease; inversely, it may increase when the prey is very small (<1 g insects).

Irrespective of the year and season, the invertebrates particularly insects were the most common prey in the diet of Spotted Owlets. Our results were similar to those documented for Spotted Owlet in India (Jain and Advani, 1984; Kumar, 1985; Naranthiran, 1989; Jadhav and Parasharya, 2003; Pande *et al.,* 2004, 2007) and Pakistan (Beg *et al.,* 1990; Shah *et al.,* 2004; Mahmood-ul-Hassan *et al.,* 2007a).

Among insects prey, Coleopterans were the most eaten by Spotted Owlet in our study area. Several previous studies also identified to Coleopterans as one of the most important food resources of this owl species in India (Jain and Advani, 1984; Kumar, 1985; Jadhav and Parasharya 2003; Pande *et al.,* 2004). Tariq *et al.* (2003), Shah *et al.* (2004) and Mahmood-ul-Hassan *et al.* (2007a) also reported that coleopterans also are the most consumed insects prey by Spotted Owlets in Pakistan. After Coleopterans, Orthopterans were also important insects prey. Kumar (1985) and Naranthiran (1989) recorded that Orthopterans accounted for 72 per cent and 67 per cent of all individual prey taken by Spotted Owlets. A number of other studies have also recorded the occurrence of Orthopteran in the diet of this owl species (Ali and Ripley, 1983; Jain and Advani, 1984; Jadhav and Parasharya, 2003; Pande *et al.,* 2004). In our study area, Coleopterans and Orthopterans were frequently recorded in agricultural fields and most often attracted to the light source during night hours. In many occasions Spotted Owlets were observed on electrical power lines to feed on the above insects which are usually attracted by the light sources. Those insects that were taken in negligent proportions possibly constituted a supplementary prey resource or it may have been captured by chance or availability.

In Madurai, Spotted Owlets also consumed scorpions (arachnids). Similarly, Kumar (1985) and Pande *et al.* (2004, 2007) found that scorpions in the diet of this owl species. Consumption of scorpions also has been documented for Collared Scops Owl (*Otus bakkamoena*) at Pondicherry, India (Verzhutskii and Ramanujam, 2002).

Consumption of reptiles by Spotted Owlets in Madurai was comparable to that documented for the species in India (Jadhav and Parasharya 2003, Pande *et al.,* 2007) and Pakistan (Shah *et al.,* 2004, Mahmood-ul-Hassan *et al.,* 2007a). Reptilian prey remains have also been commonly recorded in the food habits of other owl species in India (Verzhutskii and Ramanujam, 2002; Srinivasalu and Srinivasalu, 2007).

Composition of small mammals prey (*Rattus rattus, Tatera indica, Bandicota bengalensis* and *Mus* spp. and an insectivore *Suncus murinus* and unidentified bat) in the diet of Spotted Owlets in Madurai was similar to the results obtained by other authors in India (Jain and Advani, 1984; Kumar, 1985; Naranthiran, 1989; Pande *et al.,* 2004). Recent studies at Maharashtra, India detected 11 species of rodents in the diet of Spotted Owlet (Pande *et al.,* 2007). Mahmood-ul-Hassan *et al.* (2007a) found five species of rodents, with *Mus* sp. accounting for >20 per cent of all prey items, in the diet of Spotted Owlet in Pakistan. The frequent occurrence of mice, rats and shrews in the pellets of Spotted Owlet suggest that the owlets hunt around human settlements as well as in the agricultural lands.

The Spotted Owlet breeds during January to April as reported by Mahmood-ul-Hassan *et al.* (2007b) and Pande *et al.* (2007) elsewhere. Spotted Owlet produced only one brood per year and no second brood or replacement of brood was noticed in the present study.

The Spotted Owlet used mostly tree cavities and holes in buildings and artificial wooden nest boxes for nesting. Kumar (1985) recorded that nearly 57 per cent of Spotted Owlet nests in tree holes in Andhra Pradesh, India. Pande *et al.* (2006) found that 26.8 per cent of nests in tree holes, who have also recorded nests in holes of abandoned buildings, wall under a bridge and wells. Pande *et al.* (2007) stated that in human related area, Spotted Owlet used cavities in human-made structures, while in open country they nest in existing cavities in trees. Naik (2004) reported that Spotted Owlet readily accept and occupy nest boxes for nesting. In the present study intra-specific competition between Spotted Owlet and other cavity nesters, *viz.*, Common Myna (*Acridotheres tristis*) and Indian Roller (*Coracias benghalensis*) were recorded.

Trees used by Spotted Owlet for nesting were *Ficus benghalensis, Enterolobium saman, Cocas nucifera* and *Tamarindus indicus* and 96 per cent of all trees were alive. These trees were common, large, well branched, densely foliaged and therefore provide more suitable nesting cavities or holes to Spotted Owlet. Pande *et al.* (2006) reported that six tree species *viz., Ficus religiosa, Tamarindus indicus, Mangifera indica, Ficus benghalensis, Moringa oleifera* and *Melia azaderach* were used by Spotted Owlet in Maharashtra, India. Mahmood-ul-Hassan *et al.* (2007b) observed that Spotted Owlet nests in *Dalbergia sissu, Acacia nilotica, Butea monosperma, Ficus benghalensis, Mangifera indica* and *Eucalyptus citriodora* in Pakistan. The variation in species composition of nest trees in other region suggests that the preference is primarily based on the availability of suitable cavity to accommodate their young ones rather than tree species. Mean nest-site height, nest height and tree DBH in the present study were greater than other studies (Pande *et al.,* 2006; Mahmood-ul-Hassan *et al.,* 2007b). The considerable variation in these parameters in different regions suggests that size of the tree or heights of the nest are relatively less important in nest-site preference than the presence of a suitable cavity or hole.

Rendell and Robertson (1989) reported that avoidance of predators is one of the most important factors involved in the nest-site preference of owls and other cavity nesting species. Tome *et al.* (2004) stated that presence of predators as the main factor to nest-site preference by Little Owl (*Athene noctua*). In the present study no predators were identified except some human disturbances. The study inferred that habitat such as agricultural lands, groves, human habitations, buildings and light sources were closer to the nest-sites. These habitats provide more profitable high energy foods such as rodents, insects, lizards, etc. during breeding season to feed the growing young ones.

In the study area, mean cavity/hole height, width and depth were 34.5 cm, 24.9 cm and 80.9 cm respectively. In Maharashtra, Pande *et al.* (2006) observed that the horizontal depth of nest tunnels in earth cuttings in wells or in banks were about

45 cm and vertical depth of holes in trees was up to 60 cm. No other data in relation to nest dimensions of Spotted Owlet is available in India to compare our results.

In the present study morphometric measurements of Spotted Owlet eggs were comparatively less when compared to the descriptions of Ali and Ripley (1983) and Jadhav and Parasharya (2003). In contrast to length and width, the weight of Spotted Owlet egg was significantly greater values than in Maharashtra (Jadhav and Parasharya, 2003) and lower than the Andhra Pradesh (Kumar, 1985).

The mean clutch size of Spotted Owlet in the present study is similar to the values reported by Mahmood-ul-Hassan *et al.* (2007b). Several factors might contribute to clutch size variability *viz.*, condition of the breeding female, availability of resources necessary to produce eggs, time of laying in the season, anticipated future availability of food for feeding nestlings and natural calamities (Stinson and Stinson, 1994; Asokan *et al.,* 2009b, 2010b; Ali *et al.,* 2010c). The mean brood size of Spotted Owlet in this study is lower than in Pakistan (Mahmood-ul-Hassan *et al.,* 2007).

The overall hatching success in the present study was 83 per cent, and this is more or less similar values obtained by Mahmood-ul-Hassan *et al.* (2007b). During the study period, out of 32 nests studied, 20 nests yielded 100 per cent fledging success with an overall fledging success of 71 per cent. The unusual rainfall during April 2008 and 2009 caused 100 per cent nesting failure in three sites. However, various factors like suitable nesting-sites, prey availability, temperature, moisture and predators may also influence the nesting success of Spotted Owlet.

Artificial wooden nest boxes have been readily used by owls and are widely used in long-term studies of owls (Southern 1970; Marti *et al.,* 1979). Several studies that compared the breeding success of birds using nest boxes and cavities have demonstrated larger clutch sizes (Purcell *et al.,* 1997), lower predation rates (Purcell *et al.,* 1997; Robertson and Rendell, 1990) and more young's fledged (Purcell *et al.,* 1997) for birds using nest boxes. Because of such results, authors have recommended the use of nest boxes as a conservation tool to help augment populations of cavity-nesting birds (Koenig *et al.,* 1992; Purcell *et al.,* 1997).

The Spotted Owlet highly used more frequently the nest boxes at a height of 6-8m level. It is already established that Spotted Owlet preferred to select the roosting, nesting and foraging sites mainly at this height level. One of the main reasons of less use of nest boxes in the study area was due to owlet preferred natural sites than others. For example, nearly 79 per cent of the roosting and 84 per cent of the nesting-sites were recorded in natural trees with dense foliage. Another reason is competitions between other cavity nesters. The Common Myna (*Acridotheres tristis*), Indian Roller (*Coracias benghalensis*) and Rose-ringed Parakeet (*Psittacula krameri*) are common cavity nesters in the study area and often compete for nest boxes. During the study period, the Common Myna and Indian Roller occupied four and one nest boxes respectively and breed successfully. However, Spotted Owlet by selecting the nest boxes in trees and man-made structures are adapting to an adequate substitute for loss of natural sites.

In today's world we have learned that most of the owls superstitious are just stories, born in a time when people were fearful and trying to find answers to their lives and environment. However, many of these legends survived over time. Their wide starring eyes give them a wise appearance, while the ability to turn their head around makes them fascinating and mysterious creature. Tufts of feathers on the top of an owl's head give them the appearance of horned devils and their piercing cries add to the spook effect found in the ancient folklore of many countries.

Although considerable information is available on the cultural values of owls across the world no attempt has been made to list out them in India. In the present study, we have just studied the general knowledge about owls with regard to 183 interviews in relation to various target groups like students, public, farmers, temple workers and land owners. More studies on this aspect at various regions on various owls distributed all over India perhaps shall be very helpful to make their check list and on the basis of that awareness programmes and extension activities could be sort-out.

Acknowledgements

As authors we are thankful to the Ministry of Environment and Forests, Government of India, for providing financial support for this study. We are grateful to the Principal and Man-agement of Saraswathi Narayanan College, Madurai, for facilities and encouragement. We are thankful to Mr. U. Anbarasan and Mr. P. Muthukumar for accompanying us during field trips.

References

Agoramoorthy, G. and Minna, J.H., 2001. Predatory attack on Bats by Barn Owl *Tyto alba* and Shikra *Accipiter badius* in Tamil Nadu state, South India. *Journal of the Bombay Natural History Society*, 93(1): 90.

Ali, A.M.S., Asokan, S. and Manikannan, R., 2010a. Habitat related density and activity patterns of the White–breasted Kingfisher *Halcyon smyrnensis* in Cauvery delta, Southern India. *Podoces*, 5(1): 54–62.

Ali, A.M.S., Asokan, S., Manikannan, R. and Nithiyanandam, G. Thomas, 2010b. Population density and diurnal-activity patterns of Black Drongo *Dicrurus macrocercus* (Aves: Passeriformes) at Cauvery Delta, Tamil Nadu, India. *Global Journal of Environmental Research*, 4(3): 168–174.

Ali, A.M.S., Asokan, S., Manikannan, R. and Nithiyanandam, G. Thomas, 2010c. Nest–site characteristics and breeding biology of Black Drongo *Dicrurus macrocercus* in Cauvery Delta, Southern India. *World Applied Sciences Journal*, 9(11): 1280–1285.

Ali, S. and Ripley, S.D., 1983. *Handbook of the Birds of Indian and Pakistan*. Oxford University Press, London.

Asokan, S., 1995. Ecology of the Small Green Bee-eater, *Merops orientalis* Latham 1801 with special reference to its population, feeding and breeding in Mayiladuthurai, Tamil Nadu, South India. *Ph.D. Dissertation*, Bharathidasan University, Trichy, India.

Asokan, S. and Ali, A.M.S., 2010. Foraging behaviour of selected insectivorous birds in Cauvery delta region of Nagapattinam District, Tamil Nadu, India. *Journal of Threatened Taxa,* 2(2): 690–694.

Asokan, S., Ali, A.M.S. and Manikannan, R., 2009a. Diet of three insectivorous birds in Nagapattinam District, Tamil Nadu, India: A preliminary study. *Journal of Threatened Taxa,* 1(6): 327–330.

Asokan, S., Ali, A.M.S. and Manikannan, R., 2009b. Preliminary investigations on dietary composition and breeding biology of Indian Roller *Coracias benghalensis* in a portion of Cauvery delta, Tamil Nadu, India. *World Journal of Zoology,* 4(4): 263–269.

Asokan, S., Ali, A.M.S., Manikannan, R. and Nithiyanandam, G. Thomas, 2010a. Population densities and diurnal activity pattern of the Indian Roller *Coracias benghalensis* (Aves: Coraciiformes) in Nagapattinam District, Tamil Nadu, India. *Journal of Threatened Taxa,* 2(10): 1185–1191.

Asokan, S., Ali, A.MS. and Manikannan, R., 2010b. Breeding biology of the Small Bee-eater *Merops orientalis* in Nagapattinam District, Tamil Nadu, India. *Journal of Threatened Taxa,* 2(4): 797–804.

Asokan, S., Thiyagesan, K. and Nagarajan, R., 2003. Studies on *Merops orientalis* Latham 1801 with special reference to its population in Mayiladuthurai, Tamil Nadu. *Journal of Environmental Biology,* 24 (4): 477–482.

Balant, J.L., 1993. Nest-site selection and reproductive biology of roof and island nesting Herring Gulls. Trans. *58th N.A. Wildlife and Natural Resources Conferences,* pp. 78–86.

Balasubramanian, P., 1990. Behaviour of Southern Spotted Owlet *Athene brama brama* (Temminck) and Jungle Crow *Corvus macrorhynchos* at Point Calimere, Tamil Nadu. *Journal of the Bombay Natural History Society,* 87: 145.

Barrows, C. and Barrows, K., 1978. Roost characteristics and behavioural thermoregulation in the Spotted Owl. *Western Birds* 9(1): 1–8.

Barrows, C.W., 1981. Roost selection by Spotted Owls: An adaption to heat stress. *Condor,* 83: 302–309.

Beg, M.A., Maqbool, M. and Mushtaq-ul-Hassan, M., 1990. Food habits of the Spotted Owlet, *Athene brama. Pakistan Journal of Agriculture and Science,* 27(2): 127–131.

Bell, H.L. and Ford, H.A., 1990. The influence of food shortage on inter-specific niche overlap and foraging behaviour of three species of Australian warblers (Aanhizid). *Studies on Avian Biology,* 13: 381–388.

Block, M.W. and Brennan, L.A., 1993. The habitat concept in ornithology. Theory and applications. In: *Current Ornithology,* (Eds.) M.D. Power. Plenum Press, New York, 11: 35–91.

Buckland, S.T., Anderson, D.R., Burnham, K.P. and Laake, J.T., 1993. *Distance Sampling: Estimating Abundance of Biological Populations.* Chapman and Hall, London.

Butler, S.J., Whittingham, M.J., Quinn, J.L. and Cresswell, W., 2005. Quantifying the interaction between food density and habitat structure in determining patch selection. *Animal Behaviour,* 69: 337–343.

Carey, A.B., Horton, S.P. and Biswell, B.L., 1992. Northern Spotted Owls: influence of prey base and landscape character. *Ecological Monograph,* 62: 223–250.

Chamberlin, D.E., Wilson, J.D. and Fuller, R.J., 1999. A comparison of bird populations on organic and conventional farm systems in Britain. *Biological Conservation,* 88: 307–320.

Chandrasekhara, S. and Nameer, P.O., 2003. Short–Eared Owl (*Asio flammeus*) in Kerala, India. *Zoos' Print Journal,* 18(10): 1235.

Dabert, J., 1987. Breeding biology of the Feral Pigeon in Poznan, Poland. *Acta Ornithologica,* 23: 177–195.

Hamilton, K.L., 1980. A technique for estimating Barn Owl prey biomass. *Raptor Research,* 14(2): 52–55.

Iyer, G., 1997. Encounter with Brown Fish Owl *Bubo zeylonensis* at Rishi Valley, Chittoor district, Andhra Pradesh. *Mayura,* 14: 69.

Jadhav, A. and Parasharya, B.M., 2003. Some observations on the nesting behaviour and food of the Spotted Owlet *Athene brama. Zoos' Print Journal,* 18(8): 1163–1165.

Jain, P. and Advani, R., 1984. Winter food of Spotted Owlet *Athene brama indica. Journal of the Bombay Natural History Society,* 80(2): 415–416.

Kanakasabai, R., Neelanarayanan, P. and Nagarajan, R., 1996. Sexual dimorphism in Barn Owl (*Tyto alba*). *Newsletter for Birdwatchers,* 36: 55.

King, B.F. and Rasmussen, P.C., 1998. The rediscovery of the Forest Owlet *Athene* (*Heteroglaux*) *blewitti. Forktail,* 14: 53–55.

Koenig, W.D., Adair, G.P. and Dickinson, J.L., 1992. Boxes, barns, and bridges: confounding factors or exceptional opportunities in ecological studies? *Oikos,* 63: 305–308.

Kumar, T.S., 1985. *The Life History of the Spotted Owlet* (*Athene brama brama, Temminck*) *in Andhra Pradesh.* Raptor Research Centre, Hyderabad, India.

Lammers, W. and Collopy, M.W., 2007. Effectiveness of avian predator perch deterrents on electric transmission Lines. *Journal of Wildlife Management,* 71(8): 2752–2758.

Lloyd, H., 2003. Population densities of some nocturnal raptor species (Strigidae) in southeastern Peru. *Journal of Field Ornithology,* 7(4): 376–380.

Lynch, P.J. and Smith, D.G., 1984. Census of eastern screech owls (*Megascops asio*) in urban open–space areas using tape–recorded song. *American Birds,* 38: 388–391.

Mahmood-ul-Hassan, M., Beg, M.A., Mushtaq-ul-Hassan, M. and Rana, S.A., 2007a. Food habits of the spotted owlet *Athene brama* in central Punjab, Pakistan. *Acta Zoologica Sinica,* 53(1): 64–68.

Mahmood-ul-Hassan, M., Beg, M.A., Mushtaq-ul-Hassan, M. and Rana, S.A., 2007b. Nesting and breeding habits of the Spotted Owlet (*Athene brama*) in Punjab, Pakistan. *Journal of Raptor Research,* 41(1): 50–52.

Manakadan, R. and Pittie, A., 2002. Standardized English and Scientific names of the birds of the Indian subcontinent. *Newsletter for Birdwatchers,* 42(3): 1–35.

Marti, C.D., Wager, P.W. and Denne, K.W., 1979. Nest boxes for the management of barn owls. *Wildlife Society Bulletin,* 7: 145–148.

Nagarajan, R., 1998. Nest characteristics, productivity and nestling growth pattern of the Barn Owl *Tyto alba stertens* Hartert, 1929 in an agro-environment of Tamil Nadu, Southern India. *Ph.D. Dissertation*, Bharathidasan University, Trichy, India.

Nagarajan, R., Thiyagesan, K., Natarajan, R. and Kanakasabai, R., 2002. Patterns of growth in nestling Barn-owls. *Condor,* 104: 885–890.

Nagarajan, R., Neelanarayanan, P. and Kanakasabai, R., 1995. Descriptions of nesting sites of Barn Owl (*Tyto alba*) in man–made structures. *Zoos' Print Journal,* 10(5): 15.

Naik, S.A., 2004. Building new bird homes. *Hornbill,* Jan–Mar, pp. 38–39.

Naranthiran, G., 1989. Ecology of Southern Spotted Owlet *Athene brama brama* (Temminck) in Thanjavur District, Tamil Nadu with special reference to its food habit and ecotoxicity. *M.Phil. Dissertation,* Bharathidasan University, Trichy, India.

Neelakantan, K.K., 1971. Calls of the Malabar Jungle Owlet (*Glaucidium radiatum*). *Journal of the Bombay Natural History Society,* 68: 830–832.

Neelakantan, K.K., 1979. The voice of the juvenile Brown Hawk-Owl (*Ninox scutulata*) *Journal of the Bombay Natural History Society,* 76: 363–364.

Neelanarayanan, P., Nagarajan, R. and Kanakasabai, R., 1995. Scope for the utilization of man-made nest boxes to Barn Owl, *Tyto alba. Zoos' Print Journal,* 8(11): 3.

Pande, S., Pawashe, A., Bastawade, D.B. and Kulkarni, P.P., 2004. Scorpions and molluscs: Some new dietary records for Spotted Owlet *Athene brama* in India. *Newsletter for Ornithologist,* 1(5): 68–70.

Pande, S., Pawashe, A., Mahajan, M.N. and Mahabal, A., 2006. Changing nest site preference for holes in earth cuttings in Spotted Owlet *Athene brama. Indian Birds,* 2(1): 7–8.

Pande, S., Pawashe, A., Murlidhar, M.N., Joglekar, C. and Mahabal, A., 2007. Effect of food and habitat on breeding success in Spotted Owlets (*Athene brama*) nesting in villages and rural landscapes in India. *Journal of Raptor Research,* 41(1): 26–34.

Prasanth, J.J., Rao, V.V. and Nagulu, V., 1994. Nesting, egg size, incubation and factors affecting clutch size in Little Egret, *Egretta garzetta* at Nellore Andhra Pradesh. *Pavo,* 32 (1 and 2): 67–72.

Purcell, K.L., Verner, J. and Oring, L.W., 1997. A comparison of the breeding ecology of birds nesting in boxes and tree cavities. *Auk,* 114: 646–656.

Ramanujam, M.C., 2000a. An attempt to rationalize on the vocalizations and displays of captive Indian Eagle Owls *Bubo bubo benghalensis* Franklin. *Zoos' Print Journal,* 15(6): 269–274.

Ramanujam, M.C., 2000b. Food consumption and pellet regurgitation rates in a captive Indian Eagle Owl (*Bubo bubo benghalensis*). *Zoos' Print Journal,* 15(7): 289–291.

Ramanujam, M.C., 2003a. On the "long call" of the Indian Great Horned or Eagle–owl *Bubo benghalensis* (Franklin). *Zoos' Print Journal,* 18(7): 1131–1134.

Ramanujam, M.C., 2003b. Inter-specific intimidatory behaviour in nestling Indian Eagle Owls, *Bubo bengalensis* (Franklin). *Zoos' Print Journal,* 18(10): 1213–1216.

Ramanujam, M.C., 2007. A catalogue of auditory and visual communicatory traits in the Indian Eagle Owl *Bubo benghalensis* (Franklin, 1831). *Zoos' Print Journal,* 22(8): 2771–2776.

Ranazzi, L., Manganaro, A. and Salvati, L., 2002. Density fluctuations in an urban population of Tawny Owl *Strix aluco*: A long-term study in Rome, Italy. *Ornis Svecica,* 12: 63–67.

Ranazzi, L., Manganaro, A., Ranazzi, R. and Salvati, L., 2000. Distribution, population density and spacing of Tawny Owls (*Strix aluco*) in a Mediterranean urban area: the influence of urbanization gradient, woodland cover and fragmentation. *Biota,* 1: 83–92.

Rasmussen, P.C. and Collar, N.J., 1998. Identification, distribution and status of the Forest Owlet *Athene* (*Heteroglaux*) *blewitti. Forktail,* 14: 43–51.

Redpath, S.M., 1995. Habitat fragmentation and the individual: tawny owls *Strix aluco* in woodland patches. *Journal of Animal Ecology,* 64: 652–661.

Rendell, W.B. and Robertson, R.J., 1989. Nest-site characteristics, reproductive success and cavity availability for Tree Swallows breeding in natural cavities. *Condor,* 91: 875–885.

Robertson, R.J. and Rendell, W.B., 1990. A comparison of the breeding ecology of a secondary cavity-nesting bird, the tree swallow *Tachycineta bicolor,* in nest boxes and natural cavities. *Canadian Journal of Zoology,* 68: 1046–1052.

Rohner, P. and Doyle, J., 1992. Resources use and foraging characteristics of great horned owls. *Behaviour,* 25: 23–25.

Santhanakrishnan, R., 1987. Studies on population, food habits and nesting of Barn Owl, *Tyto alba* (Scopoli) in a portion of Cauvery river basin. *M.Phil. Dissertation,* Bharathidasan University, Trichy, India.

Santhanakrishnan, R., 1995. Ecology of Barn Owl, *Tyto alba* (Scopoli) with special reference to its population, feeding and breeding in Mayiladuthurai, Tamil Nadu, South India. *Ph.D. Dissertation,* Bharathidasan University, Trichy, India.

Schueler, F.W., 1972. A new method of preparing owls pellets boiling in NaOH. *Bird Banding,* 43: 142.

Shah, Z.A., Beg, M.A. and Khan, A.A., 2004. Prey preferences of the Spotted Little Owl (*Athene brama*) in the croplands near Faisalabad–Pakistan. *International Journal of Agriculture and Biology,* 6(2): 278–280.

Smith, G.D., Devine, A. and Walsh, D., 1987. Censusing screech owls in southern Connecticut. In: *Biology and Conservation of Northern Forest Owls*, (Eds.) B.W. Nero, R.J. Clark, R.J. Knapton and R.H. Hamre. USDA Forest Service General Technical Report, RM 142.

Sokal, R.R. and Rohlf, F.J., 1981. *Biometry*. W.H. Freeman and Company, New York.

Southern, C.J., 1970. The natural control of a population of tawny owls (*Strix aluco*). *Journal of Zoology* (*London*), 162: 197–285.

Sparks, E., Belthoff, J. and Ritchison, G., 1994. Habitat use by eastern screech owls in central Kentucky. *Journal of Field Ornithology*, 65: 83–95.

Srinivasulu, B. and Srinivasulu, C., 2007. Diet of Short-Eared Owl *Asio flammeus* (Pontopiddan, 1763) wintering in Rollapadu Wildlife Sanctuary and its vicinity in Andhra Pradesh, India. *Zoos' Print Journal*, 22(9): 2829–2831.

Sugathan, R. and Jacob, K.C., 1995. Further records of the Bay Owl from Kerala. *Newsletter for Birdwatchers*, 35: 77–78.

Tariq, S.A., Ghazi, R.R. and Nisa, N., 2003. Study on the food of Spotted Owlet (*Athene brama*) at an Agricultural Farm Gharo, Lower Sindh, Pakistan. *Journal of Natural History and Wildlife*, 2(1): 31–33.

Tome, R., Bloise, C. and Korpimaki, E., 2004. Nest-site selection and nesting success if little owls (*Athene noctua*) in Mediterranean woodland and open habitats. *Journal of Raptor Research*, 38(1): 35–46.

Trango, L.A., Valdez, R., Clemente, F. and Mendoza, G., 2001. Roost-site characteristics of Mexican Spotted Owls in Sierra Fria, Agua Scalientes, Mexico. *Journal of Raptor Research*, 35(2): 165–168.

Vanitha, V. and Kanakasabai, R., 2009. Prey selection by the Barn Owl *Tyto alba* (Scopoli, 1769) in captivity. *Journal of Threatened Taxa*,1(7): 361–365.

Verzhutskii, B. and Ramanujam, M.C., 2002. On the prey of the Collared Scops Owl *Otus bakkamoena* (Pennant) at Auroville, Pondicherry. *Zoos' Print Journal*, 17(11): 939–940.

Ward, J.P. Jr., Gutierrez, R.J. and Noon, B.R., 1998. Habitat Selection by Northern Spotted Owls: the consequences of prey selection and distribution. *Condor*, 100: 79–92.

Yalden, D.W., 2003. *The Analysis of Owl Pellets*. The Mammal Society, London.

Young, K.E., Valdez, R., Zwank, P.J. and Gould, W.R., 1998. Density and roost-site characteristics of Spotted Owls in the Sierra Madre Occidental, Chihuahua, Mexico. *Condor*, 100: 732–736.

2013, Animal Diversity, Natural History and Conservation, Vol. 3 *Pages* **323–362**
Editors: **V.K. Gupta and Anil K. Verma**
Published by: **DAYA PUBLISHING HOUSE, NEW DELHI**

Chapter 18

Natural History of the Little Known Endemic Mammal *Herpestes palustris* in Southern West Bengal, India

Jayanta Kumar Mallick*
Wildlife Wing, Directorate of Forests,
Government of West Bengal, Bikash Bhawan, North Block,
Third floor, Salt Lake City, Kolkata – 700 091, W.B., India

ABSTRACT

The little known Bengal Mongoose *Herpestes palustris* Ghose, 1965 is a wetland-dependent peri-urban mammalian species of India, which was previously considered a subspecies of *H. javanicus* or conspecific with *H. auropunctatus.* In the country, it is the only endangered mongoose species endemic to West Bengal, where it is restricted to a tiny native range spread over three southern districts (Howrah, North and South 24-Parganas) on both banks of the River Hooghly (a distributary channel of the Ganges) in its lower course. This species is well adapted to the non-saline, natural or manmade wetland habitat and lead a semi-aquatic life. But this landscape has been much converted for the agriculture, industries and human settlements but degraded, fragmented and greatly shrunk during the last four decades. The available literature does not provide an exclusive account of its natural history. Hence, a study was conducted in 2010–11 through literature review, questionnaire-based interview and field survey in this region, in order to build up a new knowledge-base on the natural history of *H. palustris.*

Keywords: *Herpestes, Behaviour, Diet, Diurnal, Habitat, Reproduction, Stress, Wetlands.*

---

* *Corresponding author.* E-mail: jayantamallick2007@rediffmail.com

Introduction

About 11 per cent of the Indian mammalian species are endemic, with the Western Ghats being the prime endemic zone, followed by the Deccan Peninsula, and also the Andaman and Nicobar Islands. For the majority of these endemic species, the detailed information regarding their natural history is not available. Many species are known by the type series or even by a single type specimen, and have not been recorded since their original description. Similarly, endemism in the wetland fauna is poorly known. Among the vertebrates, only one specimen of mammal species, namely *Herpestes palustris,* 44 species of water birds, one species of Himalayan Newt *Pleurodeles verrucosus* and 223 species of fish are known to be endemic to the Indian wetlands (Alfred and Nandi, 2000). In fact, the mammal endemicity is only marginal in West Bengal (Mallick, 2007).

There appears to be 37 recognised species (18 genera) of mongooses (with the meerkat included) in the world. Traditionally, the mongooses were included in the family Viverridae (the family of civets and genets), but now they fall into their own family Herpestidae Bonaparte, 1845. In fact, the Herpestids are differentiated from the Viverrids by their rounded ears and the presence of anal scent glands (not perianal glands) and associated sac.

The phylogeny of the Herpestidae shows that the Asian herpestids form a monophyletic clade nested within an African family. The age of the clade encompassing Asian Herpestidae (the genus *Herpestes*) is not well established, but the analysis of Gaubert and Cordeiro (2006) suggests that it is less than 10 Mya, and thus younger than the African clades of Viverridae. On the other hand, the family Herpestidae, as a whole, goes back to the Early Oligocene (about 34 million to 23 million years before the present).

The Ethiopean family Herpestidae is represented in India by one genus *Herpestes* Illiger, 1811. This genus is characterised by (a) a longish body; (b) short and rounded ears, mostly concealed by hairs; (c) the contour hairs banded with pale and darker annulations, giving a speckled appearance and (d) post-dental portion of palate extending far back half way over the mesopterygoid fossa.

There are thirteen species of the genus *Herpestes* in the world, out of which seven are found in India. The latest of them is *H. palustris* Ghose, 1965. But, there were "grey" areas in its taxonomy. Earlier, most of the popular Indian mammals' guide books did not record it as a valid species. Internationally, *H. palustris* is considered a least known and data deficient species, even if it was discovered in the mid-1960s by the scientists of the Zoological Survey of India. However, at the national level, the proposal of a new species of *Herpestes* has been accepted by two topmost research organisations of the Government of India, *viz.* Zoological Survey of India, Kolkata and Wildlife Institute of India, Dehradun as well as many other authorities. Unfortunately, there was considerable lack of motivation for due updation or attempting a further systematic study to explore the unknown facts about the species in its entire native range.

Since adequate data were not forthcoming at the time of taxonomic assessment, the species status of the Bengal Mongoose (*H. palustris*) was omitted by *The IUCN Red*

List of Threatened Species (where it is treated as a synonym of the widespread and non-threatened Small Asian Mongoose, *H. javanicus*) (Wozencraft *et al.*, 2008). This has resulted in a conservation dilemma of this exclusively wetland species: "… if Bengal Mongoose is a valid taxon, then it is probably highly threatened, but if it is not valid, then these southern (West) Bengal mongooses are irrelevant to conservation at all but the most local scale" (Mallick, 2011).

Out of seven species under Herpestidae in India, Molur *et al.* (1998) considered only *H. palustris* as 'Endangered', based on five criteria, *viz.* B1: restricted distribution, 2a: limited location, 2b: continuing decline observed in extent of occurrence, 2c: area of occupancy and 2d: quality of habitat and number of mature individuals, while other two herpestids (*H. fuscus* and *H. urva*) are 'Vulnerable' and rest of the four (*H. edwardsii, H. javanicus, H. smithii* and *H. vitticollis*) 'Least Concern'. Walker (1999) and Srinivasulu and Nameer (2003) also recorded *H. palustris* as 'Endangered'. Further, in a news entitled "Nature India Informatics", published on 1 December 2010 by the Commission on Education and Communication, IUCN, Samik Gupta, while reporting killing of the critically endangered *H. palustris* by considering *H. edwardsii,* urged that "Let us know the species first and then to conserve it".

The expert committees recommended undertaking taxonomic and morphological genetic studies, survey, monitoring, limiting factor research and life history studies on *H. palustris* (Molur *et al.*, 1998). Hence, at this stage, dissemination of authentic knowledge on the natural history of *H. palustris* is of utmost importance from the conservation point of view.

The original document (Ghose, 1965) gave only hints on a few aspects of the natural history of the species. Although a few other sources have dealt with this species subsequently during the twentieth century (Ghose and Chaturvedi 1972; Agrawal *et al.*, 1992), those pertain to its distribution and range-extension than detailing the other aspects of its life history. However, during the twenty first century, a number of studies have been conducted on the species. Dey (2007) and Deuti (2008) gave a brief account of the habits, food preference and family life of the Bengal Mongoose. The status of this species was also documented (Mallick, 2009). Recently, *H. palustris* is included in a handbook of the mammals of South Asia (Bahuguna and Mallick, 2010). The species was last reviewed by Mallick (2011).

Study Area

The study area covers the non-saline wetlands in Greater Kolkata comprising many large and small water bodies in North and South 24-Parganas districts and a few in Howrah district. The area surveyed is shown in Figure 18.1. This area falls in the 'Lower Gangetic Plains' (7B) bio-geographic zone (Rodgers and Panwar 1988) and is part of the lowland alluvial plains. It was entirely a marshy wetland in times past. Later, many of the wetland areas were reclaimed for agriculture, urbanization and industrialisation. Today, it is a mixed wetland-grassland ecosystem having patches of the typical habitat of Bengal Mongoose (Figures 18.2 and 18.3), which harbour numerous vertebrate and invertebrates needed for sustenance of this species.

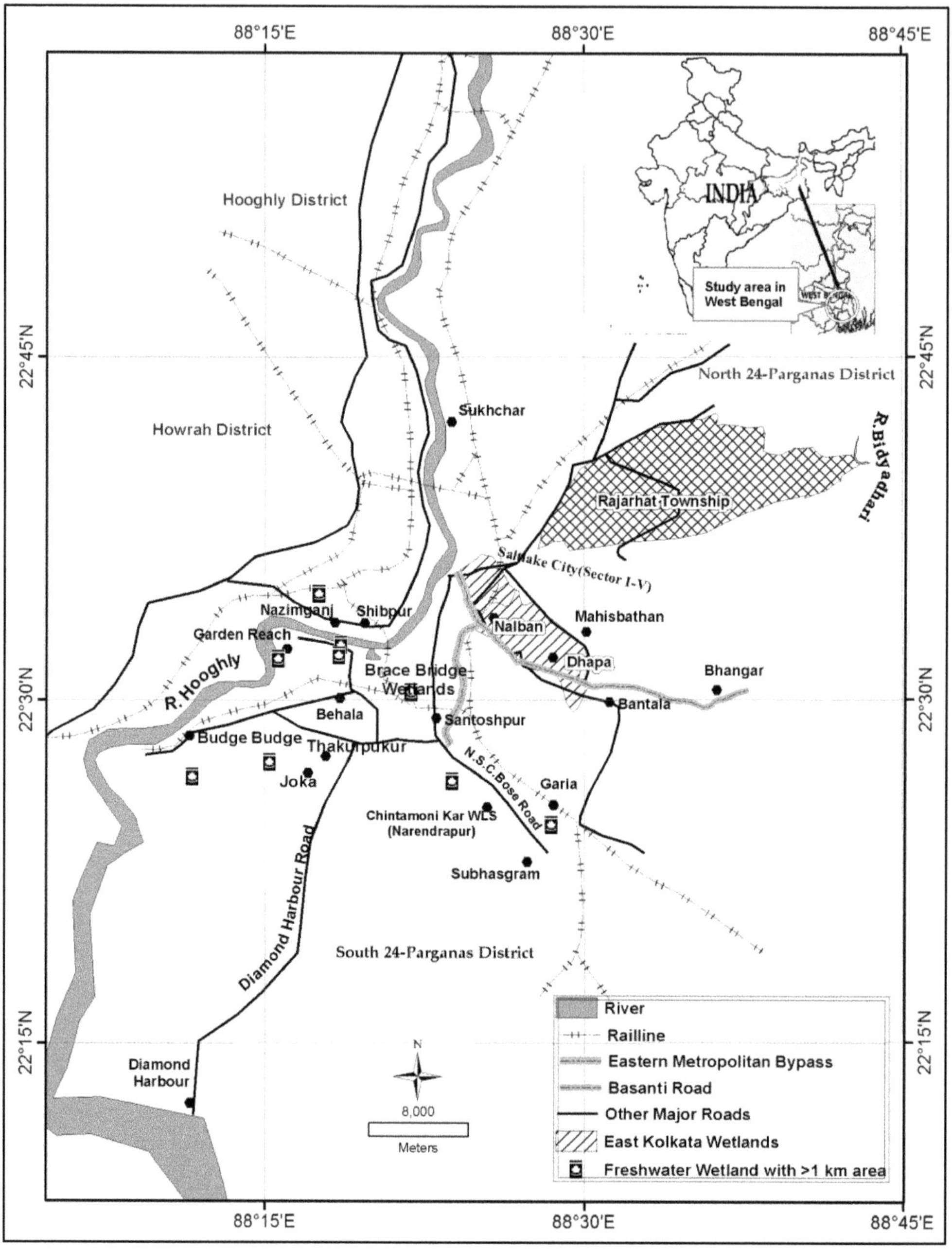

Figure 18.1: Map Shows the Surveyed Area

In the study area, the entire East Kolkata Wetlands (22°25'–22°40'N, 88°20'–88°35'E) was designated a "wetland of international importance" under the Ramsar Convention (n° 1208) on August 19, 2002. This funnel-shaped area extends over 125

Figure 18.2: *Herpestes palustris* (Indian Marsh Mongoose)

Figure 18.3: Another Specimen of *Herpestes* Showing Characteristic Morphometric Details

km^2 with 268 water bodies on both sides of a dry-weather flow channel discharging into the Kulti Gong (the waste-water outfall), including about 188 *bheries* (pisciculture area) in Bidhannagar, Bhangar, Tiljala and Sonarpur areas under North and South 24-Parganas districts. This marsh is composed of an inter-distributary (encased by the distributary and tidal channel) wetland in the mature delta of the River Hooghly. Originally, it was a vast region of tributaries and distributaries extending between the River Hooghly on the west and the River Bidyadhari (presently a derelict canal) on the east (Kundu, 2010). These wetlands sustain the world's largest and oldest integrated resource recovery practice that combines both agriculture and aquaculture by using the refuse and waste-waters of Kolkata metropolis (Kumar, 2010).

The climate is hot (temperatures ranging from 40°C to 10°C), humid and monsoonal, with the average yearly rainfall at about 1600 mm (mainly between mid-June and mid-October). January is the coolest month, while May is the hottest. Only those wetlands, which are known habitats of the Bengal Mongoose, are described below.

Nalban, North 24-Parganas District

Nalban is the largest wetland of East Kolkata Wetlands (Deuti, 2008), covering 167.14 ha and composed of a few large *bheries* (22°33′N, 88°25′E). It is bounded by the Eastern Metropolitan Bypass on the west, Basanti Road on the south and Salt Lake City (Bidhannagar) on the north and east. It is owned by the Fisheries Department of Government of West Bengal and used for pisciculture, where a cooperative was started in mid-1980s. Compared with the vastness of open water, few patches of aquatic plants have been left. The periphery of the *bheries* is covered with aquatic vegetation like Water hyacinth. Most of the recent studies of Bengal Mongoose were conducted in N° 4 bheri (Dey, 2007; Deuti, 2008; Mallick, 2009).

Sahebmara, North 24-Parganas District

The privately-owned Sahebmara is the second largest bheri of East Kolkata Wetlands, spread over 147.18 ha (22°33′N, 88°25′E). Its ecology closely resembles that of Nalban, but the habitat has been much degraded.

Rajarhat, North 24-Parganas District

This part of East Kolkata Wetlands is nearer to the erstwhile spill-over basin of the River Bidyadhari on the east (22°37′N, 88°25′E). Keshtopur, Bagjola and three other canals pass through this area. Only a few small wetlands with aquatic vegetation remain, including Khari Bari *bheri*, Diller *bheri*, Narkeldanger *bheri* and Mollar *bheri*. Due to development of New Town at Rajarhat during the recent years, the Bengal Mongoose habitat has further reduced.

Acharya Jagadish Chandra Bose Indian Botanic Garden, Howrah District

This Government-owned National Garden, covering an area of 1.09 km^2, is located at Shibpur (22°33′N, 88°18′E) on the west (right) bank of the River Hooghly. The garden has many exotic tropical plants and many indigenous species. It has 27 large and medium ponds, a few connecting with the River Hooghly. These ponds contain more than two dozen fish species. The garden is rich in molluscs, aquatic bugs and beetles, dragonfly (Odonata) nymphs, grasshoppers and crickets (Orthoptera), millipedes/centipedes (Myriopoda) and crabs (Decapoda) (A. Chatterjee *in litt.* 2010).

Santragachhi, Howrah District

Out of 24 ha area of Santragachhi (= Santragachi) (22°34′N, 88°16′E), the 'Makal *jheel*' (lake) is spread over 12.8 ha, roughly rectangular in shape (length about 915 m, width 305 m, perimeter 2,418 m and depth 4–7 ft), at an altitude of 8 m. It lies south of Santragachhi railway yard, about 20 km east of the East Kolkata Wetlands, and is surrounded by dense human habitations, *khatals* (cow sheds) and small-scale industries. The *jheel* has several small islands covered with prolific Water hyacinth and large trees along its banks. *Typha latifolia* is moderately common, but *Phragmites communis* is rare. The land is owned by the South-Eastern Railways and has been managed by the State Forest Department since 1992. The *jheel* receives various domestic and industrial sewages and refuse-dumping is engulfing part of the wetland. It supports thousands of resident and migratory birds, and a wide variety of zooplanktons, mollusks and fishes.

Joka, South 24-Parganas District

Joka (22°26′N, 88°17′E) is a semi-urban area at the edge of south-west Kolkata, where the Indian Institute of Management, Calcutta (IIMC) is located to the east of the arterial Diamond Harbour Road (NH 117). Just behind IIMC lies a government-owned tract of fallow mixed wetland-grassland (54.63 ha), rich in wild plants and animals (Raghu Ram, 2005). The area was entirely a marsh until much was converted for agriculture and towns, mainly during the early 1970s. The east–west Charial *khal* (canal) forms the northern boundary of the wetland. The grassland starts a short way down the Yani Sarani (dirt road) and is dominated by *Phragmites, Saccharum, Erianthus, Imperata, Desmostachya* and *Achyranthes*. The non-saline wetland is seasonally covered by *Typha, Sagittaria, Cyperus, Cryptocoryne, Acrostichum, Ipomoea, Eleocharis* and others. There is another wetland (area <1 ha), called Green View, near Joka Goddess Kali temple (about 3 km from IIMC) with some reed beds, grasses and aquatic plants.

Kadamtala, South 24-Parganas District

Kadamtala is part of the satellite township of Behala (22°29′N, 88°18′E). A few privately-owned lakes, including Katapukur, Harirpukur, Majhi's pond (totaling an area <1 ha), surrounded by aquatic vegetation, survive.

Survey Park, South 24-Parganas District

Survey Park (22°29′N, 88°23′E), Ajaynagar, lies at the eastern end of Santoshpur, bounded by Eastern Metropolitan Bypass on the west, about 5 km south of Salt Lake City. Of late, a satellite township has been developed here. The area has many big ponds and lakes (privately-owned) including Kalibaripukur, Nilpukur, Haludpukur, Kaanchpukur, surrounded by aquatic vegetation. There are many big *bheries* a few kilometers away at Kalikapur and Mukundapur on the eastern side of the E.M. bypass (22°29′32"N, 88°23′48"E).

Formerly, this area was a part of the Brace Bridge Wetlands, which were distributed over 12.35 km^2 (a 14 km stretch from Santoshpur on the south to Majherhat railway station on the north). Regular inflow of water from the River Hooghly at high tide with very low salinity and outflow during low tide transformed 70 per cent of the original swamp into a low vegetation marshland. This vast swamp witnessed the impact of human activities in the post-independence period (1947 onwards). Systematic removal of *Typha* and *Phragmites* and cutting of the native flora like *Aeschynomene,* and insanitary land-filling with refuse and solid waste reduced the quality of water drastically and gradually changed the area to a human-made ecosystem over the years (Ghosh and Nandi, 2009).

Subhasgram, South 24-Parganas District

Subhasgram (Changripota) is located 2.8 km south-east of Sonarpur on the left side of Netaji Subhas Chandra (N.S.C.) Bose Road (22°24′N, 88°26′E). A few privately-owned *jheels* (small, well-covered with aquatic vegetation) and gardens survive nearby, particularly west of the railway lines, where human habitations are scattered. But, the newly developed satellite townships have almost destroyed the wetland habitat on the eastern side.

Material and Methods

To initiate the study, the stuffed rolls of specimens collected by Zooogical Survey of India, Kolkata, were observed to understand the major morphological differences between the Bengal Mongoose and the Small Indian Mongoose. To explore the different aspects of natural history of Bengal Mongoose in the freshwater wetland habitats in three districts (Howrah, North and South 24-Parganas) of south Bengal, known to support it, the previous records were collated through a literature survey in addition to sorting out the unpublished data in the archive of the Wildlife Wing (Headquarters) in the Forest Department of Government of West Bengal and conducting a questionnaire-based interview with the resource persons and local inhabitants as well as field survey during 2010 and 2011 (spending about 400 person-days, on an average 12 hours/day) with the help of S. Mallick, S. Bhattacharya, S. Jha and some others, forest department staff and members of the non-government organisations working in the locality.

The total area surveyed was about 200 km^2 in extent, including the known sites at Sukhchar (22°43′N, 88°22′E) and East Kolkata Wetlands, North 24-Parganas (north-eastern part of the Bengal Mongoose range), Shibpur in Howrah (western part), Diamond Harbour, Budge Budge (Bajbaj) and Patiatala (exact locations referred to in the previous literature could not be traced) as well as Joka, South 24-Parganas (southern part). The new sites with similar marshes within the native range of Bengal Mongoose beyond East Kolkata Wetlands, which had not been surveyed for the species earlier, and the habitats were also identified. These habitats are located around: (1) Rajarhat wetlands, east of East Kolkata Wetlands next to Salt Lake City; (2) the Eastern Metropolitan Bypass up to Garia; (3) the State Highway (N.S.C. Bose Road) to Baruipur via Kamal Gazi, Narendrapur and Sonarpur; (4) Diamond Harbour Road (National Highway 117) via Behala up to Joka; and (5) Santragachhi wetlands, 20 km west of East Kolkata.

The narrow mud-banks on shallow wetlands or fisheries (*bheries*) were specially monitored for locating the burrows, dug or occupied by the Bengal Mongoose along these banks and noting the morning exit of mongooses and their evening return. Since the Bengal Mongoose is a known diurnal species (Dey, 2007; Deuti, 2008), the present survey was scheduled from 06:00 h to 18:00 h. The presence of the species was recorded by direct sighting and indirect evidences.

As because the Bengal and Small Indian Mongooses are considered sibling species in the study area, periodical trapping was done in two phases- (i) during 2005–06 and (ii) 2010–11 by Nature, Environment and Wildlife Society (NEWS) for proper identification of the species. Initially, trapping permission for 480 trap hours was given by the State Forest Department. Random trapping had been done using the wire traps having partition (Figure 18.4). Chicks of the country chickens were used as baits and put inside the smaller inside compartment of the trap. The traps were placed near the identified burrows of the Bengal Mongoose under the camouflage of leaves of Water hyacinth *Eichhornia crassipes*. After laying the traps, the field workers took position behind a hide and observed the traps from a distance of at least 20 feet and used binocular, whenever necessary. In every month, six to eight days were spent for the trapping purpose. The areas of trapping were mainly located in the North Salt lake *bheri* (Nalban) areas and partially in Bhojerhat and Bhangar area.

For measuring the exact length and weight, the trapped mongoose were immobilized by Chloroform (this practice was discontinued during the second phase for safety). Then its weight and length were measured mechanically. Before releasing the trapped animal in its natural habitat, it was marked with black hair dye to avoid measuring it again, if trapped later. The whole operation from setting the trap until release usually took about three hours.

Figure 18.4: Wire Traps Used for Capturing the Bengal Mongoose

Figure 18.5: Burrows Constructed for Nesting by the Bengal Mongoose during Present Investigation

To estimate the population status of Bengal Mongoose in the sample site at N° 4 bheri, Nalban, the burrow count method was followed. For this purpose, prior field observations were made to determine the differences between the burrows of Large Bandicoot Rat (*Bandicota indica*) and Bengal Mongoose. It may be mentioned that the Bengal Mongoose do not form a dense colony unlike the Bandicoots. Whereas piles of earth surrounding the burrows (8 to 15 cm in diameter), resembling large molehills, was an indication of presence of this large rat, the burrows of the Bengal Mongoose were devoid of any such pile. Moreover, the tunnel of the Bandicoot's burrow is of a fairly constant diameter throughout, the tunnel of the Bengal Mongoose's burrow is short-mouthed and strategically wider inside but not compartmentalized (Figure 18.5).

Based on the observed morning exit and evening return of the Bengal Mongoose, the team was able to select their burrows only on the slopes of canal banks and *bheries*, and closed the mouths of the burrows with clay particles in the late evening, when all mongooses were assumed to have entered the burrows. Next day, in the early morning (7:30 h–8:00 h), the team again observed the same stretch and counted the number of burrows that were already opened.

Results and Discussion

Taxonomy and Nomenclature

Out of thirteen species of the genus *Herpestes* (Illiger, 1811), literally meaning the "creeping thief" in Greek, in the world, six species were known to occur in India up to early 1960s–(1) Small Indian Mongoose *H. auropunctatus* Hodgson, 1836 (synonym *javanicus* É. Geoffroy Saint-Hilaire, 1818), (2) Grey Mongoose *H. edwardsi* É. Geoffroy Saint-Hilaire, 1818, (3) Indian Brown mongoose *H. fuscus* Waterhouse, 1838, (4) Ruddy Mongoose *H. smithii* Gray, 1837, (5) Crab-eating Mongoose *H. urva* Hodgson, 1836 and (6) Stripe-necked Mongoose *H. viticollis* Bennett, 1835. Of them, three species (serial numbers 1, 2 and 5) represent West Bengal.

During the course of a bird and mammal survey in the Salt Lakes (when it was an extensive swampland on the eastern fringe of Calcutta= Kolkata; later converted to a city), commenced since 1962, a new (seventh) species of mongoose having distinctive morphological characteristics was discovered in the non-saline and peri-urban wetlands of southern West Bengal, east India, that was named *Herpestes palustris* by Ghose (1965) of ZSI. He gave a systematic account of this new species as follows:

Holotype: Z.S.I. Reg. No. 17143: an adult male (in the light form or colour), Nalban (= Nalbani), *c.* 8 km east of Calcutta, North Salt Lake, 24-Parganas district (North), collected on 21 March 1965 by B. Biswas, the skin and skull of which were deposited to the National Zoological Section, ZSI, Calcutta.

Paratypes: n= 32: 20♂ and 8♀ were collected from Sukhchar, Salt Lake, Nalban, Bantala, Duttabad and Hederhat (= Hediar Hat), 1♂ from Shibpur by B. Biswas as well as 1♂ from Bhasna by Y. Chaturvedi, 1♂ and 1♀ from Nazimganj (south of Shibpur) by R.K. Ghose during the period 21 March 1964 to 27 December 1965.

It is the only species of the genus, which is endemic to India or, more precisely, a tiny range in Howrah, North 24-Parganas and South 24-Parganas districts tied to the fragmented and degraded non-saline wetlands, as referred to by the Indian authorities (Ghose, 1965, Soota and Chaturvedi, 1970; Ghose and Chaturvedi, 1972; Agrawal, 1998; Agrawal *et al.*, 1991, 1992; De, 1994; De *et al.*, 1998; Molur *et al.*, 1998; Hussain, 1999; Sanyal, 1999; Walker, 1999; Walker and Molur, 1999; Alfred and Nandi, 2000; Chattopadhyay, 2001; Alfred and Chakraborty, 2002; Alfred *et al.*, 2002; Chaudhuri and Sarkar, 2003; Srinivasulu and Nameer, 2003; Anonymous, 2004; Raghu Ram, 2005; Nandy, 2006; Dey, 2007; Mallick, 2007; Sanyal *et al.*, 2007; Deuti, 2008; Kundu *et al.*, 2008; Basak, 2009; Deb, 2009; Ghosh, 2009; Mallick, 2009; Sahajpal *et al.*, 2009; Bahuguna and Mallick, 2010; Chakraborty and De, 2010; Mallick, 2011; Sharma, 2011). However, Nandi *et al.* (1993) did not report this wetland species.

Besides, the validity of the species is also recognised in many international sources (Ewer, 1973; Schreiber *et al.*, 1989; Wozencraft, 1993; Van Rompaey and Colyn, 1996; Bininda-Emonds *et al.*, 1999; Veron *et al.*, 2004; Feldhamer *et al.*, 2007; Isaac *et al.*, 2007; Veron *et al.*, 2008).

On the contrary, a few authors (Prater, 1971; Nellis, 1989; Taylor and Matheson, 1999; Veron *et al.*, 2006) did not even mention it, at neither the specific nor subspecific level. Previously, *i.e.*, before conducting the microscopic and macroscopic analysis of the dorsal guard hairs of *H. palustris* and comparison with those of the extant species of the genus *Herpestes* (De *et al.*, 1998), its taxonomic status was controversial because some doubted it as a subspecies of *H. javanicus* (Wenzel and Haltenorth, 1972; Menon, 2003), a synonym of or conspecific with *H. auropunctatus* (Honacki *et al.*, 1982; Corbet and Hill, 1992; Wozencraft, 2005; Wozencraft *et al.*, 2008; Gilchrist *et al.*, 2009), without detailing any cogent reason for doing so.

Several lines of plausible evidences support its specific status and there seems to be no credible, let alone compelling, refutation of its status as a species or any published case that it is not a valid taxon. Though *H. auropunctatus* is the nearest relative of *H. palustris,* the typical morphological characters of the Bengal Mongoose, its distinctive profile of the dorsal guard hairs, obnoxious odour, occurrence in different ecological niche, the short duration of foraging, selective food habits, skill in semi-aquatic hunting and apparent absence of hybridisation with *H. auropunctatus* tend to indicate that the two species are siblings (Ghose, 1965).

In two separate studies conducted in 2005–2006 (Dey, 2007; Deuti, 2008), the originally identified characteristics of the holotype (Ghose, 1965) and paratype (Agrawal *et al.,* 1992) have been reestablished. Further, on the basis of the recent DNA-based comparative laboratory analysis of the guard hairs of five Indian mongoose species by ZSI, the entity of Bengal Mongoose as a distinct species has been proved (De *et al.*, 1998). Following Ghose (1965), Soota and Chaturvedi (1970), Ghose and Chaturvedi (1972), Agrawal *et al.* (1992), De *et al.* (1998), Dey (2007), Deuti (2008), Mallick (2007, 2009, 2011), and Bahuguna and Mallick (2010), *H. palustris* is treated here as a taxonomic reality at the level of species.

Locally, *H. palustris* is known by different names, of which 'Bengal Mongoose' is considered the most appropriate one (Mallick, 2011). It is called *Jalabhumir Neul* (in Bengali). Hussain (1999) named it 'Indian Marsh Mongoose' perhaps to highlight that it inhabits the marshy land just like another herpestid of the sub-Saharan Africa, having a different genus and species name (taxon), *Atilax paludinosus* Cuvier (1829), with similar preference for the permanent freshwater habitat bordered by dense vegetation, such as marshes, reed beds and estuary. It is also called 'Salt Lake Mongoose' because it is a native species of the erstwhile Salt Lake, presently its prime habitat, which was an extension of the brackish Sundarbans, the area was being fed through the River Bidyadhari up to the nineteenth century; but, due to siltation, it ceased to be functional since 1930s and, gradually, converted to a sewage-cum-rainwater fed freshwater habitat.

Museum Specimens

ZSI, Kolkata, collected as many as 32 specimens during 1960s (Agrawal *et al.,* 1992), which are preserved by them in their National Zoological Section. Specimens (skin, skull and incomplete postcranial skeleton) of a male and two unsexed Bengal Mongooses are reportedly held by the Museum Victoria, Australia (registration nos 4388/4389; 4838/4839/4840; 4920/4921), which were supplied by the Royal Melbourne Zoo (N.W. Longmore, Collections Manager, Ornithology and Mammalogy, *in litt.* 2011), but their origin is not known to the authorities. However, the Bombay Natural History Society (BNHS) confirmed that it holds no specimens of *H. palustris* or of *H. (j) auropunctatus* from within the range of *H. palustris* (S. Bajaru, Research Assistant, BNHS, *in litt.* 2011). Similarly, the Bengal Natural History Museum, Darjeeling, does not hold or display any specimen of the Bengal Mongoose (Saha and Singhal, 1996). No other information on the museum (national or foreign) specimens of *H. palustris* is readily available.

Range/Distribution

Since its discovery, *H. palustris* has been reported from over a dozen sites in southern West Bengal (*i.e.*, its entire world range) (Mallick, 2011). Since the proposal of *H. palustris* as a valid species, for which materials were examined from Duttabad (= Dattavad; north of present East Kolkata Wetlands), Hederhat (= Hedearhat; between Ka-likapur and Mukundapur, south of East Kolkata Wetlands) and Nalban (= Nalbani), all east of the Eastern Metropolitan Bypass, in erstwhile undivided 24-Parganas district, there have been several exten-sions of its known range. Agrawal *et al.* (1992) recorded collection of speci-mens from the Botanical Gardens at Shibpur and from Nazim-ganj (both in Howrah [= Haora] district) and from Bantala (= Ban-tal; south-east of the East Kolkata Wetlands) and Sukhchar (both in present North 24-Parganas district). In addition, three reports of its occurrence in South 24-Parganas district are available, but only that of Ghose and Chaturvedi (1972) was supported by a specimen, or by any detail of the record(s). Molur *et al.* (1998) and Raghu Ram (2005) simply stated that the species occurred in a few new areas (Budge Budge, Patiatala, Diamond Harbour and IIM, Joka) without any sighting record or examining any species; so these are considered as only unconfirmed reports. Moreover, two reports have been published on its prime habitat (Nalban) in the East Kolkata Wetlands. The first study was undertaken by ZSI in 2005 (Deuti, 2008) and the second by a Kolkata-based non-governmental organisation (NGO), the Nature, Environment and Wildlife Society (NEWS), in 2005–2006 (Dey, 2007). Although many other sources have mentioned the species, these are the only five presenting original records of distribution or natural history.

In addition, Mallick (2009) compiled the records available since Bengal Mongoose's original description (Ghose, 1965), including more details on NEWS' survey results (2005–06) than is avail-able in Dey (2007). No further field study has been undertaken during 2006–2009 to investigate the current global status of Bengal Mongoose and assess its conservation needs.

Of late, a few local bird-watchers and photographers like S. Mallick, S. Jha, S. Bhattacharya, A. Chatterjee, H.G. Mukho-padhyay, P.K. Biswas, L. Barman, S. Sen and A. Das have photographed a number of mongoose in the surroundings of a few fragmented wetlands of Howrah, North and South 24-Parganas districts, all within Greater Metropolitan Kolkata, though some of them misidentified the species as the Small Indian Mongoose.

During 2010–11, the Bengal Mongoose is recorded from three sites with previous records or claims: Acharya Jagadish Chandra Bose Indian Botanic Garden, Shibpur in Howrah (1.09 km^2); Nalban fisheries of East Kolkata Wetlands in North 24-Parganas (1.67 km^2); and Joka (south-west Kolkata) in South 24-Parganas (about 0.55 km^2) and also found in five new localities, all peri-urban habitats within Greater

Kolkata, three in South 24-Parganas (Survey Park, Ajaynagar [Santoshpur], close to the Eastern Metropolitan Bypass; Kadamtala [Behala]; and Subhasgram, next to Sonarpur), one in North 24-Parganas (Rajarhat)- the wetland in none of these newly recorded habitats is >1 ha in extent, but, in one case, *i.e.*, Santragachhi *jheel* in Howrah, it is spread over a comparatively large area (12.8 ha) (Mallick, 2011). Mallick (2012) also recorded a new habitat of Basail, North 24-Parganas.

The Bengal Mongoose was not found in 27 other fisheries (covering about 11 km^2 in total) surveyed in the East Kolkata Wetlands close to Nalban (Table 18.6), but it may be safely concluded that it is very rare in these areas. In addition, neither the Bengal Mongoose nor its signs were found at Dhapa (East Kolkata Wetlands), which extends over 34.2 ha, the dumping ground of solid wastes of Kolkata and its neighbourhood. No other mongoose species was seen in or reported from any wetland surveyed so far in the East Kolkata wetlands.

In Hooghly (21°56′N, 88°04′E), on the right bank of the River Hooghly, north of Howrah, the native place of the author, which supports large marshes, he found no *H. palustris*: all animals (about 100) sighted (opportunistically; there were no systematic observations) during last 40 years, including in marshes, were Small Indian Mongooses.

The Bengal Mongoose was also not reported around eight major wetlands within the Kolkata Municipal Corporation, situated in between East Kolkata Wetlands on the east to the River Hooghly on the west, where the aquatic vegetation has been destroyed due to various anthropogenic stresses. They are- Brace Bridge *Jheel* (65 ha), Rabindra Sarobar (29.5 ha), Victoria Memorial Hall Pond (24.2 ha), Subhas Sarobar (16 ha), Alipore Zoo Garden Pond (4 ha), Eden Gardens *Jheel* (4 ha), Lal Dighi (1.2 ha) and College Square Pond (1 ha).

From 1964 to 2005, the records of Bengal Mongoose ranged from Sukhchar on the north to Bantala (on the left side of Basanti Road) on the east, Shibpur on the west (with provisional reports from Budge Budge [22°28′N, 88°10′E] and Joka [(22°26′N, 88°17′E]) and Bhasna (exact location not known) on the south. Bhasna is 47 km south-east of Diamond Harbour and 110 km south of Salt Lake, whereas Sukhchar is about 12 km north of Salt Lake. So the total north–south range of *H. palustris* extends over 122 km.

During the present survey and record collation, the northern known limit is Nalban, the eastern limit is extended to the Rajarhat Wet-lands, the western limit to Santragachhi, about 5 km north-west of Shibpur (Howrah), and the southern limit to Subhasgram. Subhasgram is well north of the previous southern limit: it is 31 km north of Diamond Harbour, compared with 47 km to the south-east. Therefore, the Bengal Mongoose range may recently have contracted both in the northern and southern parts, although neither Sukhchar nor Diamond Harbour was thoroughly surveyed recently. So there is probability that the species may still occur in a few secluded and undisturbed habitat in those areas.

During 2007–11, the Bengal Mongoose was found in five new peri-urban habitats within its previously documented geographic range but no published documents indicate that any previous survey was conducted in these areas by the forest department, ZSI or any NGO and there is no reason to invoke a range expansion because these areas fall well within the previously known native range of *H. palustris*. Similarly, it is possible that the species might be found in further yet-to-be-surveyed wetlands in the region.

During the above period, the Bengal Mongoose was sighted only in the inland non-saline wetlands. No records of the Bengal Mongoose from the salty or even brackish habitats, such as the present Sundarban mangroves, were traced, but very little suitable survey has been conducted in them, so the possibility remains that the species uses such habitat. An unconfirmed report of its occurrence in Gosaba islands (22.16°N, 88.80°E) (specific location not known), the last inhabited area on the northern fringe of Sundarban Tiger Reserve (Sajnekhali), is, however, available.

In addition, although the Bengal Mongoose visits the dry land areas adjacent to some occupied wetlands, it has not yet been found at Chintamoni Kar Bird Sanctuary (22°42′N, 88°40′E), spread over 6.956 ha to the right of N.S.C. Bose Road, near Adi (old) Ganga, 12 km south of the East Kolkata Wetlands. This was formerly an orchard and holds a small population of the Grey Mongoose (direct sighting by the author and others), but not, apparently (based on considerable observations), any Bengal Mongoose or Small Indian Mongoose.

Morphology (Tables 18.1–18.4)

H. palustris has a vermiform body with small head, long face, a pointed snout, small and rounded ears, mostly concealed by hairs, short legs and long muscular tail, thick at the base, and tapering at the end, but without any band. Its fur is rough and coarsely grizzled, but not striped (a conspicuous white stripe is present on the sides of neck of *H. urva*).

The Bengal Mongoose has five toes on each of its four foot. The semi-plantigrade feet have short compressed claws, which are non-retractile and are used for digging the burrow or at the time of hunting. Its eyes are small, but glittering and snake-like, with linear erect pupils. The tongue is rough as a cat's. The female has three pairs of mammae (1 abdominal and 2 inguinal) (Ghose and Chaturvedi, 1972).

It is very similar to *H. auropunctatus* in external and skull morphology and often misidentified if it is not closely and minutely observed in the field by those who are not conversant with the distinguishing features of the two species. The original description of *H. palustris* distinguished it from Small Indian Mongoose by a number of mostly somewhat subtle features (Ghose, 1965), although he did not specify how, on the information available, this could safely be concluded. The blackish patch on the muzzle above the nose of Bengal Mongoose is its common and most prominent external distinguishing mark (not seen in any of the three mongoose species occurring in West Bengal, so identification in the field is easier). But this feature is also found among a few mongoose species elsewhere in India and other countries, as in *H. vitticollis* Bennett from South India and Ceylon (Sri Lanka), *H. siamensis* Kloss from Thailand and *H. rubrifrons* Allen from Hainan, China (Ghose, 1965). However, Ellerman and Morrison-Scott (1951: 295) treat *siamensis* and *rubrifrons* as subspecies of *H. auropunctatus*.

Sex could be determined only for the animals handled by ZSI and NEWS. Photograph of a male exposing the genital organ, while scratching the dorsal part by raising the left hind paw, was also taken at Survey Park. *H. palustris* is intermediate in size between *H. edwardsii* and *H. auropunctatus.* It is smaller than *H. edwardsii* and larger than *H. auropunctatus.* The species is sexually dimorphic. The Bengal Mongoose females are smaller (body length 30–32 cm, tail length 25–27 cm, weight 500–625 g) than males (body length 32–36 cm, tail length 26–28 cm, weight 625–900 g) (Agrawal *et al.*, 1992, Dey, 2007). The morphometrics of the Bengal Mongoose are shown in Tables 18.1 and 18.2.

Table 18.1: Morphometrics of of *H. palustris* Specimens Collected by ZSI (in mm)

Body Parts*	Male			Female		
	Mean	Max	Average	Mean	Max	Average
H&B	284.5	365.0	335.0	303.0	315.0	309.0
T	224.0	235.0	263.0	240.0	258.0	249.0
Hf	54.5	63.5	58.0	51.0	54.0	52.7
E	17.7	26.5	24.0	21.1	24.0	22.5
Cranial						
Ls	62.7	72.3	68.0	61.8	63.5	62.9
Cb	62.3	71.5	67.1	60.9	62.2	62.0
Mw	10.4	12.9	11.5	10.1	10.9	10.6
IW	10.3	12.1	11.2	9.8	11.0	10.5
Pow	8.6	13.8	10.3	9.6	13.1	11.1
Pm^4	5.7	6.3	6.0	5.6	6.1	5.8
MI	4.8	5.8	5.5	5.0	5.8	5.4
ZW	29.8	34.6	32.2	29.5	32.6	30.6

* Legend: H&B: Length of head and body; T: Length of tail; Hf: Length of hindfoot; E: Length of ear; S: Length of skull; Cb: Condylobasal length; Mw: Maxillary width; Iw: Least interorbital width; Pow: postorbital width; Pm^4: Length of fourth upper premolar; Ml: Mandibular length; Zw: Zygomatic width.

An intra-specific colour variation is reported (Ghose, 1965). In fact, it is dimorphic in colour, having both the dark and light forms. In the dark form, individual contour hair is alternately banded with black and buff-yellow, while in the light form, it is blackish-brown and straw yellow (Alfred and Chakraborty, 2002). Although the degree of darkness and paleness varies individually, both the forms occur in the same season and in individuals of similar age, and do not appear to be related to season, age or sex. During the present study, this was found to be true at Nalban as both the dark and light forms were sighted. Generally, the ventral side of the body is slightly lighter than the dorsal part.

Towards the nape the number of bands in the individual contour hairs is five, while it is generally five, but in some specimens six, in the contour hairs on the rump

(Ghose, 1965). According to Jerdon (1874: 136) and Pocock (1937: 240; 1941: 29), the number of bands in the contour hairs of *H. auropunctatus* is five. But in the specimens of the ZSI collection examined by Ghose (1965), this number on both the anterior and posterior aspects of the upper surface of a few specimens is six; and in one specimen, it has been found to be eight on the rump.

Table 18.2: Morphometrics of the Specimens of *H. palustris* trapped by NEWS

Sex	*Body Weight*	*Tip to Tail*	*Body*	*Tail Length*
Female (1 year)	450 g	560.5 mm	310.5 mm	250.0 mm
Adult female	550 g	570.5 mm	300.5 mm	270.0 mm
Adult male	625 g	610.0 mm	320.5 mm	270.5 mm
Adult male	725 g	620.0 mm	360.0 mm	260.0 mm
Adult male	900 g	630.5 mm	340.0 mm	190.5 mm
Adult female	625 g	540.0 mm	300.5 mm	230.5 mm

Hands and feet of the Bengal Mongoose are concolorous with the body, but the upper surface of the manus and pes are paler. The sole of the hind foot is always naked up to the heel. Hodgson (1836) and Blanford (1891) state that in *H. auropunctatus* the feet are not naked up to the heel, whereas in the ZSI material this character is variable, the sole being naked up to the heel in some specimens, but not so in all.

As in *H. auropunctatus,* the tail of the Bengal Mongoose is over two-third the length of, and concolorous with, the body. However, on closer examination, the tip of 8 out of 19 specimens appears darker (more clearly seen in the light form), thus differing from *H. auropunctatus,* which does not have such black tip, but approaching *H. smithi* Gray and *H. vitticollis* Bennett. One individual sighted at Santragachhi during the recent survey has such a distinguishing mark at the tail tip.

The skull of *H. palustris* is close to *H. auropunctatus* in respect of size, but is slightly longer and stouter. The forehead of Bengal Mongoose is bulging and the dorsal area behind the orbit is depressed as in *H. edwardsii.* The sagittal and occipital crests are strongly developed. In some specimens of ZSI (10 out of 18 adult skulls examined), the occipital crest is produced backward and upward forming some sort of a 'roof' over the occipital condyles, due to which the occipital plane forms an obtuse angle with the basicranial axis.

Unlike *H. auropunctatus,* the cranium of *H. palustris* does not narrow gradually towards the orbit, but narrows abruptly behind the orbit so that the postorbital region appears as a constriction between the frontals and the cranium.

Furthermore, in comparison with *H. auropuctatus, H. palustris* has the basioccipital broader, the inner margins of bullae subparallel, the teeth robust, the molars and premolars with less pointed cusps and the posterior chamber of the bulla slightly more reduced and inflated in many cases (in 12 out of 18 adult skulls examined) (Ghose, 1965). It may be mentioned that except for the size and condition of the bulla, the skull of *H. palustris* has greater similarity with that of *H. edwardsii* than with that of *H. auropunctatus.*

Table 18.3: Morphological and Behavioural differences Among Bengal Mongoose *H. palustris*, Small Indian Mongoose *H. (J.) auropunctatus* and Grey Mongoose *H. edwardsii*

Criterion	*Bengal Mongoose*	*Small Indian Mongoose*	*Grey Mongoose*	*Source*
Head and body length	280–360 mm	250 mm	330.5–350.8 mm	Agrawal *et al.*, 1992
Tail length	220–260 mm	260 mm	330.2–340.2 mm	
Weight	450–900 g	305–662 g	1,400–1,700 g	NEWS *in litt.* 2006, Dey 2007, Bahuguna and Mallick 2010
Length of hind foot	50.1–60.4 mm	40.8 mm	60.6–60.9 mm	Agrawal *et al.*, 1992
Length of ear	10.8–20.7 mm	09 mm	10.5–20.8 mm	
Cranial length	61.8–72.3 mm	58.4 mm	72.7–73.4 mm	
Condylobasal length	60.9–71.5 mm	57.2 mm	71.4–72.5 mm	
Maxillary width	10.1–12.9 mm	10.7 mm	12.5–13.7 mm	
Least interorbital width	9.8–12.1 mm	11.6 mm	12.5–13.8 mm	
Postorbital width	8.6–13.8 mm	17.0 mm	11.2–12.8 mm	
Length of fourth upper premolar	5.6–6.3 mm	5.5 mm	6.4–7.2 mm	
Length of first lower molar	4.8–5.8 mm	3.2 mm	6.2–6.3 mm	
Zygomatic width	29.5–34.6 mm	29.0 mm	34.8–38.0 mm	
Coat	Rough	Smooth and silky	Usually coarse	Ghose 1965, Bahuguna and
Dimorphism in colour	Dimorphic (both dark and light forms), not correlated with season, age or sex	Much individual variation in colouration, but no dichromatism reported	Colour varies due to age, season and locality factors	Mallick 2010
Grizzling of pelage	Coarse	Fine	Coarse	
Muzzle	Black patch	Dark brown, no patch	Rusty brown, no patch	

Contd...

Table 18.3–*Contd...*

Criterion	*Bengal Mongoose*	*Small Indian Mongoose*	*Grey Mongoose*	*Source*
Tail	Darker tip in 8 out of 19 specimens originally examined (more clearly seen in the light form)	No darker tip	Whitish or yellowish-red tip, but never black	
Cranium	Narrows abruptly behind the orbit so that post-orbital region appears as a constriction between frontals and cranium	Gradually narrows from the orbit	–	
Crest	Well-developed	Not so well-developed	–	Ghose 1965
Skull	Stronger, slightly longer, and with bulging fore-head; in general more similar to *H. edwardsii* than to H. (*j.*) *auropunc-tatus*	Weaker, shorter, and without bulging forehead	–	
Basioccipital	Broader	Narrower	–	
Inner margins of the bullae	Subparallel	Not subparallel	–	
Teeth	Robust and strong; cusps of molars and premolars less pointed	Weak, cusps of molars and premolars more pointed	–	
Posterior chamber of the bullae	Reduced and more inflated	Less reduced and less inflated	–	
Odour	"Obnoxious"	No such odour	–	
Food preference	Small fish and aquatic invertebrates, among snails preferring the water snail *Pila globosa*	Mainly carnivorous, among snails preferring the land snail *Achatina fulica*	Carnivorous	Ghose 1965, Deuti 2008, Bahuguna and Mallick 2010

Contd...

Table 18.3–*Contd...*

Criterion	*Bengal Mongoose*	*Small Indian Mongoose*	*Grey Mongoose*	*Source*
Hunting	Semi-aquatic hunter, more active during the early morning and late afternoon	Terrestrial hunter, hunts throughout the day	Terrestrial hunter	Dey 2007, Deuti 2008, Bahuguna and Mallick 2010
Ecological niche	Only non-saline wetlands, mud-banks near the water's edge	Various; but not in non-saline wetlands, in southern West Bengal	Near human habitations	Ghose 1965, Agrawal *et al.*, 1992, Dey 2007, Deuti 2008

Table 18.3 shows the distinguishing morphological and behavioural differences among the three mongoose species found in southern Bengal, *i.e.*, Bengal Mongoose *H. palustris*, Small Indian Mongoose *H.* (*j.*) *auropunctatus* and Grey Mongoose *H. edwardsii*.

Subsequent microscopic examination of the structure of dorsal guard hairs of different mongoose species (including samples from seven Small Indian Mongooses from six widely spaced localities and nine Bengal Mongooses from three localities) supports the specific status of *H. palustris* (De *et al.*, 1998): one absolute difference was found, in the scaling pattern, as were various others that are either very subtle or apparently reflect difference in the range of variation. The diameter, scale type, scale count, scale margin and medullary configuration of dorsal guard hairs of the two species are distinguishable in the samples analysed by De *et al.* (1998; see Table 18.4), and while they found that in some cases the Bengal Mongoose was more similar to other mongooses than to Small Asian Mongoose, their results require corroboration. Specifically, they did not examine enough specimens to notice the colour dimorphism in the Bengal Mongoose mentioned by Ghose (1965) and others, and the sampling of *H. javanicus–H. auropunctatus* was too spatially limited to account for intra-specific variation across its large distribution range; moreover, only one specimen came from a state next to West Bengal, and none from the state itself. Thus, the possibility of an overlap in characters, perhaps even clinally, between the Bengal Mongoose and geographically overlapping/nearby Small Indian Mongoose could not be investigated. Sahajpal *et al.* (2009) wrote of the distinctiveness of Bengal Mongoose hair under the microscope but unfortunately compared it only with *H. edwardsii*, *H. smithii* and *H. urva*, and it is, therefore, not clear how, if at all, the Bengal Mongoose differs in the characters investigated from the Small Indian Mongoose.

Although previous documents have stressed the Bengal Mongoose's similarity to the Small Indian Mongoose, confusion with the Grey Mongoose can also be problematic. In the field, the Bengal Mongoose seems closer in size to the Grey Mongoose than to the Small Indian Mongoose, an impression reflecting the measurements in Agrawal *et al.* (1992; although measurements of only one Small Indian Mongoose are given). The Grey Mongoose can be distinguished by its grey coat, heavier build, thicker and bushier tail, higher stance (due to longer legs, particularly the hind ones), and the absence of any black patch on the muzzle.

Habitat: Selection and Use

Out of three *Herpestes* species found in West Bengal, *H. urva,* which is highly distinctive from *H. palustris* in morphology, occurs only in the forests of northern Bengal. Whereas *H. auropunctatus* is a widespread species in the state, *H. palustris* is a restricted range non-forest species in south Bengal, which tends to reflect its ecological "rigidity" (*i.e.* inability to opportunistically adapt to the other terrestrial environmental conditions like arid and semi-arid habitat).

In Latin, the word '*palustris*' means 'swampy' or 'marshy' and, therefore, is used in taxonomy for the species name, particularly in the scientific name, for the animal referring to the typical habitat of the species. In fact, *H. palustris* is a wetland-dependent

Table 18.4: Microscopic and Macroscopic Profile of Dorsal Guard Hairs of Bengal Mongoose *H. palustris*, Small Indian Mongoose *H. (j.) auropunctatus* and Grey Mongoose *H. edwardsii*

Character	*H. palustris*	*H. (j.) auropunctatus*	*H. edwardsii*
Physical characters			
Colour	Alternately banded with ochraceous buff and Prout's Brown with Prout's Brown tip.	In general alternately banded with clove brown and buff with clove brown tip.	Tip clove brown then alternately banded with cream buff and clove brown. In some of the hairs overall ferrugineous tinge evident.
Total length	14–25 mm (18.5 ± 3.45)	13–18 mm (14.83 ± 1.76)	20–31 mm (25.61 ± 3.34)
Apical diameter (μ)	25–50 (40.83 ± 6.31)	25–50 (41.66 ± 5.23)	50–90 (70 ± 12.4)
Middle diameter (μ)	50–100 (80 ± 7.64)	50–100 (79.16 ± 9.19)	70–100 (90 ± 8.17)
Basal diameter (μ)	50–70 (58.33 ± 4.78)	50–75 (62.5 ± 4.5)	50–90 (60 ± 5.4)
Number of bands	Variable, 96–90 per cent hairs five banded and rest seven banded.	Variable, 78–81 per cent hairs five banded and rest seven banded.	Variable, 80–85 per cent hairs seven banded and rest five banded.
Width of bands	Variable, Prout's brown bands usually wider than ochraceous buff; width of Prout's brown bands 2–6 mm (3.77 ± 0.51) and that of ochraceous buff 1–4 mm (2.75 ± 0.07)	Variable, clove brown bands usually wider than buff except at base; width of clove brown bands 2–6 mm (3.97 ± 1.05) and that of buff 2–4 mm (2.66 ± 0.16)	Variable, but clove brown bands usually wider than Cream buff except at base; width of clove brown bands 3.30–5.0 mm (4.02 ± 0.6) and that of cream buff 2–4 mm (2.88 ± 0.54)
Surface structure			
Scale pattern	Irregular wave	Flattened irregular mosaic	Irregular wave
Scale count	148–200 (160)	95–120 (104)	180–225 (220)
Scale margin	Smooth with few notches	Smooth	Crenate
Side-to-side scale length	10–20 mm (16 ± 2.02)	10–20 mm (16 ± 3.06)	20–40 (30 ± 4.58)
Proximo–distal length	4–10 mm (7.01 ± 1.16)	7–10 mm (8.31 ± 0.08)	2–7 (5 ± 0.91)
Medulla			
Medullary configuration	Narrow aeriform lattice	Unbroken with cortical intrusion	Unbroken with cortical intrusion
Medullary Index	0.75–0.76 mm (0.755 ± 0.004)	0.88–0.90 mm (0.885 ± 0.0067)	0.81–0.83 (0.82 ± 0.003)

species. Out of ten wetland habitat types found in India, the marshes and *jheels* of the Gangetic plains in southern West Bengal are the known habitats of *H. palustris*.

In this region, *H. auropunctatus* is stated to be sympatric with *H. palustris* in the non-forest areas. According to Ghose (1965: 173), "though sympatric with *H. auropunctatus*, [*H. palustris*] is confined to marshy areas, a habitat very different from that of the former". Agrawal *et al.* (1992: 106) recorded occurrence of the Small Indian Mongoose in all districts of West Bengal, "near the human habitations, particularly adjacent to paddy or sugar cane fields and burial grounds"; marshes or swamps were not mentioned. However, there is no credible record of *H. auropunctatus* or *H. edwarsii* from any non-saline wetland within the known geographical range of *H. palustris* (Mallick, 2011). In south Bengal, only one adult and one juvenile (both unsexed) specimens of the Small Indian Mongoose were known to be collected from Kolkata district (Agrawal *et al.*, 1992) without any mention of the exact locality (and so its location relative to the Bengal Mongoose range is not known).

Though the Small Indian Mongoose is listed as widespread in the East Kolkata Wetlands by Anonymous (2004) and Kundu *et al.* (2008), the latter is derived directly from the former, given the close similarity in text in the two documents. Kundu *et al.* (2008) tabulated the occurrence of both the Bengal Mongoose and Small Indian Mongoose at all five localities in the East Kolkata Wetlands that they studied (based on the abbreviations as explained in Anonymous [2004]): Chowbhaga (CB), Bantala (BT), Jhagrasisa (JS), Sahebmara (SM) and Mahishbathan (MB), but the earlier source (Anonymous, 2004) had recorded the Bengal Mongoose only at Sahebmara (with the Small Indian Mongoose recorded at all five sites). However, these two sources gave no evidence of any critical identification of the mammals, though stated that the mammal listings were derived from "field observation and secondary data ... collected from the local residents and fishermen community" (Anonymous, 2004: 10). Therefore, this apparent claim of widespread overlap between the two species is here considered at sufficient risk of error to need corroboration.

Similarly, vague reference was made to overlap in habitat use between the Bengal Mongoose and "both the species of mongoose" (apparently the Grey Mongoose and Small Indian Mongoose) by Sanyal (1999), a statement ignored here because the evidence, on which it was based, was not presented by the author. In fact, the Grey Mongoose occurs in both north and south West Bengal, but seems not to have been collected from the specific sites of the Bengal Mongoose; specimens are held at ZSI from Basirhat, located in the southern part of North 24-Parganas district and it also occurs in Alipore in South 24-Parganas district (Agrawal *et al.*, 1992); the Bengal Mongoose has not yet been recorded from these two sites. During the present study, the Grey Mongoose was photographed at Barasat and Badu in North 24-Parganas and Chintamoni Kar Bird Sanctuary (Narendrapur) in South 24-Parganas, where no Bengal Mongoose was sighted during the present survey or recorded previously.

Dey (2007) commented: "each and every trapped mongoose population in the East Kolkata (Calcutta) Wetlands is constituted of (Bengal) Mongoose." Five mongooses were captured during January–April 2011 by NEWS at Nalban, following the protocol of Dey (2007), who in 2006 had caught six of them (three adult males, two adult females and one juvenile male) in north Salt Lake *bheri* areas and partially in Bhojerhat and Bhangar area. Whereas ZSI collected comparatively many Bengal Mongooses (n= 32) during the 1960s, before promulgation of the Indian Wildlife (Protection) Act in 1972, NEWS caught only a few individuals, all from Nalban, reflecting legal restrictions (so that the habitat and the habitants are not disturbed excessively, which may cause adverse impact on their behaviour, feeding ecology and reproduction), not an inability to trap more mongooses.

Thus, it seems that within the geographic range of animals showing the gross morphological characters of Bengal Mongoose, all individuals from wetlands critically examined have fitted the Bengal Mongoose, not other mongoose species, and so, for the present review, all recent records of the Small Indian/Bengal Mongooses in the East Kolkata Wetlands are taken to be the Bengal Mongoose.

The study area is home to over 100 plant species. It has been observed in the study of CEMSAP (Ghosh, 1997) that a rich variety of macrophytic vegetation offers suitable site for laying eggs and cover all size of fishes. The interdependence also provides facilities for roosting and breeding of the aquatic birds. Here, the marshy edges, infested with the aquatic plants like cattails *Typha*, reeds *Phragmites*, sedges (Cyperaceae), grasses (Gramineae) and Water-hyacinth *Eichhornia crassipes* form the typical habitat of Bengal Mongoose (NEWS *in litt.* 2006, corroborated by records assembled here). This marshy habitat of the Bengal Mongoose differs from other Indian wetlands by their negligible surface catchment; instead, a perched aquifer lies more than 400 feet below them.

The floristic diversity of these wetlands is mediocre, even in the core area. The macrophytes include *Sagittaria montevidensis, Cryptocoryne ciliata, Cyperus* spp., *Acrostichum aureum* and *Ipomoea aquatica*. The embankment vegetation is mostly dominated by *Fimbristylis ferruginea, Suaeda maritima, Acanthus ilicifolius, Excoecaria agallocha* and *Avicennia officinalis*. Salt Lake proper, which, despite its name, has been non-saline since 1930 (Furedy, 1987), is dominated by numerous algae, clumps of the reeds like *Aegiceras majus* and the cattails like *Typha elephantina*. The marshy, shallow and large water bodies or wetland bogs, including swampy edges, are fully or partially filled with emergent hydrophytes, such as *Phragmites karka* and *Typha angustifolia* (Anonymous, 2004). These wetlands are lined by narrow mud-banks, on which grasses and a few stunted trees grow (Deuti 2008). These wetlands are rich in the food items preferred by the Bengal Mongoose, *e.g.*, the fish, freshwater crustaceans including shrimps, molluscs (Gastropoda, Bivalvia), bugs (Hemiptera), beetles (Coleoptera), amphibians (Amphibia) and turtles (Testudines) (Anonymous 2004).

During 1960s, when reclamation of wetlands for the Salt lake township started, ZSI recorded 22 species of mammals representing Shrew (1 species), Bats (9 species), Carnivores (10 species) and Rodents: Squirrel, Rat and Mouse (5 species) in the area including the Bengal Mongoose *H. palustris*, Fishing Cat *Prionailurus viverrinus*, Smooth-coated Otter *Lutrogale perspicillata*, Small Indian Mongoose *H. auropunctatus*, Indian Gray Mongoose *H. edwardsii*, Small Indian Civet *Viverricula indica*, Common Palm Civet *Paradoxurus hermaphroditus*, Jungle Cat *Felus chaus*, Jackal *Canis aureus*,

Greater Bandicoot Rat *Bandicota indica,* Lesser Bandicoot Rat *Bandicota bengalensis,* Common Indian Field Mouse *Mus booduga,* Indian Brown Spiny Mouse *Mus platythrix* and Grey Musk Shrew *Suncus murinus,* all depend on these wetlands for food and shelter (Mallick, 2009). Among them, *H. palustris, H. edwardsii, V. indica, P. hermaphroditus, F. chaus, C. aureus* and *L. perspicillata* are rare (Anonymous, 2004).

There are also about 64 species of birds, both local and migratory, which include the kingfishers, grebes, cormorants, egrets, terns, eagles and sandpipers. During 1964-69, a total of 248 bird species were recorded by ZSI from the study area. Among them 109 species (50 per cent aquatic) have become locally extinct (Ghosh, 2004). 15 fish and 112 reptilian species have also become rare here. The amphibians include Common Indian Toad (*Bufo melanostictus*), Ornate Narrow-mouthed Frog (*Microhyla ornata*), Skittering Frog (*Rana cyanophlyctis*), India Bull Frog (*Rana tigerina*) and Cricket Frog (*Rana limnocharis*). A definite prey-predator relationship between the Bengal Mongoose and other life forms (smaller vertebrates and invertebrates) is well established in this ecosystem. However, there are a very few predators of the Bengal Mongoose in the study area, *e.g.*, man and Monitor Lizards.

A single location-specific change in the habitat quality (say, re-profiling the earthworks) might have caused an equally location-specific change in the mongoose population, or distribution of burrow. However, all lines of evidence (common-sense assumptions of effects of habitat conversion; resurvey of 1960s sites; and decreased number found at N° 4 *bheri*) are consistent with a severe decline having occurred. The area occupied by the Bengal Mongoose during the late twentieth century was reported to be <500 km² (Molur *et al.*, 1998), but the present survey assessed the probable extent of such area to be <200 km², representing a major decline since the 1960s.

The Bengal Mongoose is a den-digging animal living around large but shallow non-saline water bodies, fully or partly covered with a thick growth of aquatic plants, *i.e.* bogs, marshes and swamp-edges infested with reeds (Alfred and Chakraborty, 2002; Dey, 2007; Deuti, 2008). It makes inter-connected burrows of about 45–55 cm long and 33 cm deep with 3–4 entries, mainly along the slopes of the water bodies. These permanent underground burrows are short-mouthed, whereas the tunnel is wider with conspicuous downward slope from the mouth, which prevents the predators from entering them easily. The burrows assumed to be those of Bengal Mongoose were found mainly in the secluded and undisturbed portions of the mud-banks of the wetlands. The area surrounding the occupied burrow is devoid of clay particles, but often having fresh scratch marks near its mouth. The burrows were not found in the dry land areas located far from the wetlands. It seems that the individual territory of the Bengal Mongoose is not extensive, but small, fixed, exclusive and usually spaced along the water's edge.

These burrows are multi-channelled with openings located in close proximity, presumably for easy escape, but the number of burrow-openings does not indicate the population. In a stretch of ½ km in the Nalban *bheri* (n° 4), 11 to 13 burrows were identified. It was also found foraging near the human habitations, particularly the shrubby backyard gardens. However, after thorough searching, no such burrow could be detected in these gardens. Hence, it is anticipated that the Bengal Mongooses observed here were merely visitors from the nearby marshlands. A few decades back, these areas were part of the extensive wetlands, but since human settlements have been established there, they might have shifted to a suitable habitat still left.

Group Size

H. palustris is believed to be solitary excepting the breeding season. Solitary animals were mostly sighted. Out of the 49 individuals sighted during 2007–11, 30 sightings were of singles, six of duos, one of a trio and one of a group of four. Pairing during the breeding season was also observed. A pair moving close to each other was photographed at the backyard garden in Survey Park. A mating pair was also photographed at Nalban. Three pups accompanied by an adult-sized animal, presumably the mother, were seen at Nalban, who hurriedly entered a burrow at the sign of presence of human-beings. But, occasionally, they were found gregarious. They often form small groups, particularly during the afternoon foraging period. Such association appears to be helpful in hunting.

Behaviour

Earlier, the behaviour of Bengal Mongoose was poorly known. However, it is believed to be diurnal. It exhibits a routine exitentry movement, *i.e.*, coming very cautiously out of the monitored burrow few hours after sunrise, and if the environment is found undisturbed, they go on foraging following a particular route in its territorial range and returning to the burrow just before sunset. They stay inside the burrows overnight.

Out of 20 burrow-mouths sealed with mud in the late evening after return of the foragers, five were found opened early the next morning, suggesting that at least a quarter of the burrows contained animals. It may be more, because a disturbed animal may not come out of the burrow in the morning, monitoring the surroundings, but when a bicycle passed at a distance of about 10 m, it returned to its burrow and did not emerge during the following hour's observation.

A loner in a backyard at Survey Park, Ajaynagar (Santoshpur) stood on its hind legs to monitor the surroundings, a stance also observed at Nalban.

The Bengal Mongoose is more active in the morning during 06:00 h–08:00 h in the summer and 07:00–10:00 hrs in the winter. They seem taking shelter in the grasses and shrubs during the mid-day, because they were neither seen in the open in scorching sunlight nor to return to their burrows during the hot mid-day. Deuti (2008) also observed them generally hiding in the reed-bed or long grasses and aquatic vegetation. After a gap of six to seven hours (probably resting period), it resumes the foraging activity in the afternoon during 16:00 h–17:00 h in the summer and 15:00 h–17:00 h in the winter. By contrast, *H. auropunctatus* is active throughout the day.

The Bengal Mongoose can bark, scream and purr, but these are exceptions. It is generally a silent animal and sometimes mews, varying with low yelps or growling in rage with bristling of the hair, particularly of the tail. A Bengal Mongoose, finding a trap with live bait encircled the trap a number of times, then approached close to threaten the bait with a hissing sound, hitting at it with its paw. It also often clambered about, with skill, on the cage gratings. Failing to capture the bait from outside, it had no option but to take the risk of entering the cage through the opening. Immediately after the cage door shut behind, the trapped animal began showing aggression, made an open-mouthed fierce protest or display and moved up and down within the short space of the trap. However, after struggling for some time, it calmed down. After making observations for some time by the study team, when the shutter was lifted for its escape, it usually moved backwards to come out of the trap. Creeping (reflecting the literal meaning of its genus) was particularly observed at the time of entering or leaving the trap. This is also true when emerging out of or entering the burrow. Some mongooses were seen to avoid the trap in the presence of people or other sources of disturbance.

It was seen to use the tail to balance either when standing upright, when it is laid straight on the ground, or moving fast, when it is kept almost straight in the air. The Bengal Mongoose usually communicates its mood and expressions through the tail movements, which may be variable from individual to individual. A mongoose, before entering a trap, was found to curve (angular and U-shaped) the tapering lower part of the tail outward, possibly as a gesture of threat. When it entered the trap, the tail-tip was seen to form an inward loop in defence. In another case, a mongoose fearlessly entered the trap straightway, keeping its tail in normal position, when the tip was touching the ground. No upward loop of the tail was, however, observed during the survey.

The Bengal Mongoose was sometimes encountered along a path, while shifting from one patch to another. It runs at a high speed and, astonishingly, can even turn without slowing down. When photographed, it passed by the camera a few times and moved so fast that most of the pictures were blurred, or slightly out of frame. However, it paused for a moment to look around and then quickly dashed into the nearby cover and disappeared. This type of short immobility helped focus the animal.

That the Bengal Mongoose is alert, ferocious and courageous is evinced, particularly when the young accompany the mother. If enraged, it can make its body swell by erecting the body hairs. It can squirt out a fetid fluid from its anal glands similar to the Crab-eating Mongoose, *H. urva,* which acts as a means of self-defence or territorial marking. Chemicals in the anal glands may constitute individual signatures and indicate reproductive condition, sex and/or dominance rank.

Before coming out of the burrow in the morning, the Bengal Mongoose peeps out of the burrow, the mouth and neck first and observes whether there is any imminent danger. If feels safe, it comes out of the nest and moves hither and thither in the territory.

The Bengal Mongoose, particularly the juvenile, is very inquisitive and often approaches very near to the new objects like the camera without hesitation. While coming out for hunting during daytime, the adult occasionally sits up on its haunches or stands up on its hind legs, when the forepaws remain hanging free, for surveillance of the environs by directional movements of its head. They can also make small leaps in the air, particularly when hunting.

Diet/Feeding Habits

H. palustris is generally a predator, feeding on a wide range of animals, but it seems to be an opportunistic feeder. Though it hunts both the vertebrate and invertebrate preys, it feeds largely on the aquatic preys. Amongst the Asiatic mongooses, *H. urva,* inhabiting the stream-banks in northern Bengal, has also preference for similar aquatic preys. The Bengal Mongoose is known to prey on the crustaceans, mollusks, crabs, amphibians, small reptiles, birds including their eggs, insects and larvae. It is a good rat-hunter (*Mus* and *Bandicota* spp.) and seems to keep the quick-growing population of this pest under control. Occasionally, it takes poisonous or non-poisonous snakes. The frogs (*Rana* and *Bufo* spp.) are also preferred by them. Secondarily, the Bengal Mongooses are known to eat a variety of fruits, tubers and berries as well as smaller mammalian species, which it may overpower.

Above all, the Bengal Mongoose prefers a specialized diet. It feeds mainly on the aquatic snails and small fish found in their habitats, for example, Salt Lake and Bhasna in North and South 24-Parganas respectively (Ghose and Chaturvedi, 1972).

The local farmer and fisher communities use the Water hyacinth to create a buffer between the land and water in order to minimise the erosion. The dense bed of this plant, a 3–4 m wide band, kept in captivity along the periphery of the Bengal Mongoose habitat, is beneficial for providing such specialised food resources. For example, this weed is the feed of some fishes like *Tilapia* spp. Lylontika also lays their eggs on the Water hyacinth in shallow water. These plants help protect these eggs. Water hyacinth also provides shade to the fish during summer. Thus, this weed bed is the most important hunting ground of the Bengal Mongoose for the small fishes and their eggs. Its special diet also includes snails, particularly the common aquatic snail, *Pila globosa* and the bivalve mollusks like *Lamellidens marginalis* as well as the gastropod mollusks like *Bellamya bengalensis* and *Lymnaea* spp. These mollusks are abundant at the water's edge. This observation was confirmed during the present survey because the broken shells of these mollusks were seen scattered at the mouth of some of their burrows.

Besides, various aquatic Hemipteran bugs (*Gerris spinolae, Sphaerodema annulatum, Ranatra elongata, R. varips, Laccotrephes griseus, Diplonychus annulatus* and *D.* [=*Sphaerodema*] *molestum*) and Coleopteran beetles (*Canthydrus laetabilis, Cybister tripunctatus, Hydrocoptus subvittatus, Hypoporus bengalensis, Eretes sticticus, Hydrophilus olivaceus* and *Berosus indicus*), dragonfly nymphs, terrestrial grasshoppers, crickets, centipedes, crabs, etc. are taken by the Bengal Mongoose. They were seen taking the giant water bugs and diving beetles, which possess hard elytra and wing membranes. The juveniles, who are not adapted to the aquatic hunting, have to feed on various terrestrial insects and small mollusks (*Lymnaea* spp.). Peculiarly, after the meal, the Bengal Mongooses were sometimes seen using one of the long claws of the forefeet to clean the teeth like a toothpick (Deuti, 2008). It is also a selective secondary feeder of the domestic residues (Mallick, 2012).

Hunting Technique

The Bengal Mongooses use different techniques for hunting different types of prey. The method of hunting is direct and open headlong attack. It often scampers off with its prey for consuming it conveniently. The most interesting part of its hunting techniques is that it frequently goes to the fringe area of water body, jumps on the dense bed of water hyacinth and searches for the aquatic food, astonishingly, without being submerged into the water. Only its feet become wet in this process. It also wades along the pond banks, poking its hands into the crevices and sifting through the mud to take the frogs and crustaceans out. It also fearlessly enters water to take the underwater prey.

After seizing the snail or crab, the Bengal Mongoose throws the prey against a hard base from standing position in order to break the shell open. For consuming the bird's eggs, it uses its fore-paws to hold the egg, cracking a little hole at the small end to suck out the yolk. It was also seen to hunt a cattle egret (*Bubulcus ibis coromandra* Boddaert), a common resident wader or marsh bird, weighing about 500 g, by seizing its throat. It drew attention of the inquisitive bird by waving its tail. Another hunting situation involving a Bengal Mongoose was reported and photographed at Santra-gachhi Jheel on 26 January 2010 at 09:49 h (Sen, 2010). A snipe *Gallinago* spp. was spotted on a water-hyacinth mat by the said mon-goose, which then sat down, prospecting; the snipe flew, apparently sensing the presence of the mongoose. When fighting with a snake, it takes care that the snake does not envenomate it. If the prey is a non-poisonous one, the task becomes easier for the predator.

Reproduction

In general, the Bengal Mongoose attains sexual maturity within a year or so. The females are polyoestrus. Though they usually breed twice a year, evidences showed that they do not exhibit a regular breeding time. The courtship and mating usually starts after winter (January) and continues up to early summer (March), when the pairs could be easily seen. At Nalban, it was observed that the courtship and mating had started from the first week of March and the babies were born in the first week of June. A lactating female was collected in June (Ghose, 1965). The pups were also seen in early June during the present survey at Nalban. A subadult male was trapped in June by ZSI (Ghose, 1965).

The female in heat finds out the partner in the vicinity. But the actual mating is short and takes place repeatedly after a gap of sometime. While foraging together, the male follows or chases the female. There are very little preliminaries or ceremonial, like sniffing or licking genital organs or vocalisations. The male suddenly jumps on the female's back and tries to penetrate her from behind. If she accepts, she lifts her tail, seen to bend to the right or left as much as possible and also splays the hind legs a little. After mounting, the male's forepaws rest on the female's sides or clasp her belly for support while thrusting, but griping the female's nape in jaws was not, however, seen. After a short while, they break and move around till the next copulation begins. For mating, they prefer the bushy than open areas and do not like to be disturbed in any way. At Nalban, a mating pair was found in a reed bed just on the fringe of the wetland. The forepaws of the female were placed in the water and the hind legs on the land.

Their gestation period is around 45 days and the litter size varies from two to three or rarely four. It seems that most of the birth takes place in the burrow just before the monsoon between April and June. The peak birth season correlates with the abundance of the small insects and mollusks in the wetlands during the monsoon, which are the normal diet of the young.

Sometimes the pregnant female, before giving birth, enlarges her own burrow in order to accommodate the litter comfortably. However, during the survey, a family group of four mongooses (caretaker mother and three youngsters), peeping and coming out of a burrow, was also sighted on 29th January 2011, in the late afternoon, which indicates that the breeding may also take place in the early winter.

While born, the pups are helpless and cannot protect themselves from the predators. Hence, leaving them alone in the burrow, while going out for foraging, is rather risky for the mother. Since the male does not take part in the care of the young in any way, the mother has to take the sole responsibility of rearing the neonates. Initially, the mother mongoose nourishes her young on milk produced by her mammary glands. When the young are able to come out of the burrows with the mother, they are acquainted with the habitat and learn the techniques of hunting small prey. The mother always protects her cubs ferociously. Even she does not hesitate to attack large monitor lizards (*Varanus flavescens*), which often try to attack and steal her cubs.

The cubs grow rapidly within 2-3 months and by August-September they are able to come out of the burrow for hunting with the mother. Just after the monsoon, in September-October, 2–3 cubs were often seen playing with each other in the evening between 16:00 h-17:00 h. They nibbled at each other's tail and ran helter-skelter near the mouth of the burrow. When they can start living independently, the mother again breeds.

The Bengal Mongoose usually lives up to ten years. However, they often confront premature death due to anthropogenic threats or natural diseases. While examining the specimens at Bhasna in South 24-Parganas for endoparasites, 2♂ and 1♀ examples of a strongylid nematode worm were found from the rectum of a Bengal Mongoose (Ghose and Chaturvedi, 1972). Earlier, Soota and Chaturvedi (1970) described this parasite *Herpestostrongylus palustris* sp. nov. from the lungs of its host *H. palustris,* collected from the same location.

Anthropogenic Stresses

Several potential natural and anthropogenic stresses threatening the habitat of Bengal Mongoose have been proposed (Banerjee, 1999; Chattopadhyay, 2001; Alfred and Chakraborty, 2002; Dey, 2007; Mallick, 2009; Bahuguna and Mallick, 2010). These are catastrophic events, edaphic factors, human interferences, habitat fragmentation and reclamation, particularly because the type localities (originally swamps) have become satellite townships (Molur *et al.,* 1998; Srinivasulu and Nameer, 2003). Threats to the species comprise habitat loss due to large-scale conversion for agriculture, satellite townships, industries and roads; conversion of a large portion of the natural wetlands to fisheries; clearing of aquatic vegetation; canal concretisation; dumping of municipal solid wastes; poaching and illegal trade in mongoose hair for paint-brush industry.

Many *bheries* were filled up and converted to human habitations in the north-eastern part of East Kolkata Wetlands (Tables 18.5–18.6). Due to large-scale conversion since the 1950s there were only about 200 *bheries* in East Kolkata Wetlands in the 1990s. In 1945, of the total wetlands of about 8,000 ha, nearly 4,628 ha of area were occupied by fisheries. Since 1953, after the formation of Salt Lakes Reclamation Scheme, the area under fisheries declined steadily. By 1970, the expanding city had engulfed Salt Lakes by about 2,000 ha. In the northern Salt Lake area, 26 fisheries were taken over by Salt Lake City housing complex alone. During 1969, there was a large-scale conversion of fisheries when some *bheries* turned vested.

Table 18.5: List of some Old *Bheries* and their Conversion at Salt Lake

Sl.No.	*Name of Bheries*	*Present Location*
1.	Bidyadhari Spill Cooperative Fisheries	Vidyasagar, Laboni, B.D. Market
2.	Knakrimari *Bheri*	Bhaha Atomic Research Center
3.	Boro *Bheri*	Baisakhi, Digantika
4.	Daser *Bheri*	Government office premises
5.	Nortala Khas *Bheri*	Salt Lake Stadium
6.	Kansar *Bheri*	Baisakhi Houshing
7.	Bager *Bheri*	Jhilmil (Now Nicco Park)
8.	Kajar *Bheri*	Industrial Estate
9.	Hansar *Bheri*	Industrial Estate

The major parts of wetlands are held either by Government or influential people and the people living on the wetland mostly do not have property rights on the wetland. The land, considered as owned land, is actually the vested land distributed amongst the marginal farmers and fishers through the process of land reforms under West Bengal Land Reforms Act of 1955, but these latter people do not have selling rights of these vested lands. Saha Ghatak (2010) found that the local people do not really care for the conservation of environment or are not aware of the ecological importance of the East Kolkata Wetlands, and most of them prefer the non-wetland-based livelihood to the wetland-based livelihood.

Under political patronage, these vested lands were distributed among the landless people and some co-operatives were formed. A large part was converted into paddy fields: the total conversion around this time was about 60 km^2. Cultivation also started on the dried-up bed of the River Bidyadhari. Although local fishery owners and legal actions halted further encroachment of the wetlands, the process was not stopped. Throughout the 1970s, creeping conversion continued. Starting from the northern hub of Ultadanga flyover to Garia in the south, the Eastern Metropolitan Bypass, operating since the late 1970s, runs over a stretch of 21 km, cutting through the East Kolkata Wetlands and then extending southwards, where many satellite townships have developed during the last four decades. New townships like Kasba, Vaishnabghata, Patuli and others came up and spread over about 8 km^2. The remaining wetlands were under threat from real estate interests in 1991–92, when a "Science

City" and an amusement park (named Nicco Park), covering an area of 16.18 ha, were built here.

Table 18.6: List of the *Bheries* Surveyed in East Kolkata Wetlands, where No Sighting was Recorded

Name of the Bheri	*Area (ha)*	*Remarks*
Chinta Singh *Bheri*	75.10	Dumping of refuse by Bidhannagar Municipality, filled up over the years and being converted to paddy fields
Sardar *Bheri*	66.79	Drying up due to reduced flow of waste water.
Nater *Bheri*	71.18	Drying up due to reduced flow of waste water
Munshir *Bheri*	69.25	Located in Mahishbathan area; degraded and encroached by Nabadiganta (New Town) Industrial township
Mollar *Bheri*	54.99	Degraded and encroached
Narkeltala *Bheri*	75.88	Degraded and encroached
Patrabad *Bheri* Nos. I-IV	132.94	Destroyed by Rajarhat New Town Project
Baro Paresh *Bheri*	98.89	Degraded and encroached
Choto Paresh *Bheri*	37.35	Degraded and encroached
Goltolla *Bheri*	42.72	Reduced in size
Barochaulari *Bheri*	27.38	Degraded and encroached
Chotochaulari *Bheri*	23.15	Degraded and encroached
Jhagra Sisha *Bheri*	62.32	Degraded and encroached
Gonpotta *Bheri*	50.73	Degraded and encroached
Uttar Gorumera *Bheri*	57.52	Degraded and encroached
Durga Bhasan *Bheri*	56.98	Under cultivation
Heder *Bheri*	NA*	Both under cultivation and fishery
Chaker *Bheri*	27.29	Degraded and encroached
Gopeshwar *Bheri*	34.18	Degraded and encroached
Chachari *Bheri*	NA	Degraded and encroached
Hena Khali *Bheri*	NA	Under cultivation and fishery
Har. Kara *Bheri* Nos. I-V	22.22	Degraded and encroached
Eani Jheel *Bheri*	NA	Degraded and encroached
Diller *Bheri*	17.17	Degraded and encroached
Danir *Bheri*	14.79	Degraded and encroached
Ban *Bheri*	17.67	Degraded and encroached
Garumara *Bheri* (South)	39.40	Degraded and encroached

*NA: Not available.

Of those (semi-)natural wetlands that have not been drained, many have been converted to fisheries, with fragmentation and degradation of those parts not yet converted. Pisciculture involves clearing of the aquatic vegetation used by the foraging

Bengal Mongooses. In the past, the reed beds were dominant features of the habitat. At present, there is almost absence of the reed beds. So, compared with the vastness of the open water, only a few patches of aquatic plants remain. The animals are thus deprived of the shelter, nesting site, roosting site and food resources. The drastic reduction of plant diversity and biomass in most of the *bheries* had resulted in loss of the Bengal Mongoose foraging habitat. Besides, concretisation of both sides of the main canals in East Kolkata Wetlands, where the burrows of Bengal Mongoose are located, has occurred. This has reduced their burrowing opportunities, and, presumably, foraging habitat and invertebrate populations.

Municipal solid wastes (domestic and commercial organic-rich sources) of Kolkata are dumped (2,500 tons/day) for production of vegetables in the landfill sites, and untreated industrial effluents at Dhapa (East Kolkata Wetlands) have been released over an area of 21.4 ha. As this dumping site was being filled up, an alternative site was selected in 2010 by the side of the main dumping ground. Much methane is discharged from this site, needing immediate remediation because runoff into nearby water bodies creates environmental hazards for wetland-dependent species (Purkait and Chakraborty, 2011).

East Kolkata Wetlands receives a huge amount of the city's composite industrial effluent (about 600 million liters per day) throughout the year and the waste-element-contaminated water flows down through a major storm water flow (SWF) canal into the wetland for nearly 40 km farther eastward (Chatterjee *et al.*, 2010). The stress on urban pollution on this largest natural wetland ecosystem is also discussed by Saha *et al.* (2003), Mukherjee (2011) and others. Under the circumstances, some of the sites surveyed in the East Kolkata Wetlands, where no Bengal Mongooses *Herpestes palustris* or their signs were found (Table 18.6).

Mongoose poaching to meet the demands of the paintbrush industry is on the rise in south Bengal (K. Ghosh, Statesman News Service, Kolkata, *in litt.* December 2002). Trapped mongooses are beaten to death, and the hair (about 10 g/animal, fetching Rs. 25.00) is plucked by hand (K. Ghosh, 2002). K. Ghosh (2002) identified a few villagers in North and South 24-Parganas districts engaged in this occupation.

The Forest Department of West Bengal seized mongoose hairbrushes (508 in 2002–03, 56 in 2004–05, and 887 in 2007–08) and mongoose hair (2 kg in 2004–05) in Kolkata and suburbs (Annual Report 2009–10). For a species with such a small and fragmented range, even modest levels of offtake could heighten the risk of extinction. Although Nandy (2006) considered that the Bengal Mongoose might be threatened by capture for the pet trade, this seems unlikely to be so.

Legal Status

The species was upgraded by the Indian Wildlife Protection Act (1972), as amended in 2002, from originally Schedule IV to Part II of Schedule II, providing absolute protection, the offences under which are prescribed the highest penalties (Bahuguna and Mallick, 2010). In such cases, the minimum imprisonment prescribed is three years, which may extend to seven years with a minimum fine of INR 25,000 (see Section 51 of the said Act). However, it is not listed in CITES, RDB, National

(1994) and RDB, International (1996). Given enough proof that Bengal Mongoose is a valid and highly threatened species, the global Red List should review this stance and at least consider listing it, perhaps as Data Deficient reflecting the taxonomic uncertainty, pending extensive survey of the habitats scattered in the native range of *H. palustris* in south Bengal.

Conclusions

In sum, it seems that the Bengal Mongoose has become rare in its entire range except at Nalban (Dey, 2007; SWREJUK, 2007; Deuti, 2008; this survey), its largest remaining habitat-block. Though Nalban has a number of *bheries*, only the little-disturbed core area of N° 4 bheri is considered the most viable habitat (NEWS *in litt.* 2006). The Bengal Mongoose is rare in Sahebmara bheri on the west, its second largest habitat-block. It was not sighted at four locations, Bantala (south-east), Choubhanga (south) [opposite Bantala], Jhagrasisa (east) and Mahishbathan (north) areas within the East Kolkata Wetlands, despite survey during October 2010 at levels adequate to find the species if present and anything other than extremely rare. Moreover, the East Kolkata Wetlands have reportedly been omitted from the latest Indian National Wetland Atlas prepared by the Ahmedabad-based Space Application Centre (SAC), a part of the Indian Space Research Organisation (ISRO) (Times of India, 18 July 2011). So, there is a risk of derecognition of this Ramsar site.

Survival of the restricted-range Bengal Mongoose depends on existence of non-saline wetlands. Therefore, Nalban (a government-owned fishery used by the communities) is recommended for declaration as a Conservation Reserve as per provisions under the Wildlife (Protection) Amendment Act of 2002. Active conservation of the habitats and inhabitants should be implemented earnestly by the nodal agency (East Kolkata Wetlands Management Authority) involving the local people and NGOs. Captive breeding of the Bengal Mongoose has already been recommended (Walker, 1999; Srinivasulu and Nameer, 2003), but no follow up action has yet been undertaken.

Clarification of the current distribution and status of Bengal Mongoose require an in-depth survey of all non-saline wetlands remaining in southern West Bengal, and adjacent areas of potentially suitable habitat. This would allow a cohesive species action plan to be developed and implemented. But most importantly, the taxonomic validity of Bengal Mongoose urgently requires assessment, through a critical re-evaluation of all purported morphological, behavioural and odour characters, and DNA-based investigation. This should use many individuals of *H. javanicus–H. auropunctatus*, from within West Bengal and across the rest of their range. Depending on the result, the Bengal Mongoose may be anything from one of the most threatened and high-conservation priority small carnivores in the world, to merely an indicator of pervasive habitat problems of only local significance.

The records on which these conclusions are based have not been subject to opinions on their identification from people other than the original observers and/or authors. And the number of animals examined in-hand from southern West Bengal remains small. Thus, as stated in the introduction, this study has assumed that the small mongooses of wetlands in this part of West Bengal are all Bengal Mongooses.

Other than by a major trapping programme, the validity of this assumption could be tested by the collation of many photographs of Small Indian Mongoose, Grey Mongoose and Bengal Mongoose from localities across their range in India and neighbouring countries, and taken from a range of lighting conditions and angles showing the animals in varied postures and behaviours, and of varying ages and sexes, and the group discussion of what these images show and how useful are the purported identification characters. Many cases with birds have shown that this approach (pioneered by Porter *et al.* (1974), and now widespread particularly since the advent of the internet) gives immensely valuable perspective, addition to what can be gained from direct study of skins, on how to use distinguishing characters in the field. The continuing lack of clear guidelines on field identification (the species is not in the two most commonly-used identification guides to Indian mammals, Prater (1971) and Menon (2003) mean that people who have not the opportunity to study skins lack clear guidance on how to identify Bengal Mongoose. This knowledge gap is bridged in this updated note.

Acknowledgements

I am specifically grateful to all those persons and institutions, who helped in conducting this study at various locations or provided field information to build up the database on Bengal Mongoose – ZSI; NEWS; IIM, Kolkata; staff of Forest Department; respondent villagers as well as the birdwatchers and photographers like Sumanta Mallick, Soma Jha, Shantanu Bhattacharya, Indranil Mitra, Debarati Bose, A. Chatterjee, H.G. Mukhopadhyay, P.K. Biswas and L. Barman for their records that stimulated the search in the new areas. I am also indebted to Indranil Mitra for providing the map of the study area.

References

Agrawal, V.C., 1998. Faunal diversity in India: Mammalia. In: *Faunal diversity in India,* (Eds.) J.R.B. Alfred, A.K. Das and A.K. Sanyal. Zoological Survey of India, Kolkata, India, pp. 459–469.

Agrawal, V.C., Das, P,K. and Ghose, R.K., 1991. Mammals. In: *Animal Resources of India: Protozoa to Mammalia: State of the Art,* (Ed.) Director, Zoological Survey of India. Zoological Survey of India, Kolkata, India, pp. 659–678.

Agrawal, V.C., Das, P.K., Chakraborty, S., Ghose, R.K., Mandal, A.K., Chakraborty, T.K., Poddar, A.K., Lal, J.P., Bhattacharyya, T.P. and Ghosh, M.K., 1992. Mammalia. In: *Fauna of West Bengal, part I, overview, Mammalia, Aves and wildlife,* (Ed.) Director, Zoological Survey of India. Zoological Survey of India, Kolkata, India, pp. 105–107.

Alfred, J.R.B. and Chakraborty, S., 2002. Endemic mammals of India. *Records of Zoological Survey of India,* Occasional Paper, 201: 1–37.

Alfred, J.R.B. and Nandi, N.C., 2000. Faunal diversity in Indian wetlands. *ENVIS Newsletter,* 6(2): 1–3.

Alfred, J.R.B., Sinha, N.K. and Chakraborty, S., 2002. Checklist of mammals of India. *Records of Zoological Survey of India,* Occasional Paper, 199: 1–289.

Anonymous., 2004. *Final report on preliminary study on biodiversity of sewage [fed] fisheries of East Kolkata Wetland ecosystem*. Institute of Wetland Management and Ecological Design, Salt Lake, Kolkata, India, p. 40.

Bahuguna, N.C. and Mallick, J.K., 2010. *Handbook of the mammals of South Asia*. Natraj Publishers, Dehra Dun, India, xxxiv+541 p. + plates 31 p.

Banerjee, S., 1999. Heavy metal in waster water ponds in and around Calcutta and their effect on mammalian system contaminated through fish raised in waste water improvement. *Annual Report (1998–99)*. Department of Zoology, University of Calcutta.

Basak, A., 2009. *Environmental studies*. Dorling Kindersley (India) Pvt. Ltd. Delhi, India, p. 320.

Bininda–Emonds, O.R., Gittleman, J.L. and Purvis, A., 1999. Building large trees by combining phylogenetic information: a complete phylogeny of the extant Carnivora (Mammalia). *Biological Reviews of the Cambridge Philosophical Society,* 74(2): 143–175.

Blandford, W.T., 1891. *The Fauna of British India including Ceylon and Burma,* Part II: *Mammalia.* Taylor and Francis, Red Lion Court, Fleet Steet, London.

Chakraborty, R. and De, J.K., 2010. *Atlas on hairs of Indian mammals, Part I, Carnivora*. Zoological Survey of India, Kolkata, India, pp. 79–80.

Chatterjee, S., Chattopadhyay, B. and Mukhopadhyay, S.K., 2010. Monitoring waste metal pollution at Ganga Estuary via the East Calcutta Wetland areas. *Environmental Monitoring and Assessment,* 170(1–4): 23–31.

Chattopadhyay, K., 2001. *Environmental conservation and valuation of East Calcutta Wetlands.* Indira Gandhi Institute of Developmental Research, Goregaon, Mumbai, India, p. 67.

Chaudhuri, A.B. and Sarkar, D.D., 2003. *Megadiversity conservation: flora, fauna and medicinal plants of India's hotspots*. Daya Publishing House, Delhi, India, pp. xx+300.

Corbet, G. B. and Hill, J. E., 1992. *The mammals of the Indomalayan region: a systematic review.* Natural History Museum Publications and Oxford University Press, Oxford, U.K, pp. vi+ 488.

De, B. (Ed.) 1994. *West Bengal District Gazetteers, 24–Parganas.* Government of West Bengal, Calcutta.

De, J.K., Chakraborty, S. and Chakraborty, R., 1998. Identification of dorsal guard hairs of five Indian species of mongoose, *Herpestes* Il–tiger (Mammalia: Carnivora). *Mammalia,* 62: 285–296.

Deb, S.B., 2009. Urban wildlife: status in the Kolkata metropolis. *West Bengal,* 51(9&10): 22–28.

Deuti, K., 2008. Habits, food preference and family life of the Marsh Mongoose (*Herpestes palustris* Ghose, 1965) in the Salt Lake Swamps near Calcutta, West Bengal. *Records of Zoological Survey of India,* 108(2): 47–50.

Dey, P., 2007. Study on Marsh Mongoose (*Herpestes palustris*, Ghose) in the East Kolkata Wetland. *Environ*, 10: 31–33.

Ellerman, J.R. and Morrison–Scott, T.C.S., 1951. *Checklist of Palaearctic and Indian mammals.* British Museum (Natural History), London, U.K.

Ewer, R. F., 1973. *The carnivores.* Cornell University Press, Ithaca, New York, U.S.A., p. 544.

Feldhamer, G.A., Drickamer, L.C., Vessey, S.H., Merritt, F. and Krajewski, C., 2007. *Mammalogy: adaptation, diversity, ecology*. 3rd Edn. Johns Hopkins University Press, Baltimore, Maryland, U.S.A., p. 672.

Furedy, C., 1987. From waste land to waste–not land: the role of the Salt Lakes, East Calcutta, in waste treatment and recycling, 1845–1930. In: *Calcutta: the urban experience. Essays in honour of Prof. Nisith R. Ray*, (Ed.) P. Sinha. Riddhi–India, Calcutta, India, pp. 145–153.

Gaubert, P. and Cordeiro–Estrela, P., 2006. Phylogenetic systematics and tempo of evolution of the Viverrinae (Mammalia: Carnivora: Viverridae) within feliformians: implications for faunal exchanges between Asia and Africa. *Molecular Phylogenetics and Evolution*, 41: 266–278.

Ghose, R.K., 1965. A new species of mongoose (Mammalia: Carnivora: Viverridae) from West Bengal, India. *Proceedings of the Zoologi–cal Society of Calcutta*, 18(2): 173–178.

Ghose, R.K. and Chaturvedi, Y., 1972. Extension of range of the mongoose, *Herpestes palustris* (Mammalia, Carnivora, Viverridae), with a note on its endoparasitic nematode. *Journal of the Bom–bay Natural History Society*, 69(2): 412–413.

Ghosh, A., 1997. *Management of East Calcutta Wetlands and canal system*. Department of Environment, Government of West Bengal assisted by the UK Overseas Development Administration.

Ghosh, A., 2009. *Biodiversity conservation*. A.P.H. Publishing Corporation, New Delhi, India, p. 320.

Ghosh, A.K., 2004. Avian diversity in East Calcutta wetlands. *Environ.*, 9(1): 8–13.

Ghosh, A.K. and Nandi, N.C., 2009. Community–based wetland management: A case study of Brace Bridge Nature Park (BBNP), Kolkata, India. In: *Sustaining the world's wetlands: setting policy and resolving conflicts*, (Ed.) R.C. Smardon. Springer Science, New York, U.S.A., pp. 125–152.

Gilchrist, J.S., Jennings, A.P., Veron, G. and Cavallini, P., 2009. Family Herpestidae (mongooses). In: *Handbook of the mammals of the world, vol. 1. Carnivores*, (Eds.) D.E. Wilson and R.A. Mittermeier. Lynx Edicions, Barcelona, Spain, pp. 262–328.

Hodgson, B.H., 1836. *Mangusta auropunctata. Journal of the Asiatic Society of Bengal*, 5: 235–236.

Honacki, J.H., Kinman, K.E. and Koeppl, J.W. (Eds.) 1982. *Mammal species of the world: a taxonomic and geographic reference*. Allen Press Inc. and the Association of Systematics Collections, Kansas, U.S.A., p. 694.

Hussain, S.A., 1999. Mustelids, viverrids and herpestids of India: spe–cies profile and conservation status. In: *Mustelids, viverrids and herpestids of India*, (Ed.) S. A. Hussain. Wildlife Institute of India, Dehradun, India, pp. 27–28.

Jerdon, T.C., 1874. *The mammals of India.* John Wheldon, London, U.K.

Isaac, N.J.B., Turvey, S.T., Collen, B, Waterman, C. and Baillie, J.E.M., 2007 . Mammals on the Edge: Conservation Priorities Based on Threat and Phylogeny. *PLOS One,* 2(3): e296.

Kumar, R., 2010. Integrated management planning for East Kolkata Wetlands. *East Kolkata Wetlands,* 1: 7–15.

Kundu, N., 2010. East Kolkata Wetlands: an introduction. *East Kolkata Wetlands* 1: 1–6.

Kundu, N., Pal, M. and Saha, S., 2008. East Kolkata Wetlands: a resource recovery system through productive activities. In: *Proceedings of Taal 2007: the 12th World Lake Conference,* (Eds.) M. Sengupta and R. Dalwani. International Lake Environment Committee, Jaipur, India, pp. 868–881.

Mallick, J.K., 2007. Endemic mammals of West Bengal. *West Bengal,* 49(10&11): 53–55.

Mallick, J.K., 2009. Endemic Marsh Mongoose *Herpestes palustris* (Carnivora: Herpestidae) of East Kolkata Wetlands, India: A status report. *Journal of Threatened Taxa,* 1(4): 215–220.

Mallick, J.K., 2011. Records and conservation status review of the endemic Bengal Mongoose *Herpestes palustris* Ghose, 1965 in southern West Bengal, India. *Small Carnivore Conservation* 45: 31–48.

Mallick, J.K., 2012. New records and extension of north-eastern range of endeimic Bengal Mongoose *Herpestes palustris* Ghose, 1965 in southern West Bengal, India. *Journal of New Biological Reports,* 1(2): 47–60.

Menon, V., 2003. *A field guide to Indian mammals*. Dorling Kindersley (India) Pvt. Ltd, Delhi, India, p. 200.

Molur, S., Nameer, P.O. and Walker, S. (Eds) 1998. *Report of the workshop "Conservation Assessment and Management Plan for Mammals of India" (BCCP – Endangered Species Project)*. Zoo Outreach Organisation and Conservation Breeding Specialist Group– India, Coimbatore, India.

Mukherjee, D.P., 2011. Stress of urban pollution on largest natural wetland ecosystem in East Kolkata–causes, consequences and improvement. *Archives of Applied Science Research,* 3(6): 443–461.

Nandi, N.C., Das, S.R., Bhuinya, S. and Dasgupta, J.M., 1993. Wetland faunal resources of West Bengal: 1. North and South 24–Parganas districts. *Records of the Zoological Survey of India,* Occasional Paper No.150: 10. Zoological Survey of India, Kolkata, India.

Nandy, H., 2006. Threatened mammals of West Bengal. *West Bengal,* 48(10): 25–30.

Nellis, D.W., 1989. *Herpestes auropunctatus. Mammalian Species,* 342: 1–6.

Pocock, R.I., 1937. The mongooses of British India, including Ceylon and Burma. *Journal of the Bombay Natural History Society,* 39: 211–245.

Pocock, R.I., 1941. *Fauna of British India,* Mammalia, 2. Tailor and Francis, London, U.K.

Porter, R.F., Willis, I., Christensen, S. and Nielsen, B.P., 1974. *Flight identification of European raptors.* T. and A.D. Poyser, Calton, U.K., p. 184.

Prater, S.H., 1971. *The book of Indian animals.* Bombay Natural History Society and Oxford University Press, Mumbai, India, p. 348.

Purkait, N.N. and Chakraborty, D.K., 2011. Pattern of methane emission from a garbage dump. *Current Science,* 101(4): 528–531.

Raghu Ram, T.L., 2005. *Biodiversity register Indian Institute of Management Calcutta: birds–snakes–lizards–mammals.* CMDS Project, Centre for Development and Environment Policy, IIM, Kolkata, India.

Rodgers, W.A. and Panwar, H.S., 1988. *Planning wildlife protected area network in India: the report, Vol. 1&2.* Wildlife Institute of India, Dehradun, India, p. 267.

Saha Ghatak, D., 2010. *Trade–off between conservation of environment and economic development? A case study of East Kolkata Wetland.* M.A. thesis, Graduate School of Development Studies, Interna–tional Institute of Social Studies, The Hague, Netherlands, pp. x+55.

Saha, R., Chattopadhyay, A. and Naidu, B.R., 2003. Pollution potential of leachate from East Kolkata solid waste dumpsite. *Workshop on sustainable landfill management,* 3–5 December 2003, Chennai, India, pp. 303–307.

Saha, S.S. and Singhal, N., 1996. *Mammalian specimens of Bengal Natural History Museum, Darjeeling.* Darjeeling: Department of Forest (Wildlife Wing), Government of West Bengal, 69 pp.

Sahajpal, V., Goyal, S.P., Raza, R. and Jayapal, R., 2009. Identification of mongoose (genus: *Herpestes*) species from hair through band pattern studies using discriminate functional analysis (DFA) and microscopic examination. *Journal of the Forensic Science Society,* 49: 205–209.

Sanyal, A.K., Ramakrishna and Alfred, J.R.B., 2007. Faunal resources in West Bengal. *West Bengal,* 49(10&11): 38–48.

Sanyal, P., 1999. Common mustelids, viverrids and herpestids of West Bengal. In: *Mustelids, viverrids and herpestids of India,* (Ed.) Hussain, S.A. Wildlife Institute of India, Dehradun, India, pp. 59–61.

School of Water Resources Engineering Jadavpur University Kolkata (SWREJUK)., 2007. *Monitoring and modeling of discharge of dry weather flow (DWF) to East Kolkata Wetlands (EKW): Final report.* Jadavpur University, Kolkata, p. 62.

Schreiber, A., Wirth, R., Riffel, M. and Van Rompaey, H., 1989. *Weasels, civets, mongooses, and their relatives: an action plan for the conservation of mustelids and viverrids.* IUCN, Gland, Switzerland, p. 99.

Sen, S.K., 2010. The hunter and the hunted and the effectiveness of camouflage. Delhibirdpix. [*https: //groups.google.com/group/delhibirdpix*]. Downloaded on 20 April 2011.

Sharma, U., 2011. Effect of global warming on biodiversity in India. In: *National Conference on Forest Biodiversity: Earth's living treasure, 22nd May 2011*, (Ed.) Uttar Pradesh State Biodiversity Board Uttar Pradesh State Biodiversity Board. Lucknow, India, pp. 76–84.

Soota, T.D. and Chaturvedi, Y., 1970. On five new species of nematodes from vertebrates. *Proceedings of the Indian Academy of Sciences, Section B*, 71(3): 100–108.

Srinivasulu, C. and Nameer, P.O., 2003. Report on the training workshop on reintroduction, conservation welfare and conservation breeding with special emphasis on non–volant small mammals. *Rat–a–tattle – RILSCINSA Newsletter*, 3(1): 7–12.

Taylor, M.E. and Matheson, J., 1999. A craniometric comparison of the African and Asian mongooses in the genus *Herpestes* (Carnivora: Herpestidae). *Mammalia*, 63: 449–464.

Van Rompaey, H. and Colyn, M., 1996. An annotated bibliography of the Herpestidae and Viverridae (Carnivora, Mammalia). *Musée Royal de l'Afrique Centrale, Annales Sciences Zoologiques*, 279: 1–411.

Veron, G., Colyn, M. and Dunham, A. E., 2004. Molecular systematics and origin of sociality in mongooses (Herpestidae, Carnivora). *Molecular Phylogenetics and Evolution*, 30(3): 582–598.

Veron, G., Patou, M.L., Pothet, G., Simberloff, D. and Jennings, A.P., 2006 (for 2007). Systematic status and biogeography of the Javan and Small Indian Mongooses (Herpestidae, Carnivora). *Zoologica Scripta*, 36(1): 1–10.

Veron, G., Patterson, B.D. and Reeves, R., 2008. Global diversity of mammals (Mammalia) in freshwater. *Hydrobiologica*, 595(1): 607–617.

Walker, S., 1999. Endemic and non–endemic mammals in Indian zoos– from Central Zoo Authority Data Base 1998 and BCPP CAMP Workshop Output. Part III – civets and mongooses of India. *Zoos' Print*, 14(8): 7–18.

Walker, S. and Molur, S., 1999. Conservation breeding and Indian species of mustelids, viverrids and herpestids: a perspective. In: *Mustelids, viverrids and herpestids of India*, (Ed.) S.A. Hussain. Wildlife Institute of India, Dehra Dun, India, p. 108.

Wenzel, E. and Haltenorth, T., 1972. System der Schleichkatzen (Viverridae). *Säugetierkundliche Mitteilungen*, 20: 110–127.

Wozencraft, W.C., 1993. Order Carnivora. In: *Mammalian species of the world. A taxonomic and geographic reference*, 2nd edition, (Eds.) D.E. Wilson and D.M. Reeder. Smithsonian Institution Press, Washington D.C., U.S.A., pp. 279–348.

Wozencraft, W. C., 2005. Order Carnivora. In: *Mammalian species of the world. A taxonomic and geographic reference*, 3rd edition, (Eds.) D.E. Wilson and D.M. Reeder. The John Hopkins University Press, Baltimore, U.S.A., pp. 532–628.

Wozencraft, W.C., Duckworth, J.W., Choudury, A., Muddapa, D., Yonzon, P., Kanchanasaka, B., Jennings, A. and Veron, G., 2008. *Herpestes javanicus*. In: *IUCN Red List of Threatened Species*, version 2010.4, IUCN 2010. <*www.iucnredlist.org*>. Downloaded on 29 May 2011.

2013, Animal Diversity, Natural History and Conservation, Vol. 3 *Pages* **363–387**
Editors: **V.K. Gupta and Anil K. Verma**
Published by: **DAYA PUBLISHING HOUSE, NEW DELHI**

Chapter 19

Animal Diversity in Kuldiha Wildlife Sanctuary, Odisha, India

S.D. Rout*, Sunit K. Das, B. Baruah, P. Medhi and R.K. Mishra
Department of Wildlife and Biodiversity Conservation, North Orissa University, Takatpur, Baripada – 757 003, Odisha, India

ABSTRACT

Kuldiha Wildlife Sanctuary is an undocumented area of Balasore district, Odisha and no previous literature is available on it. Therefore, this study was conducted to get some baseline information on animal diversity of the sanctuary and it is highlighted in this paper. Like other forests in the region, forest of the sanctuary are relatively preserved, maintaining naturalness of the area. Habitat surveyed includes tropical semi evergreen forest, sal tree forest, moist mix deciduous forest, roadside vegetation, water bodies and their surroundings. A total of 86 species of invertebrates, 9 species of fishes, 5 species of amphibian, 39 species of reptiles, 141 species of birds and 25 species of mammals were recorded during the study. The study reflects an overview on faunal diversity and conservation need of the sanctuary. Sloth bear and Elephant are the key species of the area and study reveals that the sanctuary is a good habitat for Leopard. Other mammals have low population due to hunting and food gathering habit of tribal and loss of habitat. The presence of giant squirrel also indicates the good health of the upper canopy of the sanctuary. The Sanctuary is also connected to Simlipal Biosphere Reserve of Mayurbhanj district where Sukhpada hills and Nato hills in the west act as a corridor. Due to high temperature, low rainfall and inadequate number of perennial water bodies, Sanctuary faced the high shortage of water.

---

* *Corresponding author.* E-mail: srusti_d_rout@rediffmail.com

Forest fire is also occurred rapidly during dry months, which are the main issues of wildlife management point of view. Considering these facts, we suggested the conservation and management of animals which were identified must be taken as essential element of sustainable wildlife biodiversity.

Keywords: *Kuldiha, Biological diversity, Conservation, Wildlife biodiversity, Odisha.*

Introduction

Variations is the law of nature. It occurs everywhere and every moment, which makes the nature beautiful. On the other hand the variety and variability of organism and ecosystems is referred to as biological diversity or biodiversity (Kumar and Ajija, 2002).Relating the distribution of animal species with spatial patterns of environmental features is one of the major goals of landscape ecology. It carries considerable significance in terms of conservation planning. The biodiversity of State Odisha reflects constituents from four Biogeographic provinces namely Eastern plateau, Chhotnagpur plateau, the lower gangetic plain and the Eastern coast (Wildlife scenario in Orissa, 2008). The region comes under Eastern Ghat and act as a connecting link between the Himalayas and the Western part of India. Kuldiha Wildlife Sanctuary comes under Biogeographic region of Deccan peninsular which has a tremendous genetic and ecological importance with variety of flora and fauna. The area is a home to several endangered species like Sloth bear, Leopard, Mouse deer, etc. and provides a good habitat for Asian elephants. Though the Kuldiha Wildlife Sanctuary is rich in biodiversity but no systematic account is available on faunal diversity. Therefore, this study was designed for understanding the distribution of fauna in the Kuldiha Wildlife Sanctuary in relation to identify animal species and their habitat, and to identify the threats to the sanctuary.

Study Area

The Kuldiha Wildlife Sanctuary is located in the Nilgiri Civil Sub division of Balasore (Wildlife) division bordering to Mayurbhanj District and the area lies between 210-20/N to 210-30/N (Latitude) and 860-30/E to 860-45/E (Longitude) (Figure 19.1). The Sanctuary encompasses an area of 272.75 km^2 which comprises of five ranges. Interestingly Kuldiha Wildlife Sanctuary is connected to Similipal Biosphere Reserve of Mayurbhanj district through Sukhpada hills and Nato hills in the west, which is typical representative of mixture of peninsular (coastal) Sal forest and moist deciduous forests (Champion and Seth, 1968). It comes in the biogeographic region of Deccan peninsular zone which is of tremendous genetic and ecological importance with diversity of flora and fauna.

Biogeographically the sanctuary forms a part of 6-Deccan Peninsula Biogeographic Zone and 6-C "Eastern-plateau"–biotic province. It forms north eastern tip of 6 C and North eastern ghat sub division or region (Rodger and Panwar, 1988). Udvardy (1969) classifies the area under "Chotanagpur plateau biotic province (6D) of Deccan Peninsula Biogeographic Zone". It falls under "tropical deciduous forest biome". The area exhibits a high level of heterogeneity in vegetation both at the local and landscape level, strongly influenced by the topographic complexity.

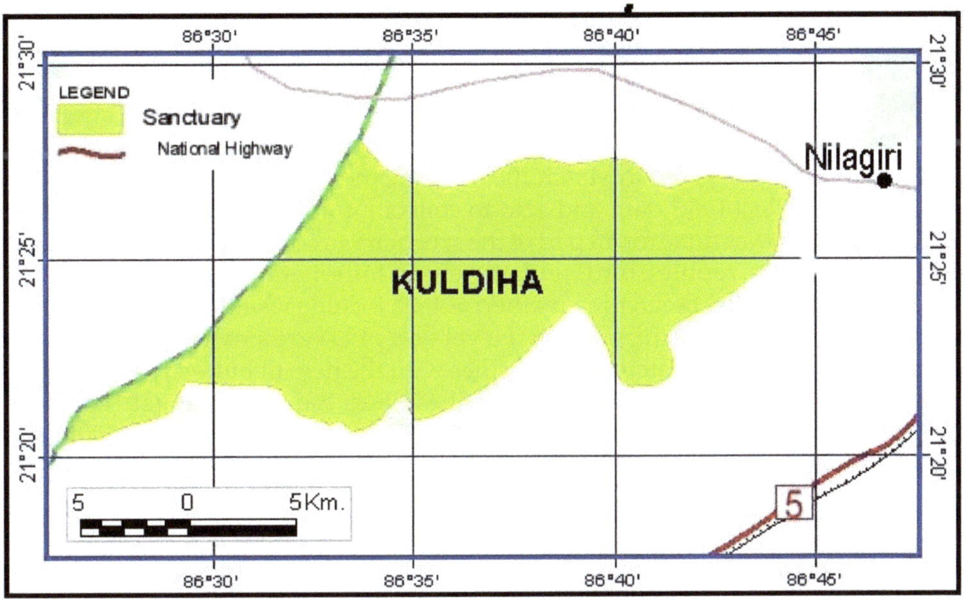

Figure 19.1: Map Shows the Study Area

The range of altitudes varies from 169 mtrs. to 682 mtrs. The river Sono, a tributary of Budhabalanga, flows in the western and northern portion of the forest with several seasonal streams. The climate is tropical with three prominent seasons which are summer, monsoon and winter. Summer is the dominant season with high temperature from March to mid July. The maximum temperature at times touches beyond 42°C. Kuldiha Wildlife Sanctuary faces acute water shortage during the dry months. Water remains in very few places and the wildlife in the area face acute shortage in summer especially during drought.

The area is home of several endangered species like leopard, mouse deer, pangolin, python, monitor lizard, peafowl, etc, elephant being the flagship species. Past record reveals that the 'Makhanas' in the elephant population which are rare found in this sanctuary. The area has got vast potency for ecotourism, where community based eco-tourism will be a viable proposition giving source of alternate livelihood.

No special documentation on animal diversity has been recorded from the sanctuary so far. However, the preliminary documentation for establishment of the reserve as a wildlife sanctuary the forest official recorded the presence of 30 species of snakes, 14 species of lizards, 3 species of turtles and 4 species of anurans from the sanctuary (Kuldiha Wildlife Division, 1984). This study outlines the significance of dry forest habitats to the overall species richness of the Kuldiha. Due to the proximity of dry forest habitats to human habitation and their inclusion in the buffer zone of the reserves, these habitats are severely threatened by anthropogenic activities. The need for bringing greater extent of these habitats under current protected area network for an effective conservation of the fauna of Kuldiha Wildlife Sanctuary has been highlighted and discussed in the light of current human pressure.

Material and Methods

The data on faunal diversity inside the sanctuary area was systematically collected from October 2008 to March 2009. During the study 453 technical man-days was spent to collect field data and also to collect information from other sources. Extensive survey was done for all part of the sanctuary to collect data on invertebrates, fishes, amphibians, reptiles, birds and mammals. Information on invertebrates was obtained through direct searching method, which include visual encounter surveys, especially in care of butterflies which (Heyer *et al.,* 1994 are small insects. So for the identification we had to catch the butterflies with the help of butterfly net and after photography and identification released these into the same habitat. For better identification of the butterflies used the book of Haribal (1992), Kunte (2000), and Kehimkar (2008).

For fishes trapping method was used, which is one of the most versatile fish censusing method (Sutherland, 1996). Like that of invertebrates taxonomic and ecological data on herpetofauna in the area was collected systematically through direct searching method. Avifauna of the study area was surveyed through visual encounter by line and road transect method (Buckland *et al.,* 1993). In addition to these methods vocal individuality method (Saunders and Wooler, 1988) was also used to identify some bird species, because many species of birds are detected by their calls and songs (Rappole *et al.,* 1993).

In case of mammalian fauna modified transect method (Burnham *et al.,* 1980) was used and transects were laid randomly to find out abundance and rarity of mammals. Animals on both side of the transect were listed. Direct and indirect evidences were collected from primary and secondary sources and people were also interviewed to collect data from them (Das, 2010). The indirect evidences such as scats/droppings, feathers, scales or hairs were collected to get species level confirmation in doubtful cases. Permanent preserved hair materials were used as reference materials for comparison.

Results and Discussion

Mammalian Fauna

A perusal of Table 19.1 reveals that a total of 25 species of mammals were recorded from the sanctuary area of which 12 species were directly sighted. They are elephant, bison, giant squirrel, small Indian otter, barking deer, cheetal, common langur, rhesus macaque, Indian wild boar, common mongoose, Indian field mouse, Indian Flying Fox. Indirect evidences have been obtained for 14 other species. Sloth bear is the key species of the area. But no direct sighting was recorded. Indirect evidences were seen a number of times in form of dung (fresh and dried), pugmarks, scratch marks, ant and termite hills with feeding hairs, pits dug by the sloth bear and compressed grass in the area, where animal could have sat or moved. Elephant is the flagship species of the sanctuary and the area comes under Mayurbhanj Elephant Reserve. The elephant population during 2005 census was estimated to be 59. Direct sighting occurred during the survey period and indirect evidences were obtained in form of dung (fresh and dried), feeding signs, their resting place found near Badagohalo hills. Elephant calls were also clearly heard and their footprints were also found at different places of the sanctuary.

Table 19.1: List of Mammalian Fauna of Kuldiha Wildlife Sanctuary

Common Name	*Scientific Name*	*Family*	*Site Record*	*Schedule*
Barking deer	*Muntiacus muntjak*	*Cervidae*	Several places	III
Mouse deer	*Moschiola meminna*	*Tragulidae*	KLD	I
Sambar	*Cervus unicolor*	*Cervidae*	Several places	III
Cheetal	*Axis axis*	*Cervidae*	Several places	III
Indian bison	*Bos gaurus*	*Bovidae*	PU,KLD,DH	I
Wild boar	*Sus scrofa*	*Suidae*	KLD	II
Indian fox	*Vulpes benghalensis*	*Canidae*	DH	II
Jackel	*Canis aureus*	*Canidae*	JO	IV
Jungle cat	*Felis chaus*	*Felidae*	Several places	II
Royal Bengal tiger	*Panthera tigris*	*Felidae*	KLD	I
Leopard	*Panthera paradus*	*Felidae*	Several places	I
Common mogoose	*Herpestes edwardsi*	*Harpestidae*	JO, KLD	IV
Striped hyaena	*Hyaena hyaena*	*Hyaenidae*	JO	III
Small Indian otter	*Lutra perspicillata*	*Mustelidae*	CH,NIL	
Sloth bear	*Melursus ursinus*	*Ursidae*	JO,KLD.NIL	II
Small Indian civet	*Vivericula indica*	*Viverridae*	DH,KLD	II
Indian flying fox	*Pteropus giganteus*	*Pteropodidae*	DH,KLD	III
Indian hare	*Lepus nigricollis*	*Leporidae*	Several places	IV
Common langur	*Semnopithecus entellus*	*Cercopithecidae*	Several places	II
Rhesus macaque	*Macaque mullata*	*Cercopithecidae*	KLD	II
Indian elephant	*Elephus maximus*	*Elephantidae*	Several places	I
Indian giant squirrel	*Ratufa indica*	*Sciuridae*	Several places	II
Indian porcupine	*Hystrix indica*	*Hystricidae*	DH	IV
Indian field mouse	*Mus booduga*	*Muridae*	Several places	IV
Rat	*Rattus rattus*	*Muridae*	Several places	IV

Kuldiha is hilly and suitable for peaceful existence of wild animals. Tiger and leopard population was estimated during 2004, which was 6 in numbers (Tiger census report, 2004). The pugmarks of leopards and Royal Bengal Tiger were found near Borogohalo and Debigarh hills during the survey period. A good number of bison is present inside the protected area and sighted at Jodachua salt lick during the survey period. Buffalos are also present but they are feral in nature. Mammalian species of the sanctuary were identified following Prater (1998).

Avifauna

A total of 141 species of birds of 52 families were reported from the sanctuary (Table 19.2). Out of 141 species 85.8 per cent (121) were resident forms and 14.2 per cent (20) migrants. Among the birds Black-bellied Tern (*Sterna acuticauda*) is considered nearly threatened according to the Red Data Book of Birdlife International (2006). No avifaunal survey work had previously been conducted in this Sanctuary. Due to the lack of published information on birds in this region, the main aim of this research was to investigate the diversity of birds in the study area. This checklist is totally new for the area as there no any earlier reports are available. Analysis of encounter rates revealed the fact that Black Drongo, Black Hooded Oriole, Black Napped Oriole, Plum Headed Parakeet, Rose Ringed Parakeet, Alexandrine Parakeet, Rufous Tree Pie, Crimson Breasted Barbet, Large Tailed Nightjhar, Red Vented Bulbul, Spotted Dove are the main birds of the sanctuary with high encounter rates. From the present survey it is also clear that Kuldiha is a good habitat for raptors and a total number of 15(10.6 per cent) species of raptors were recorded. Indian grey hornbill was also observed in the sanctuary.

The numbers of water birds are very less inside the sanctuary due to the absence of perennial water sources. Artificial water reservoirs should be constructed to fill up the shortage of water.

Herpetofauna

Among the total 5 amphibian species, 4 species documented previously in this area were observed in the present study (Table 19.3). In addition, another 1 species (*Polypedates maculatus*) as added, for a total of 5 species of recorded amphibians. Of the lizard species, out of 15 species 2 species (*Cosymbotus platyurus* and *Cyrtodactylus nebulosa*) have not been recorded previously from this area and remaining 13 species documented previously were observed in this study (Table 19.3). In case of snakes *Ptyas mucosum, Xenochropis piscator* and *Lycodon aulicus* were common throughout the sanctuary. During the project work it is also clear that Kuldiha provides a good habitat for King cobra (*Ophiophagus hanna*), which is the largest venomous snake in the world (Aggard, 1983). Although 24 species of snake previously noted was observed during this study, an additional 2 new species (*Ahaetulla nasutus issabalinus* and *Lycodon aulicus*) were documented from this area which has not recorded in the preliminary survey; we recorded a total of 26 species of snakes from this area. Among the anurans *Duttaphrynus melanostictus, Hoplobatrachus tigerina* show the dominancy in the entire forest areas of the sanctuary. *Psammophilus dorsalis* show the dominancy among the other lizards. In case of snakes *Ptyas mucosum* and *Xenochropis piscator* and *Lycodon aulicus* common throughout the sanctuary. But the chelonian diversity is poor in compare to the other reptiles groups.

No turtles were directly observed or captured in the wild during the surveys. The information was collected about the turtles and tortoises of the sanctuary from the secondary source through the interaction with villagers. However, it appears that the freshwater turtles in the area are heavily harvested by the local peoples. Two species were identified as *Chitra indica* (Gray, 1831) and *Lissemys punctata* (Lacépède,

Table 19.2: List of Avifauna of Kuldiha Wildlife Sanctuary

Sl.No.	*Common Name*	*Family/Scientific Name*	*Site Record*	*Status*
	Grebes	**Podicipedidae**		
1.	Little Grebe (5)	*Tachybaptus ruficollis* (Pallas, 1764)	KLD	R
	Cormorants/Shags	**Phalacrocoracidae**		
2.	Little cormorant (28)	*Phalacrocorax niger* (Vieillot, 1817)	Several places	R
3.	Indian Shag (27)	*Phalacrocorax fuscicollis* Stephens, 1826	PU	R
	Herons, Egrets and Bitterns	**Ardeidae**		
4.	Little Egret (49)	*Egretta garzetta* (Linnaeus, 1766)	CH	R
5.	Large Egret (45-46)	*Casmerodius albus* (Linnaeus, 1758)	JO	R
6.	Median Egret (47, 48)	*Mesophoyx intermedia* (Wagler, 1829)	Several places	R
7.	Cattle Egret (44)	*Bubulcus ibis* (Linnaeus, 1758)	Several places	R
8.	Indian Pond Heron (42-42a)	*Ardeola grayii* (Sykes, 1832)	Several places	R
9.	Chestnut Bittern (56)	*Ixobrychus cinnamomeus* (Gmelin, 1789)	JO	R
	Storks	**Ciconiidae**		
10.	Asian Openbill-Stork (61)	*Anastomus oscitans* (Boddaert, 1783)	CH, NIL	R
	Hawks, Eagles, Buzzards, Old world Vultures, Kites, Harriers	**Accipitridae**		
11.	Oriental Honey-Buzzard (129-130)	*Pernis ptilorhynchus* (Temminck, 1821)	DH	R
12.	Black-shouldered Kite (124)	*Elanus caeruleus* (Desfontaines, 1789)	DH, KLD	R
13.	Black Kite ((132-134)	*Milvus migrans* (Boddaert, 1783)	Several places	R
14.	Short-toed Snake Eagle (195)	*Circaetus gallicus* (Gmelin, 1788)	BH	R
15.	Crested Serpent Eagle (196-200)	*Spilocornis cheela* (Latham, 1790)	BH, DH, PLR	R
16.	Pallied Harrier (190)	*Circus macrourus* (S.G.Gmelin, 1770)	JO	M
17.	Shikra (137-140)	*Accipiter badius* (Gmelin, 1788)	Several places	R
18.	White-eyed Buzzard (157)	*Butastur teesa* (Franklin, 1832)	KLD	R

Contd...

Table 19.2–*Contd...*

Sl.No.	*Common Name*	*Family/Scientific Name*	*Site Record*	*Status*
	Osprey	**Pandionidae**		
19.	Osprey (203)	*Pandion haliaetus* (Linnaeus, 1758)	BH	M
	Falcons	**Falconidae**		
20.	Common Kestrel (222-224)	*Falco tinnunculus* Linnaeus, 1758.	JO	M
21.	Peregrine Falcon (209-211)	*Falco peregrinus* Tunstall, 1771.	BH	R
	Pheasants, Partridges, Quails	**Phasianidae**		
22.	Jungle Bush –Quail (255-258)	*Perdicula asiatica* (Latham, 1790)	JO	R
23.	Red Junglefowl (299-300)	*Gallus gallus* (Linnaeus, 1758)	KLD, JO, PU	R
24.	Indian Peafowl (311)	*Pavo cristatus* Linnaeus, 1758.	KLD	R
	Rails, Crakes, Moorhens, Coots	**Rallidae**		
25.	White BreastedWaterhen (343-345)	*Amaurornis phoenicurus* (Pennant, 1769)	PU	R
26.	Common Moorhen (347-347a)	*Gallinula chloropus* (Linnaeus, 1758)	PU	R
	Jacanas	**Jacanidae**		
27.	Pheasant-tailed Jacana (358)	*Hydrophasianus chirurgus* (Scopoli, 1786)	PU	R
28.	Bronze-winged Jacana (359)	*Metopidus indicus* (Latham, 1790)	KLD	R
	Ploveres, Dotterels, Lapwings	**Charadridae**		
29.	Little ringed Plover (379-380)	*Charadrius dubius* Scopoli, 1786.	JO	R
30.	Red-wattled Lapwing (366-368)	*Vanellus indicus* (Boddaert, 1783)	PLR	R
	Sandpipers, Stints, Snipes, Godwits and Curlews	**Scolopacidae**		
31.	Common Sandpiper (401)	*Actitis hypoleucos* Linnaeus, 1758.	PU	M
	Gulls, Terns and Noddies	**Laridae**		
32.	Black-headed Gull (455)	*Larus ridibundus* Linnaeus, 1766.	CH	M
33.	River Tern (463)	*Sterna aurantia* J.E.Gray, 1831.	CH	R
34.	Black-bellied Tern (470)NT	*Sterna acuticauda* J.E.Gray, 1831.	CH	R

Contd...

Table 19.2–*Contd...*

Sl.No.	*Common Name*	*Family/Scientific Name*	*Site Record*	*Status*
	Pigeons and Doves	**Columbidae**		
35.	Blue Rock Pigeon (516-517)	*Columba livia* Gmelin, 1789.	JO	R
36.	Europeaan Turtle Dove (529)	*Streptopelia turtur* (Linnaeus, 1758)	KLD, JO	R
37.	Oriental Turtle-Dove (530-533)	*Streptopelia orientalis* (Latham, 1790)	KLD, PLR, SGH	R
38.	Spotted Dove (537-540)	*Streptopelia chinensis* (Scopoli, 1786)	Several places	R
39.	Eurasian Collared-Dove (534)	*Streptopelia decaocto* (Frivaldszky, 1838)	JO, PU	R
40.	Emerald Dove (542-544a)	*Chalcophaps indica* (Linnaeus, 178)	NIL, DH	
	Parakeets and Hanging-Parrots	**Psittacidae**		
41.	Alexandrine Parakeet (545-548)	*Psittacula eupatria* (Linnaeus, 1766)	KLD, JO	R
42.	Rose-ringed Parakeet (549-550)	*Psittacula krameri* (Scopoli, 1769)	Several places	R
43.	Plum headed Parakeet (557-558)	*Psittacula cyanocephala* (Linnaeus, 1766)	Several places	R
	Cuckoos, Malkohas and Coucals	**Cuculidae**		
44.	Pied Crested Cuckoo (570-571)	*Clamator jacobinus* (Boddaert, 1783)	BH	M
45.	Brainfever Bird (573-574)	*Hierococcyx varius* (Vahl, 1797)	DH, KLD	
46.	Indian Cuckoo (576)	*Cuculus micropterus* Gould, 1838.	PLR	R
47.	Asian Koel (590-592)	*Eudynamys scolopacea* (Linnaeus, 1758)	PLR, JO	R
48.	Sirkeer Malkoha (596-598)	*Phaenicophaeus leschenaultii* (Lesson, 1830)	BH	R
49.	Greater Coucal (600-602)	*Centropus sinensis* (Stephens, 1815)	PLR, KLD, PU	R
	Barn Owls	**Tytonidae**		
50.	Barn Owl (606-607)	*Tyto alba* (Scopoli, 1769)	JO	R
	Owls	**Strigidae**		
51.	Collared Scops-Owl (619-624)	*Otus bakkamoena* Pennant, 1769.	BH	R
52.	Jungle Owlet (636-637)	*Glaucidium radiatum* (Tickell, 1833)	JO, SGH	R
53.	Spotted Owlet (650-652)	*Athene brama* (Temminck, 1821)	NIL	R

Contd...

Table 19.2–*Contd...*

Sl.No.	*Common Name*	*Family/Scientific Name*	*Site Record*	*Status*
	Nightjars	**Caprimulgidae**		
54.	Indian Jungle Nightjar (670-672a)	*Caprimulgus indicus* Latham, 1790	JO, KLD	R
55.	Common Indian Nightjar (680-681)	*Caprimulgus asiaticus* Latham1790	JO	R
	Swifts	**Apodidae**		
56.	Asian Palm Swift (707-708)	*Cypsiurus balasiensis* (J.E.Gray, 1829)	BH	R
57.	House Swift (702-706)	*Apus affinis* (J.E.Gray, 1830)	SGH, NIL	R
	Tree-Swifts	**Hemiprocnidae**		
58.	Crested Tree-Swift	*Hemiprocne coronata* (Tickell, 1833)	DH	R
	Kingfishers	**Alcedinidae**		
59.	Small Blue Kingfisher (722-724)	*Alcedo atthis* (Linnaeus, 1758)	JO, PU	R
60.	White Breasted Kingfisher (735-738)	*Halcyon smyrnensis* (Linnaeus, 1758)	PLR, JO, PU	R
	Bee-eaters	**Meropidae**		
61.	Small Bee-eater (749-752)	*Merops orientalis* Latham, 1801	KLD	R
62.	Blue-tailed Bee-eater (748)	*Merops philippinus* Linnaeus, 1766	JO, KLD	M
	Rollers	**Coraciidae**		
63.	Indian Roller (755-757)	*Coracias benghalensis* (Linnaeus, 1758)	PLR, KLD, NIL	R
	Hoopoes	**Upupidae**		
64.	Common Hoopoe (763-766)	*Upupa epops* Linnaeus, 1758.	SGH, NIL, KLD	R
	Hornbills	**Bucerotidae**		
65.	Indian Grey Hornbill (767)	*Ocyceros birostris* (Scopoli, 1786)	BH	R
	Barbets	**Capitonidae**		
66.	Brown-headed Barbet (780-782)	*Megalaima zeylanica* (Gmelin, 1788)	BH, DH, JO	R
67.	Coppersmith Barbet (792)	*Megalaima haemacephala* (P.L.S. Muller, 1776)	BH	R

Contd...

Table 19.2–*Contd...*

Sl.No.	*Common Name*	*Family/Scientific Name*	*Site Record*	*Status*
	Woodpeckers	**Picidae**		
68.	Brown-capped Pygmy Woodpecker (851-854)	*Dendrocopos nanus* (Vigors, 1832)	JO	R
69.	Lesser Golden-backed Woodpecker (818-823)	*Dinopium benghalense* (Linnaeus, 1758)	NIL, JO	R
70.	Greater Golden-backed Woodpecker (860-863)	*Chrysocolaptes lucidus* (Scopoli, 1786)	NIL, JO, KLD	R
	Larks	**Alaudidae**		
71.	Ashy-crowned Sparrow-Lark (878)	*Eremopterix grisea* (Scopoli, 1786)	PU	R
72.	Eastern Skylark (904-909)	*Alauda gulgula* Franklin, 1831	PU, DH	M
	Wagtails and Pipits	**Motacillidae**		
73.	Forest Wagtail (1874)	*Dendronanthus indicus* (Gmelin, 1789)	KLD	M
74.	White Wagtail (1885-1890)	*Motacilla alba* Linnaeus, 1758.	KLD, NIL	M
75.	Large Pied Wagtail (1891)	*Motacilla maderaspatensis* Gmelin, 1789	SGH, PU	R
76.	Yellow Wagtail (1875-1880)	*Motacilla flava* Linnaeus, 1758	JO	M
77.	Grey Wagtail (1884)	*Motacilla cinerea* Tunstall, 1771	KLD	M
78.	Paddyfield Pipit (1858-1860)	*Anthus rufulus* Vieillot, 1818	PLR, KLD	R
79.	Eurasian Tree Pipit (1854-1855)	*Anthus trivialis* (Linnaeus, 1758)	JO	M
	Cuckoo-Shrikes, FlycatcherShrikes, Trillers, Minivets, Woodshrikes	**Campephagidae**		
80.	Large Cuckoo-Shrike (1072-1075)	*Coracina macei* (Lesson, 1830)	BH	R
81.	Black-headed Cuckoo-Shrike (1078-1079)	*Coracina melanoptera* (Ruppell, 1839)	BH, DH, JO	R
82.	Rosy Minivet (1089)	*Pericrocotus roseus* (Vieillot, 1818)	JO	R
83.	Small Minivet (1090-1095)	*Pericrocotus cinnamomeus* (Linnaeus, 1766)	NIL	R
84.	Scarlet Minivet (1080-1083)	*Pericrocotus flammeus* (Forster, 1781)	PLR	R

Contd...

Table 19.2–*Contd...*

Sl.No.	*Common Name*	*Family/Scientific Name*	*Site Record*	*Status*
85.	Large Woodshrike (1067-1068)	*Tephrodornis gularis* (Raffles, 1822)	BH	R
86.	Common Woodshrike (1069-1071)	*Tephrodornis pondicerianus* (Gmelin, 1789)	KLD	R
	Bulbuls and Finchbills	**Pycnonotidae**		
87.	Red-whiskered Bulbul (1118-1122)	*Pycnonotus jocosus* (Linnaeus, 1758)	KLD, JO, PU	R
88.	Red-vented Bulbul (1126-1132)	*Pycnonotus cafer* (Linnaeus, 1766)	Several places	R
	Ioras, Chloropsis/Leafbird, Fairy-Bluebird	**Irenidae**		
89.	Common Iora (1097-1101)	*Aegithina tiphia* (Linnaeus, 1758)	JO	R
	Shrikes	**Laniidae**		
90.	Brown Shrike (949-950a)	*Lanius cristatus* Linnaeus, 1758	SGH, PLR, PU	M
91.	Bay-backed Shrike (939-940)	*Lanius vittatus* Valenciennes, 1826	KLD	R
92.	Rufous-backed Shrike (946-948)	*Lanius schach* Linnaeus, 1758	BH	M
	Thushes, Shortwings, Robins, Forktails, Wheaters	**Turdinae**		
93.	Blue Rock-Thrush (1725-1726)	*Monticola solitaries* (Linnaeus, 1758)	DH, KLD	M
94.	Orange-headed Thrush (1733-1736)	*Zoothera citrina* (Latham, 1790)	JO	R
95.	Oriental Magpie- Robin (1661-1664)	*Cpsychus saularis* (Linnaeus, 1758)	Several places	R
96.	White-rumped Shama (1665-1668)	*Copsychus malabaricus* (Scopoli, 1786)	JO, NIL	R
97.	Black Redstart (1671-1672)	*Phoenicurus ochruros* (Gmelin, 1774)	JO	M
98.	Indian Robin (1717-1721)	*Saxicoloides fulicata* (Linnaeus, 1776)	NIL, JO, SGH	R
99.	Pied Bushchat (1700-1703)	*Saxicola caprata* (Linnaeus, 1766)	Several places	R
	Babblers, Laughingthrushes, Babaxes, Barwings, Yuhinas	**Timaliinae**		
100.	Common Babbler (1253-1254)	*Turdoides caudatus* (Dumont, 1823)	JO, DH	R
101.	Jungle Babbler (1261-1265)	*Turdoides striatus* (Dumont, 1823)	Several places	R

Contd...

Table 19.2–*Contd...*

Sl.No.	*Common Name*	*Family/Scientific Name*	*Site Record*	*Status*
	Goldcrest, Prinias, Tesias, Warblers	**Sylviinae**		
102.	Streaked Fantail-Warbler (1498-1500a)	*Cisticola junidis* (Rafiesque, 1810)	PU, DH	R
103.	Franklin's Prinia (1502-1505)	*Prinia hodgsonii* Byth, 1844	BH, JO	R
104.	Ashy Prinia (1515-1518)	*Prinia socialis* Sykes, 1832	BH	R
105.	Plain Prinia (1510-1514)	*Prinia inornata* Sykes, 1832	PU	R
106.	Common Tailorbird (1535-1539)	*Orthotomus sutorius* (Pennant, 1769)	Several places	R
107.	Common Chiffchaff (1574-1575)	*Phylloscopus fuscatus* (Blyth, 1842)	PU, KLD, NIL	M
108.	Red-throated Flycatcher (1411-1412)	*Ficedula parva* (Bechstein, 1792)	BH	M
109.	Tickell's Blue-Flycatcher (1442-1430)	*Cyornis tickelliae* Blyth, 1843.	BH, JO	R
	Flycatchers	**Muscicapinae**		
110.	Asian Brown Flycatcher (1407)	*Muscicapa dauurica* Pallas, 1811	KLD	M
	Monarch-Flycatchers and Paradise-Flycatchers	**Monarchinae**		
111.	Black –naped Monarch-Flycatcher (1465-1469)	*Hypothymis azurea* (Boddaert, 1783)	DH	R
	Fantail-Flycatchers	**Rhipidurinae**		
112.	White-Throated Fantail- Flycatcher (1454-1459)	*Rhipidura abicollis* (Vieillot, 1818)	KLD, CH, NIL	R
113.	White-browed Fantail-Flycatcher (1451-1453)	*Rhipidura aureola* Lesson, 1830	CH, JO	R
	Tits	**Paridae**		
114.	Great tit (1790-1797)	*Parus major* Linnaeus, 1758	KLD, SGH	R
115.	Black-lored Yellow Tit (1809-1811)	*Parus xanthogenys* Vigors, 1831	JO	R
	Nuthatches and Wallcreeper	**Sittidae**		
116.	Chestnut-bellied Nuthatch (1827-1831)	*Sitta castanea* Lesson, 1830	JO, PU	R

Contd...

Table 19.2–*Contd...*

Sl.No.	*Common Name*	*Family/Scientific Name*	*Site Record*	*Status*
	Flowerpeckers	**Dicaeidae**		
117.	Thick-billed Flowerpecker (1892-1894)	*Dicaeum agile* (Tickell, 1833)	SGH	R
	Sunbirds and Spiderhunters	**Nectariniidae**		
118.	Purple-rumped Sunbird (1907-1908)	*Nectarinia zeylonica* (Linnaeus, 1766)	NIL, BH	R
119.	Purple Sunbird (1916-1918)	*Nectarinia asiatica* (Latham, 1790)	JO, PLR	R
	White-eyes	**Zosteropidae**		
120.	Oriental White-eye (1933-1936)	*Zosterops palpebrosus* (Temminck, 1824)	PLR, PU	R
	Munias (Estrildid Finches)	**Estrildidae**		
121.	Red Munia (1964)	*Amandava amandava* (Linnaeus, 1758)	JO	R
122.	White-throated Munia (1966)	*Lonchura malabarica* (Linnaeus, 1758)	JO, PU	R
123.	Spotted Munia (1974-1975)	*Lonchura punctulata* (Linnaeus, 1758)	KLD	R
124.	Black-headed Munia (1976-1978)	*Lonchura malacca* (Linnaeus, 1766)	PLR, PU	R
	Sparrows and Snowfinches	**Passerinae**		
125.	House Sparrow (1938-1939a)	*Passer domesticus* (Linnaeus, 1758)	PLR, KLD, CH	R
	Weavers	**Ploceinae**		
126.	Baya Weaver (1957-1959)	*Ploceus philippinus* (Linnaeus, 1766)	BH	R
	Starlings and Mynas	**Sturnidae**		
127.	Grey-headed Starling (987-989)	*Sturnus malabaricus* (Gmelin, 1789)	BH, DH, PU	R
128.	Brahminy Starling (994)	*Stunus pagodarum* (Gmelin, 1789)	Several places	R
129.	Asian Pied Starling (1002-1004)	*Sturnus contra* Linnaeus, 1758	Several places	R
130.	Common Myna (1006-1007)	*Acridotheres tristis* (Linnaeus, 1766)	Several places	R
131.	Bank Myna (1008)	*Acridotheres ginginianus* (Latham, 1790)	PLR, NIL, CH	R
132.	Jungle Myna (1009-1011)	*Acridotheres fuscus* (Wagler, 1827)	DH	R

Contd...

Table 19.2–*Contd...*

Sl.No.	*Common Name*	*Family/Scientific Name*	*Site Record*	*Status*
	Orioles	**Oriolidae**		
133.	Eurasian Golden Oriole (952-953)	*Oriolus oriolus* (Linnaeus, 1758)	JO, PU, KLD	R
134.	Black –napped Oriole (954, 956-957)	*Oriolus chinensis* Linnaeus, 1766	JO	R
135.	Black-headed Oriole (958-960a)	*Oriolus xanthornus* (Linnaeus, 1758)	Several places	R
	Drongos	**Dicruridae**		
136.	Black Drongo (962-964)	*Dicrurus macrocercus* Vieillot, 1817	Several places	R
137.	White-bellied Drongo (967-969)	*Dicrurus caerulescens* (Linnaeus, 1758)	JO	R
138.	Greater Racket-tailed Drongo (976-981)	*Dicrurus paradiseus* (Linnaeus, 1766)	JO, PU	R
	Crows, Jays, Treepies, Magpies	**Corvidae**		
139.	Indian Treepie (1030a-1034)	*Dendrocitta vagabunda* (Latham, 1790)	Several places	R
140.	House Crow (1048-1051)	*Corvus splendens* Vieillot, 1817	Several places	R
141.	Jungle Crow (1054-1057)	*Corvus macrorhynchos* Wagler, 1827	JO, BH	R

Numbers within brackets after the common names are the numbers given to species in Ali and Ripley (1987)

NT: Near Threatened (IUCN Red List Category, Birdlife International, 2006); R: Resident; M: Migrant.

Sight records

KLD: Kuldiha; CH: Chandipur; PLR: Panchulingeswar; JO: Jorachua; PU: Purnapani; SGH: Sergarh; BH: Borogohalo Hill; DH: Debagiri Hill; NIL: Nilgiri.

Table 19.3: List of Herpetofauna of Kuldiha Wildlife Sanctuary

Sl.No.	*Common Name*	*Family/Scientific Name*	*Site Record*	*Status*
	Amphibia	**Bufonidae**		
1.	Common Asian Toad	*Dattuphrynus melanostictus* Schneider, 1799	JO	R
		Ranidae		
2.	Indian Bull Frog	Bouleng *Hoplobatrachus tigerinus* 1918	JO	R
3.	Cricket Frog	*Fejervayra limnocharis* Gravenhorst, 1829	Several places	C
4.	Indian Skipping Frog	*Euphlyctis cyanophlyctis* Schneider, 1799	Several places	C
		Rhacophoridae		
5.	Tree Frog	*Polypedates maculatus* J.E.Gray, 1830	PU	C
	Squamata (Sauria)	**Scincidae**		
6.	Bronze Grass Skink	*Eutropis macularia* Blyth, 1853	JO	C
7.	Common Skink	*Eutropis carinata* Schneider, 1801	JO,PU, DH	C
8.	Sand Skink	*Eutropis bibroni* Gray, 1838	SGH	R
9.	White-spotted Supple Skink	*Lygosoma albopunctata* Gray, 1846	PU	C
		Agamidae		
10.	Indian Garden Lizard	*Calotes versicolor* Fitzinger, 1826	JO	R
11.	Forest Calotes	*Calotes rouxii* Duméril and Bibron, 1837	KU, JO	C
12.	Peninsular Rock Agama	*Psammophilus dorsalis* Gray, 1831	KLD	R
13.	Fan-throated Lizard	*Sitana ponticaniana* Cuvier, 1829	DH	R
		Gekkonidae		
14.	Asian House Gecko	*Hemidactylus frenatus* Schlegel, 1836	Several places	C
15.	Brook's House Gecko	*Hemidactylus brookii* Gray, 1845	JO	R
16.	Leaf-toed Gecko	*Hemidactylus leschenaultii* Duméril and Bibron, 1836	DH	R

Contd...

Table 19.3–*Contd...*

Sl.No.	*Common Name*	*Family/Scientific Name*	*Site Record*	*Status*
17.	East Indian Leopard Gecko	*Eublipharis hardwickii* Gray, 1827	PU	C
18.	Flat-tailed Gecko	*Cosymbotus platyurus* Schneider, 1792	DH	R
19.	Ground Gecko	*Cyrtodactylus nebulosa*	DH	R
		Varanidae		
20.	Bengal Monitor	*Varanus benghalensis* Daudin, 1802	SGH	C
	Squamata (Serpentes)	**Typhlopidae**		
21.	Diard's Worm Snake	*Typhlops diardii* Schlegel, 1839	PU	R
22.	Common Blind Snake	*Typhlops. acutus* Duméril and Bibron, 1844	CH	R
23.	Brahminy Blind Snake	*Ramphotyphlops braminus* Daudin, 1803	KLD	R
		Uropeltidae		
24.	Shield Tailed Snake	*Uropeltis ocellata* Beddome, 1863	KU	C
25.	Shield Tailed Snake	*Uropeltis ocellata* Beddome, 1863	JO	C
		Boidae		
26.	Indian Rock Python	*Python molurus* Linnaeus, 1758	SGH	C
		Colubridae		
27.	Common Water Snake	*Xenocrophis piscator* Schneider, 1799	KLD	C
28.	Buff Stripped Keel Back	*Amphiesma stolatum* Linnaeus, 1758	DH	C
29.	Common Indian Bronze Back	*Dendralaphis tristis* Gmelin, 1789	DH	C
30.	Tree snake	*Chrysopelea ornate* Shaw, 1802	KLD	R
31.	Common Bronzeback Tree Snake	*Ahaetulla nasutus* Lacépède, 1789	KLD	R
32.	Vine Snake	*Ahaetulla.nasutus issabalinus* Dutta, 2008	PU	C
33.	Common Wolf Snake	*Lycodon aulicus* Linnaeus, 1758	PU,JO	C

Contd...

Table 19.3–*Contd...*

Sl.No.	*Common Name*	*Family/Scientific Name*	*Site Record*	*Status*
34.	Yellow Speccled Wolf Snake	*Lycodon jara*	SGH	C
35.	Indian Rat Snake	*Ptyas mucosa* Linnaeus, 1758	NIL, SGH, KLD	C
36.	Cat snake	*Boiga trigonata* Schneider, 1802	DH	C
37.	Green Keelback	*Macropisthodon plumbicolor* Cantor, 1839	DH	R
38.	Olive Keelback	*Atretiuim schistesum*	KLD	C
39.	Smooth Water Snake	*Enhydris enhydris* Schneider, 1799	KLD	C
40.	Banded Racer	*Argyogera fasciolatus*	JO	C
		Elapidae		
41.	Banded Krait	*Bangarus caeruleus* Schneider, 1801	BH	C
42.	Monocled Cobra	*Naja kaouthia* Lesson, 1831	NIL, KLD	C
43.	Spectacled Cobra	*Naja naja* Laurenti, 1768	DH	R
44.	King Cobra	*Ophiophagus hannah* Cantor, 1836	NIL	R
		Viperidae		
45.	Bamboo Pit Viper	*Trimeresurus gramineus* Shaw, 1802	KLD	R
46.	Russell's Viper	*Vipera russelli* Shaw and Nodder, 1797	DH	C

R: Resident; C: Common

Sight records.

KLD: Kuldiha; CH: Chandipur; PLR: Panchulingeswar; JO: Jorachua; PU: Purnapani; SGH: Sergarh; BH: Borogohalo Hill; DH: Debagiri Hill; NIL: Nilgiri.

1788) as per the information of local village people. Identification was done using the keys after Smith (1931); Schleich and Kastle (2002); Das (1995); Dutta (1997); David and Vogel (1996).

Fish–Fauna

Fishes are important in ecosystem as a food for various species and as a predator for different species. They are important as Biological agents (Jhingran, 1991). Kuldiha Wildlife Sanctuary supports many indigenous fish species, such as *Labeo rohita, Catla catla, Wallago attu, Clarius batrachus, Anabus testudinius, Channa striatus, Punctius ticto, Puntius stigma* and *Heteropnestus fossilies.* Catfishes are very common in the sanctuary area. Law in wildlife sanctuaries bans commercial fishing, but local communities are catching the fishes for themselves as a food source.

Invertebrates

A total of 86 species were encountered in the sanctuary area (Table 19.4–19.5). During the study 2 species of annelida, 85 species of arthropod and 5 species of mollusca were collected.

Table 19.4: List of Butterflies in Kuldiha Wildlife Sanctuary

Sl.No.	*Common Name*	*Family/Scientific Name*
	Papilionidae	
1.	Glassy bluebottle	*Graphium antiphates*
2.	Common Jay	*Graphium doson*
3.	Common	*Graphium sarpedon*
4.	Blue bottle	*Graphium antiphates*
5.	Common jay	*Papilio demoleus*
6.	Lime butterfly	*Papilio polymnestor*
7.	Blue mormone	*Pachliopta aristolochiae*
8.	Common rose	*Pachliopta hector*
9.	Crimson rose	*Papilio clytia*
	Pieridae	
10.	Common jezebel	*Delias eucharis*
11.	Painted jezebel	*Delias hyparete indica*
12.	Common wanderer	*Pareronia valeria*
13.	Three spotted grass yellow	*Eurema blanda*
14.	Spotless grass yellow	*Eurema laeta*
15.	Common grass yellow	*Eurema hecabe*
16.	Small grass yellow	*Eurema brigitta*
17.	Common emigrant	*Catopsilia pomona*
18.	Mottled emigrant	*Catopsilia pyranthe*

Contd...

Table 19.4–*Contd...*

Sl.No.	*Common Name*	*Family/Scientific Name*
	Lycaenidae	
19.	Angled sunbeam	*Curetis acuta*
20.	Fluffy tit	*Zeltus amasa*
21.	Large oakblue	*Arhopala amantes*
22.	Common pierrot	*Castalius rosimon*
23.	Dark cerulean	*Jamides bochus*
	Nymphalidae	
24.	Common duffer	*Discophora sondiaca zal*
25.	Great duffer	*Discophora timora*
26.	Common evening brown	*Melanitis leda*
27.	Dark evening brown	*Melanitis phedima bela*
28.	Common fourring	*Ypthima huebneri*
29.	Common threering	*Ypthima aseterope*
30.	Common bushbrown	*Mycalesis perseus*
31.	Dark band bushbrown	*Mycalesis mineus*
32.	Common Indian crow	*Euploea core*
33.	Blue king crow	*Euploea klugii*
34.	Common tiger	*Danaus genutia*
35.	Blue tiger	*Tirumala limniace*
36.	Peacock pansy	*Junonia almana*
37.	Grey pansy	*Junonia atlites*
38.	Lemon pansy	*Junonia lemonias*
39.	Grey count	*Tanaecia lepidea*
40.	Common sailer	*Nyptis hylas*
41.	Baronet	*Euthalia nais*
45.	Angled castor	*Ariadne ariadne*
46.	Common lascar	*Pantoporia hordonia*
47.	Common sergeant	*Athyma perius*
48.	Great eggfly	*Hypolimnas bolina*

Out 18,000 species of butterfly in the world, India existing 1,501 species, from these 48 species are recorded from Kuldiha Wildlife Sanctuary. It was found that Nymphalidae is the largest family with 25 genera inside the sanctuary area. Overall species diversity in the study area was found to be very low. This is due to the natural climate, which is generally arid zone. Different species of butterfly were collected from the wet area of the sanctuary. The less number of butterfly diversity found due to the hot climate, as the rain has not been raining since long period. Basically the common Indian crow is the dominant species of that particular study area, followed

Table 19.5: List of Odonates in Kuldiha Wildlife Sanctuary

Sl.No.	*Family*	*Name as in Fraser, 1933-36*	*Name as in Prasad and Varshney, 1995*
I	**Coenagrionidae**		
1.		*Agriocnemis pygmea*	*Agriocnemis pygmea* (Rambur,1842)
2.		*Ceriagrion coromandelianum*	*Ceriagrion coromandelianum* (Fabr.,1798)
3.		*Ceriagrion olivaceum*	*Ceriagrion olivaceum* Laidlaw,1914
4.		*Ischnura delicata*	*Ischnura aurora* (Brauer,1865)
5.		*Ischnura senegalensis*	*Ischnura senegalensis* (Ramb.,1842)
6.		*Pseudagrion microcephalum*	*Pseudagrion microcephalum* (Rambur,1842)
II	**Lestidae**		
7.		*Lestes viridula*	*Lestes viridulus* Rambur,1842
III	**Calopterygidae**		
8.		*Neurobasis chinensis*	*Neurobasis chinensis* (Linnaeus,1758)
9.		*Vestalis apicalis*	*Vestalis apicalis* Selys,1873
IV	**Gomphidae**		
10.		*Ictinus rapax*	*Ictinogomphus rapax* Rambur,1842
V	**Aeshnidae**		
12.		*Anax guttatus*	*Anax guttatus* (Burmeister,1839)
VI	**Libellulidae**		
13.		*Acisoma panorpoides*	*Acisoma panorpoides* Rambur,1842
14.		*Aethriamanta brevipennis*	*Aethriamanta brevipennis* (Rambur,1842)
15.		*Brachythemis contaminata*	*Brachythemis contaminata* (Fabr.,1793)
16.		*Bradinopyga geminata*	*Bradinopyga geminata* (Rambur,1842)
17.		*Crocothemis servilia*	*Crocothemis servilia* (Drury,1770)
18.		*Diplacodes trivialis*	*Diplacodes trivialis* (Rambur,1842)
19.		*Neurothemis fulvia*	*Neurothemis fulvia* (Drury,1773)
20.		*Neurothemis tullia*	*Neurothemis tullia* (Drury,1773)
21.		*Orthetrum pruinosum*	*Orthetrum pruinosum* (Rambur,1842)
22.		*Orthetrum sabina*	*Orthetrum sabina* (Drury,1770)
23.		*Pantala flavescens*	*Pantala flavescens* (Fabr.,1798)
24.		*Potamaracha obscura*	*Potamarcha congener* (Rambur,1842)
25.		*Rhodothemis rufa*	*Rhodothemis rufa* (Rambur,1842)
26.		*Rhyothemis variegata*	*Rhyothemis variegate* Linn.,1763
27.		*Tholymis tillarga*	*Tholymis tillarga* (Fabr.,1798)
28.		*Tramea basilaris*	*Tramea basilaris* Kirby,1889
29.		*Trithemis aurora*	*Trithemis aurora* (Burmeister,1839)
30.		*Trithemis festiva*	*Trithemis festiva* (Rambur,1842)
31.		*Trithemis pallidinervis*	*Trithemis pallidinervis* Selys,1889

by common evening brown. The fluffy tit was found only once during the study period, the common tiger is also not rare in the area. The diversity of the butterfly is not very high but according the time and duration it is a good sign. Further study require for better information about the presence of the family.

In all, 58 odonate species are known to inhabit Odisha. (Srivastava and Das,1987). This study records distribution of 31 species of odonates in the Kuldiha Sanctuary under 25 genera and 6 families. A detailed systematic list of odonates is presented in Table 19.5. Of these odonates family Libellulidae (19) is well represented followed by Coenagrionidae (6), Calopterygidae (2), Gomphidae (2), Lestidae (1), and Ashenidae (1). Among the collected Libelluids, *Pentala flavescens* (Fabricius) and *Orthretum sabina sabina* (Drury) were more abundant during monsoon season. In case of *Zygopterans Ischnura aurora* (Brauer) was more abundant than the others inside the study area during the all seasons. The family and species level classification follows Davies and Tobin (1984, 1985) and Prasad and Varshney (1995). The prey of the adults consists mostly of the harmful insects of crops, orchards and forests and thus has a regulatory impact on the agro-forestry. Their aquatic larvae constitute a natural biological control over mosquito larvae and thus help to control several epidemic diseases like malaria, dengue, filarial diseases etc. (Mitra, 2002). But several developmental activities like construction of building, road, and stone crushers in the peripheral areas have directly impact on the population of the odonates as the habitat and food is being destroyed by such activities. It may lead to local extinction of sensitive species. Public awareness is required to conserve these odonates and their habitats.

Conclusions

Faunal diversity of the Kuldiha wildlife Sanctuary is high; also the area has very high conservation value. But in a very short period of time is not possible to document completely a large area of 272.75 km^2. In the sanctuary sloth bear and elephant are the key species, when other mammals have low encounter rates. Avifaunal and herpetofaunal diversity of the sanctuary also very high. Number of raptor species is also encouraging, which indicates the good health of the forest ecosystem. In case of cat families leopard density is high inside the sanctuary. The area is also well connected with Similipal Biosphere Reserve, where Sukhpada hills and Nato hills in the west act as corridors, but it is necessary to develop it more to facilitate better movement of animals between these two areas. But the sanctuary faced the high shortage of water during dry months which creates some limitations to the area Forest fire is an another problem which creates major damage to the sanctuary area. Kuldiha also faced the high grazing pressure by the livestock of fringe villages and shortage of grasslands. Other threats to the sanctuary are timber and other forest product collection, poaching, trapping and poisoning of wild animals. After the survey it was clear that man is directly or indirectly responsible for depletion of fauna and their habitat inside the sanctuary.

Forest fire is another problem which creates major damage to the sanctuary area. Watchtower is essential inside the sanctuary to monitor the forest fire and to minimize the damage. Due to the proximity of dry forest habitats to human habitation, these habitats are severely threatened by anthropogenic activities. Kuldiha faced the high grazing pressure by the livestock of fringe villages and shortage of grasslands. Other threats to the sanctuary are timber and collection of NTFP, poaching, trapping and poisoning of wild animals. An awareness programme is essential for school students, leaders, villagers and forest staffs to change the attitude of people towards conservation. Kuldiha is a rich biodiversity area but the high anthropogenic pressure decreases the richness of this area day by day. The sanctuary needs better management practices which are necessary to reduce the pressure on the forest. A long term study is needed to explore this undocumented area on faunal diversity of Kuldiha forest.

Recommendations

- ☆ A detailed long-term survey should be conducted to completely explore the sanctuary.
- ☆ Improved management of grasslands in the buffer zone will help decrease the pressure on reserve grassland and forest.
- ☆ Fire watchtower is essential inside the sanctuary to monitor the forest the forest fire and to minimize the damage.
- ☆ The illegal entry of domestic animals into the sanctuary for grazing should be controlled.
- ☆ Artificial water reservoirs should be constructed to fill up the shortage of water.
- ☆ Monitoring of raptors community will be conducted for relative abundance, distribution pattern and habitat use.
- ☆ It is necessary to develop corridors to facilitate movement of animals.
- ☆ An awareness programme is essential for school students, leaders, villagers and forest staffs to change the attitude of people towards conservation of forest and its wildlife.
- ☆ A short-term study to assessing turtle conservation information should be conducted involving visit to fringe villages as well as the forest habitats and inspecting and enumerating the species in midden piles. A longer term ecological study could also be undertaken, aimed at elucidating the important features of the life cycles of the different species; *i.e.*, breeding areas and periods, migrations (if any), nursery areas, diet and so on.

Acknowledgements

The authors are thankful to the Range Officer, Kuldiha Range for his co-operation and logistic arrangement in conducting this exercise. We are indebted to the Conservator of Forests and Field Director, Similipal Biosphere Reserve, Baripada for assigning the job to the North Orissa University. The authors are thankful to the officials of all ranks who have helped us to traverse many untraversable terrains and have shared risks and joys with us during this work.

References

Aagaard, C.J., 1924. Cobras and King Cobras. Natural History. *Bull Siam. Soc.*, 6: 315–316.

Anonymous, 2004. *Tiger Census Report*.Forest Department of Orissa.

Anonymous, 2008. Wildlife Census in Orissa, Govt. of Orissa, Forest and Environment Department, Bird Life International 2006. *Sterna acuticauda., 2006* IUCN Red List of Threatened Species. *IUCN*, 2006.

Buckland, S.T., Anderson, D. R., Burnham, K.P. and Laake, J.L., 1993. *Distance Sampling: Estimating Abundance of Biological Populations*. Chapman and Hall, London.

Burnham, K.P., Anderson, D.R. and Laake, J.L., 1980. Estimating of density from line transect sampling of biological populations. *Wildlife Monographs*, 72: 1–200.

Champion, H.G. and Seth, S.K., 1968. *A Revised Survey of the Forest Types of India*. Manager of Publication, Government of India, Delhi.

Das, I., 1995. The Reptiles of South Asia. Checklist and distribution summary. *Hamadryad*, 19: 15–40.

Das, Sunit K., Rout, S.D., Sahoo, D.P., Mohanta, P. and Brahma, A., 2010. A preliminary survey on faunal diversity of Kuldiha Wildlife Sanctuary with special reference to lesser-known forms, Orissa". *Indian Forester*, 136(10): 1391–1402.

Davies, D.A.L. and Tobin, P., 1984–1985. The dragonflies of the world: A systematic list of the extant species of Odonata. *Vols. I and II. Soc. Internat. Odonatol. Rapid Comm.* (*Suppl.*), Nos. 3 and 5, pp. 1–127 and 1–151.

Dutta, S. K., 1997. *Amphibians of India and Sri Lanka: Checklist and Bibiliography.* Odyssey Publishing House, Bhubeneshwar, pp. xiii + 342+ xxii.

David, P. and Vogel, G., 1996. *Snakes of Sumatra: An Annotated Checklist and Key with Natural History Notes*. Edition Chimaira, Frankfurt am Main. Germany, 260: 33 colour photo.

Jhingran,V.G., 1991. *Fish and Fisheries of India*, Third Edition. Hindustan Publishing Corporation (India) Delhi.

Haribal, M., 1992. *The Butterflies of Sikkim Himalaya and their Natural History*. Publishied by Sikkim Nature Conservation Foundation (SNCF).

Kehimkar, I., 2008. *The Book of Indian Butterflies*. Published by Bombay Natural History Society.

Heyer, W.R., Donnelly, M.A., Mcdiarmid, R.W., Hayek, L.C. and Foster, M.S., 1994. *Ensuring and Monitoring Biological Diversity: Standard Methods for Amphibians*. Smithsonian Institution Press. Washington.

Kumar, U. and Asija, M.J., 2002. *Biodiversity: Principles and Conservation*. Agrobios, India.

Kunte, K., 2002. *India: A Lifescape: Butterflies of Peninsular India*. University areas, Hyderabad.

Mitra, T.R., 2002. Geographical Distribution of Odonata(Insecta) of Eastern India. *Memories of the Zoological Survey of India*, 19(1): 171.

Prasad, M. and Varshney R.K., 1995. A checklist of the Odonata of India including Data on larval studies. *Oriental Insects*, 29: 385–428.

Prater, S.H., 1998. *The book of Indian Animals*. Bombay Natural History Society, Oxford University Press.

Rappole, J.H., McShea, W.J. and Vega-Rivera, J., 1993. Evaluation of two survey methods in upland avian breeding communities. *Journal of Field Ornithology*, 64: 55–70.

Saundars, D. A. and Wooler, R.D., 1988. Consistent individuality of voice in birds as a management tool. *Emu.*, 88, 25–32.

Schleich, H.H. and Kastle, W., 2002. *Amphibians and Reptiles of Nepal*. A.R.G. Gantner Verlage K.G., Ruggell, Liechtenstein, pp. x + 1201.

Smith, M.A., 1931. *The Fauna of British India, including Ceylon and Burma, Vol. 1: Loricata*. xxviii + 185 pp + 2 pl. Reprinted 1970 Ralph Curtis Books, Hollywood, Florida and 1980 Today and Tomorrow's Publishers, Dehradun.

Srivastava, V.K. and Das, S., 1987. Insecta: Odonata, pp. 135–150. In: *Fauna of Orissa, Part–I*, Zoological Survey of India, Calcutta.

Sutherland, W.J., 1996. *Ecological Census Techniques*. Cambridge University Press, Cambridge.

2013, Animal Diversity, Natural History and Conservation, Vol. 3 *Pages* **389–397**
Editors: **V.K. Gupta and Anil K. Verma**
Published by: **DAYA PUBLISHING HOUSE, NEW DELHI**

Chapter 20

Seasonal Movement Pattern of Great Indian One-Horned Rhinos (*Rhinoceros unicornis*) Outside Pobitora Wildlife Sanctuary, Assam, India

Ramesh Bhatta[1] and P.K. Saikia[2]*
[1] Research Scholar, Department of Zoology,
Gauhati University and presently working as Programme officer in Kaziranga-Karbi Anglong Landscape Conservation Programme, WWF-India
[2]Professor, Department of Zoology, In-charge,
Animal Ecology and Wildlife Biology Lab. and Centre for Wildlife Conservation and Management, Gauhati University, India

ABSTRACT

Pobitora Wildlife Sanctuary (38.81 sq. km.) holds 84 individual (As per the recent census record Assam Forest Department; 2009) of Great Indian One-horned Rhino with a sex ratio 1:2.19 and growth rate of 3.70 per cent compared to 2006 census. The actual rhino bearing area is approximately 16 km ^{2}, making it an area with the highest density of Great Indian Rhino within the current distribution range of this endangered species. It has been observed that some rhinos move out of the sanctuary in search of food (as intensive cattle grazing inside the sanctuary

* *Corresponding author.* E-mail: saikiapk@rediffmail.com

leads to degradation of major portion of grasslands), for preferred space or may be for their habit. The intensity of the movement varies in different seasons of the year. This seasonal movement of rhino to unprotected areas around Pobitora Wildlife Sanctuary leads to negative impact of rhino human interaction and also increase poaching pressure many folds. The present study was designed with forest department aiming to identified the movement route and understand the movement pattern in different season which will definitely help in better management of the species outside the protected area. After the survey, 27 major straying tracks of rhino have been identified and mapped of the track were used to move from one part of the sanctuary to other part of the Sanctuary and rest 20 Nos. were the actual straying track to the nearby village and agricultural field. During winter months straying incidences was recorded as maximum and during monsoon it was recorded as minimum. Out of 27 stray tracks 7-9 tracks are extensively used by the rhino throughout the year.

Keywords: *Great Indian One-horned Rhino, Rhinoceros unicornis, Straying movement, Seasonal movement pattern, Pobitora Wildlife Sanctuary, Assam.*

Introduction

Overall, the conservation of *Rhinoceros unicornis* in Assam and India has been a great success. In 1905, numbers of the species in Assam had declined to 10-20 rhino in Kaziranga National Park. Through strict protection, this population has recovered to over 2000 individuals. A smaller (about 84) but still secure and growing population of rhinos also occurs in Pobitora Wildlife Sanctuary. The total population in Assam now stands at around 2200 individuals (2009 census Assam Forest Department) in four separate populations in the protected areas of lf Kaziranga National Park (2045 nos), Pobitora Wildlife Sanctuary (84 Nos.), Orang National Park (64 Nos.) and Manas National Park (5 Nos. newly translocated)

The population of rhinos in Pobitora could be stated as healthy looking and indicating a steady trend of growth in the last decade. On the other hand this population is also seems to be beyond ideal density levels resulting in animals straying out of the sanctuary creating a negative human-rhino interaction. If this state is allowed to continue, it will spell doom for the rhinos in Pobitora. Hence the need for developing a strategic conservation action plan that will

- Reduce burden on the management for protection of stray rhinos
- Reduce of negative human rhino interaction
- Maintain a steady growth rate in the population
- Build positive attitude towards conservation among people and decision makers.

The specific objectives of the study were as follows:

- To identify and document important rhino straying track outside sanctuary.
- To document seasonal use of straying tracks by rhino.

This will help in preparation of proper strategy for controlling the straying problem of the rhino in the sanctuary.

Study Area

Pobitora Wildlife Sanctuary is situated in the south bank flood plains of river Brahmaputra at a distance of about 50 kilometers East from the capital city of Guwahati. The sanctuary is located in the Mayang Civil Circle (Former Mayang Kingdom; the land of black magic) area in the Western most part of Morigaon District bordering Kamrup district of Assam. The sanctuary has an area of about 38.81 km^2 and extends between latitude 26°12'N to 26°16'48"N and 91°58'48" and 92°05'24" East longitude (Figure 20.1).

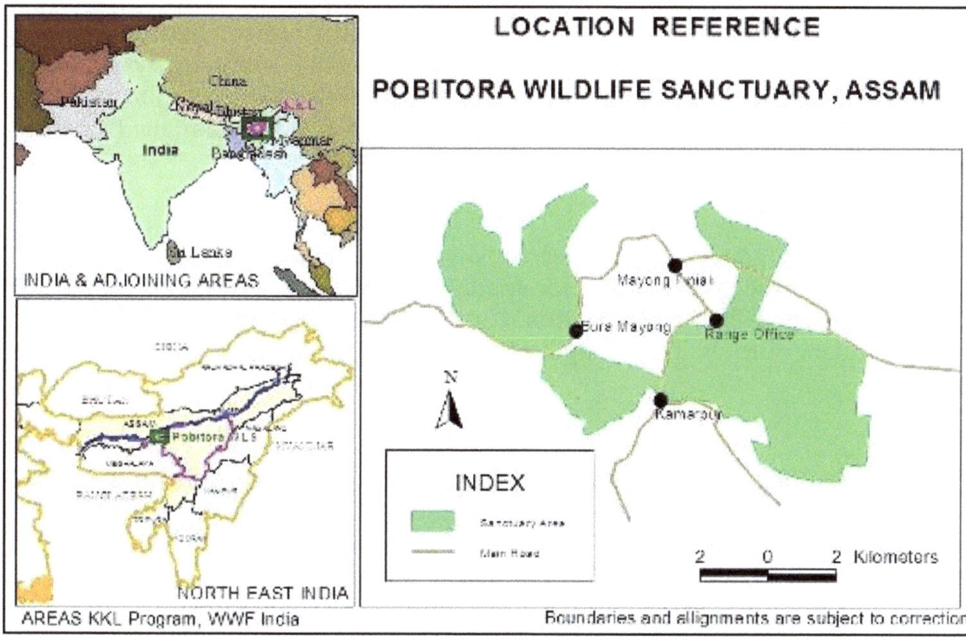

Figure 20.1: Location Map of Pobitora Wildlife Sanctuary

River Brahmaputra forms the northern boundary of the sanctuary (north of Mayang Hills), revenue villages numbering 37 (thirty-seven) lies along the east, south and western boundaries. The sanctuary is primarily a plain in nature, prone to annual floods with undulating topography in Burha-Mayang hill areas. A number of small hillocks scattered along and outside the boundary. The altitude of the area varies from about 40 meters to 350 meters. The habitat of sanctuary is mix type composing woodland, grassland and wetland with rich biodiversity. The soil of the area is mostly alluvial deposits of river (fertile clayey loam) in the low-lying area and rocky/ sandy for the hillock.

The climate of the Pobitora Wildlife Sanctuary can be treated as subtropical monsoon type with four distinct seasons (Winter, Pre-monsoon, Monsoon and

Retreating-monsoon; Borthakur 1986). Maximum and minimum temperature of the year ranges from 9°C to 35°C and average annual rainfall is approximately 2000 mm.

Materials and Methods

Linear survey following direct and indirect evidence from the sanctuary to the outside that rhino use to move has been carried out for each individual route. Latter part of retreating monsoon (September–October) and early part of pre monsoon (Feb-Mar) has been selected for survey as maximum possibility of evidence can be sighted. Historic straying records have been collected and studied prior to field survey. Experienced forest personal and local villagers accompany the survey team for better output. Secondary evidences (information from forest personal, villagers etc.) taken into consideration where route couldn't be traced/linked due to unavailability of direct evidence for bad soil type. The data for each route have been recorded in a pre-designed field format with geo-coordinates using Global Positioning System. After identification of straying route observer (FD and experienced local villagers) have been deputed for each route for monitoring rhino movement. A pre-designed datasheet has been used to record the data.

Analysis of Data

The data sets generated have been designed as a database using MS Access for integration into the GIS platform for spatial analysis and output generation. GIS software from ESRI like Arc View GIS and Spatial Analyst has been used for obtaining the preliminary outputs. The strength of the results tested using advanced statistical methods.

Results

Straying Zone-After analysis of straying incidences of rhino around Pobitora Wildlife Sanctuary during the year 2004-05 and 2005-06 it has been observed that rhinos utilized a space around 270 km^2 at 90 per cent probability level, which is 7 times compared to the area of Pobitora Wildlife Sanctuary.

Straying Track- 27 major straying tracks of rhino have been identified and mapped some (7 Nos.) of the tracks are use to move from one part of the sanctuary to other part of the sanctuary and rest 20 Nos. are the actual straying track to the nearby village and agricultural field.

Seasonal and intensity of use-daily monitoring of rhino straying in identified tracks for the year 2004-05 and 2005-06 have been analyzed as follows (*See* Figures 20.2–20.7 and Table 20.1).

Discussion

Findings of the study reveals that rhinos of the of Pobitora Wildlife Sanctuary stray out in most part of the year and utilized around 270 sqkm covering mostly agricultural areas and scattered villages, which is not a healthy situation for conservation of rhino in the area. Most of the rhino poaching incidences occur during stray. The cause of this stray behavior of rhino though to be due to the quick degradation of grassland of the sanctuary because of extensive domestic cattle grazing every day

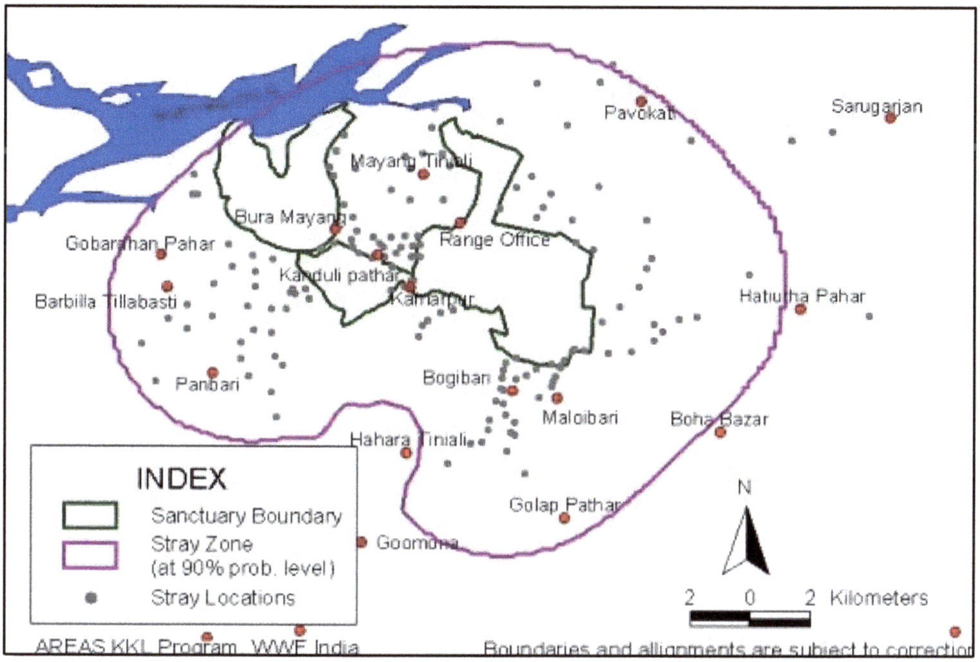

Figure 20.2: Pobitora Wildlife Sanctuary Showing Rhino Stray Zone (Area at 90 per cent probability level is about 270 kms²)

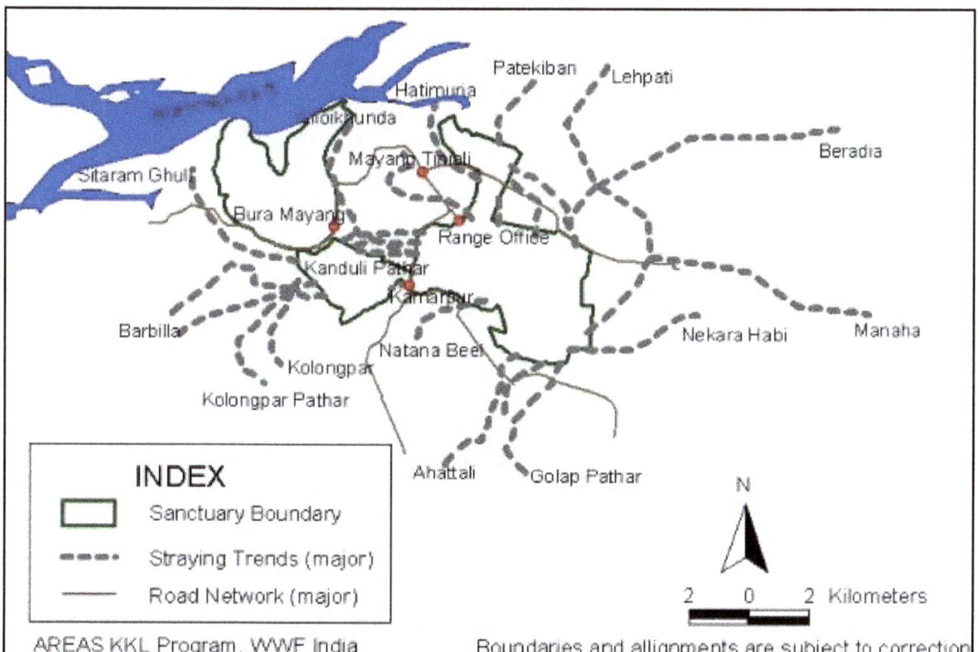

Figure 20.3: Pobitora Wildlife Sanctuary Showing Rhino Movement Pattern Outside PWLS

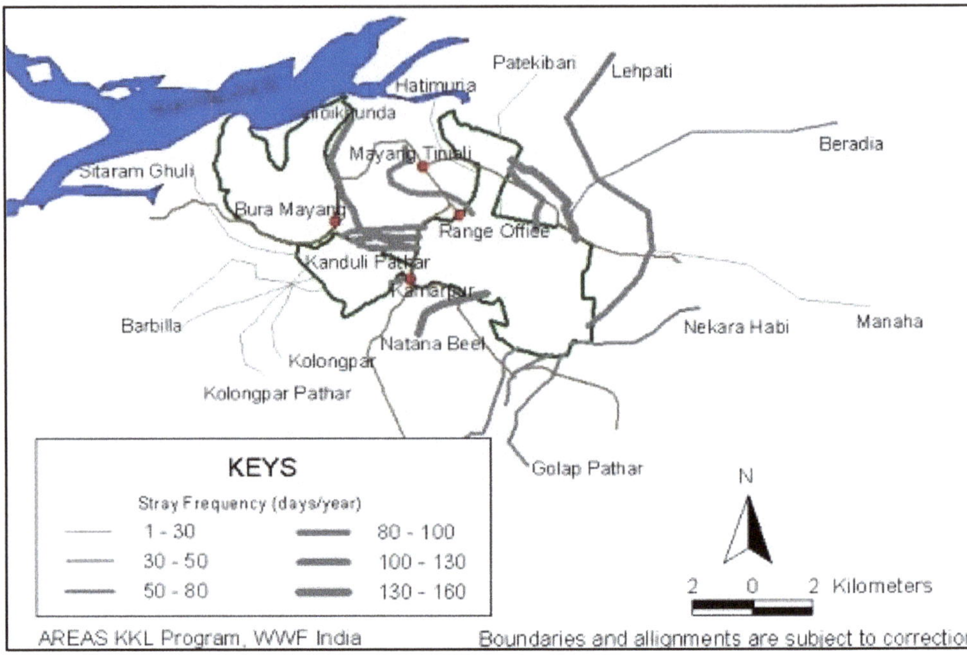

Figure 20.4: Pobitora Wildlife Sanctuary Showing Rhino Movement Intensity (2004-05) Outside PWLS

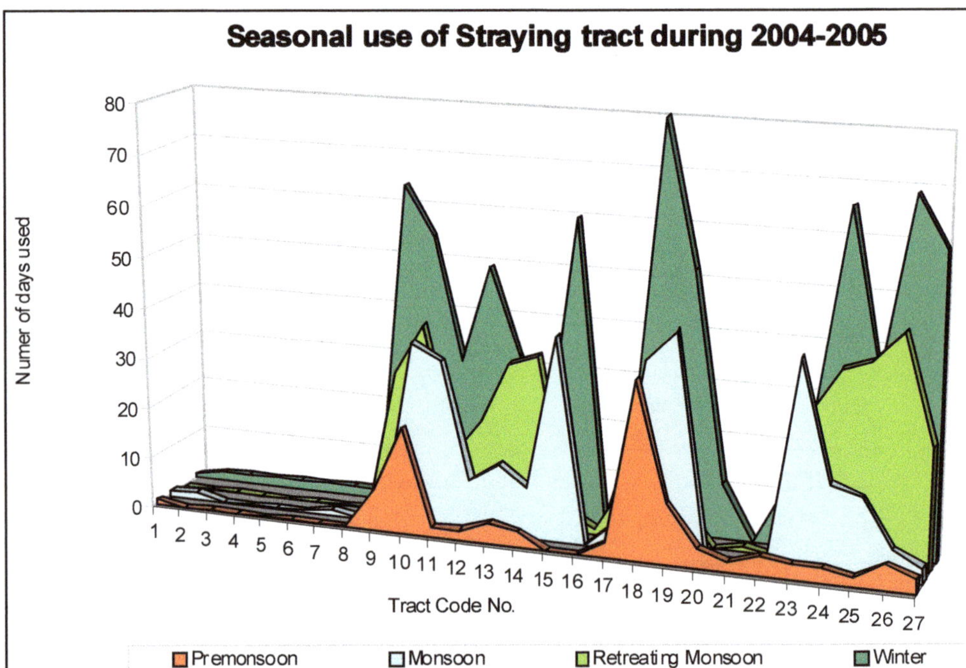

Figure 20.5: Pobitora Wildlife Sanctuary Showing Rhino-Movement Intensity (2004-05) in different Season for different Track Outside PWLS

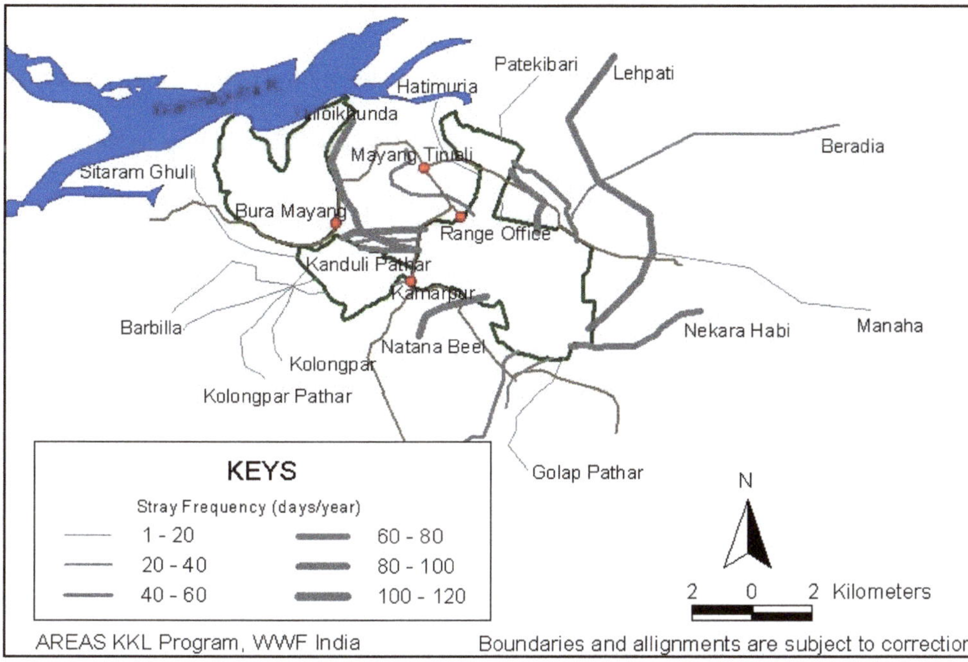

Figure 20.6: Pobitora Wildlife Sanctuary Showing Rhino Movement Intensity (2005-06) Outside PWLS

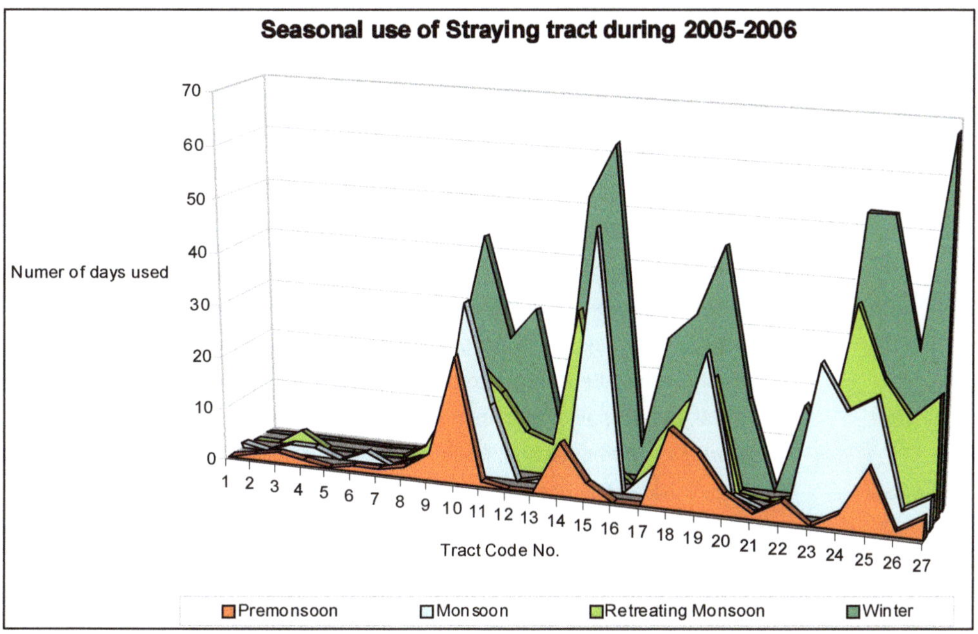

Figure 20.7: Pobitora Wildlife Sanctuary Showing Rhino Movement Intensity (2005-06) in different Season for different Track Outside PWLS

within the sanctuary, extensive collection of thatch by villagers, increased number of water buffalo grazing in the same land and effect of flood and siltation. Moreover the projected 38.81 sq km area of the sanctuary is yet not under the management controls, around 1170.24 hec. of land is occupied by villagers for agriculture. Unplanned tourism development is also one of the upcoming cause of disturbance to the animal may lead increased stray activity.

Table 20.1: Straying Tracks of Rhino around Pobitora Wildlife Sanctuary

Track Code	*Track Name*
1	Govali –Kajali Track
2	Barchapara beel–gobardhan–Barbilla Track Towards Songbeel)
3	Barchapara beel–Barbilla Track (Towards Kalbari)
4	Borchapara–Kolong- Digaru Track
5	Dhepuji–Kolong par Pathar Track (Towards Dimoria)
6	Dhepuji – Borchapara Track
7	Kardia–Barchapara beel1 Track
8	Kardia–Barchapara beel2 Track
9	Kamarpur–kamarpur Track
10	Thengbhanga- Garangamandir-Natanapathar Track
11	Thengbhanga- Bogibari–Amguri- Track (Towards Hahara).
12	Moliarpar-Thengbhanga- Bogibari Track
13	Diprang-Bogibari-Maloibari-(Towards Khetri.)
14	Dholi-Nekarahabi-bahadalani- Baha Pahar Track
15	Nekera Kukari-Jhargaon-Kwarhagi-bhurgaon garmari Track
16	Kukuri–hatiutha pahar – manaha Track
17	Kholabhuyan-pokoria – burgeon Track
18	Kholabhuyan–kukuri – dubaritali Track
19	Naltali-Sildubi-bardia-Dubaritali Track Track
20	Amaramul-Rajamayang Track
21	Dubaritali-hatibhangi–pabhakati–Patekibari Track.
22	Rajamayang-Hatimuria- Hatibhangi- Tepuri Track (towards Orang)
23	Amaramul rajamayang Kalsila- Bordal pathar Track
24	Tamlidova- Kandulipathar-Hiloikhunda Track.
25	Haduk bridge – Burhamayang Track
26	Tamulidova–Kandulipathar 1 Track
27	*Tamulidova–Kandulipathar 2 Track*

Considering all those straying activities, it could be concluded that the straying of rhino outside might be due to shortage of food and space and thus listed following recommendations to save the population from a catastrophic crash"

- ✰ All the 38.81 km^2 proposed area needs to bring under management control with proper demarcation.

- ✰ Do needful to add BARBEEL area near newly created AAMSENG Wildlife Sanctuary (Govt. khas land) with proper connecting link and protection mechanism to Pobitora Wildlife Sanctuary.
- ✰ Stop domestic cattle grazing inside the sanctuary through awareness among local villagers and by implementing community development projects.
- ✰ Reduce the number of feral buffalo grazing inside the sanctuary.
- ✰ Implement organized monitoring system for stray rhino and compensation package for affected villagers.
- ✰ Reduce the number of rhino from the sanctuary by translocation to other good habitat
- ✰ Implement proper strategy for tourism to reduce the disturbance of wild animal inside sanctuary

Acknowledgements

We thank the Department of Environment and Forests, Government of Assam for their permission and overall support to conduct the programme. Most sincere thanks to officials and the front line staff of Pobitora Wildlife Sanctuary without whom this would never have been possible. We are very much indebted for the valuable advice, guidance and support of Mr. Ravi Singh (SG and CEO) and Dr. Dipankar Ghose, WWF-India. Our heartiest thanks to Pobitora Conservation Society, Lasit Karmi Bahini Diprang, Mayang College and all other organizations/individuals that has helped us at various stages of the study.

References

Bora, C. K., 2003. Management action plane of Pobitora Wildlife Sanctuary submitted to Dept. Env. and forest Govt. of Assam.

Bairagi *et al.,* 2005. Conservation of Rhinoceros through assessment of grassland utilization by the local communities and its effect on Rhino habitat in Pobitora Wildlife Sanctuary.

Choudhary, A.U., 1996. The Greater one–horn Rhino outside the protected areas of Assam, India. *Pachyderm,* 22: 7–9.

Das, P.J., 2005. GIS mapping of Rhino stray movement from Pobitora Wildlife Sanctuary to other neighbouring civil areas.

Talukdar, B.K., 2005. Conservation of great Indian one horned Rhino at Pobitora Wildlife Sanctuary.

2013, Animal Diversity, Natural History and Conservation, Vol. 3 *Pages* **399–415**
Editors: **V.K. Gupta and Anil K. Verma**
Published by: **DAYA PUBLISHING HOUSE, NEW DELHI**

Chapter 21

Changing Habitat Considerations for Conservation of the Asian Openbill (*Anastomus oscitans*) in the Raiganj (Kulik) Bird Sanctuary, Raiganj, India

Debashis Das[1], Prithwiraj Jha[2]* and Apurba Sen[3]

[1]Department of Zoology, Tufanganj Mahavidyalaya, Tufanganj – 736 160, West Bengal, India

[2]Department of Zoology, Raiganj Surendranath Mahavidyalaya, Raiganj – 733 134, West Bengal, India

[3]Office of the Divisional Forest Officer, Raiganj Division, Raiganj – 733 134, West Bengal, India

ABSTRACT

Asian openbill (*Anastomus oscitans*) is the dominant migratory species in the Raiganj (Kulik) bird sanctuary, West Bengal, India. Being an exclusive bird sanctuary, Kulik has a good record for creating a conducive environment for openbill congregation. Over the years, many anthropogenic activities have happened in the vicinity of the sanctuary that could effect the habitat requirements. This review

---

* *Corresponding author.* E-mail: jhakingshuk@rediffmail.com

amalgamates findings of earlier papers, particularly on the openbill behaviour, nesting and feeding requirements and anthropogenic interference within and near the sanctuary, with some observations made by the present authors throughout several years with the aim of considering the changing birds' habitat that could provide effective cues to future conservation initiatives.

Keywords: *Asian openbill (Anastomus oscitans), Raiganj (Kulik) bird sanctuary, Behaviour, Nesting requirements, Feeding requirements, Anthropogenic interference.*

Introduction

The Raiganj bird sanctuary (latitude: 25°37′48.82″–25°38′21.06″ N; longitude: 88°07′09.30″–88°07′00.49″ E) is a famous heronry of eastern India. The sanctuary is also called the Kulik bird sanctuary because the river Kulik flows round a part of the sanctuary and acts as the boundary in its eastern and southern parts. The area is located in the outskirts of the town of Raiganj (district headquarters of Uttar Dinajpur, West Bengal) and was a zamindari forest till it was vested to the Government in the early 1950's. About 1.26 km^2 acre of land situated in Soharai and Bhattadighi mouzas was declared as Raiganj reserve forest with effect from 1 October 1959. Plantation activities began soon and the earliest flock of migratory Asian open-billed stork or openbill (*Anastomus oscitans*) was noticed in the year 1962. After enactment of the West Bengal wildlife protection act in 1972, the area was placed under Malda-West Dinajpur division (central circle) of the forest department. Some semblance of scientific management and conservation was initiated in 1976. Large scale plantation began in 1982 through the social forestry scheme. In 1985, an additional 0.04 km^2 of vested land in Abdulghata mouza was added and a total of 1.30 km^2 of forest land was declared as a wild life sanctuary.

Although 29 species of avifauna had been recorded in the sanctuary (Pramanik *et al.*, 2010), the area acquired the atmosphere of a bird sanctuary due to congregation of few migratory species, *i.e.*, openbill (*Anastomus oscitans*), two heron species (*Nycticorax nycticorax, Ardeola grayii*), two egret species (*Bulbulcus ibis, Egretta garzetta*) and two cormorant species (*Phalacrocorax fuscicillis, Phalacrocorax niger*) in large numbers (Pramanik *et al.*, 2010). According to Pal and Das (2008), the colonizing species were *Anastomus oscitans, Nycticorax nycticorax, Phalacrocorax niger* and three egret species (*Egretta alba, Egretta intermidia* and *Egretta garzetta*). Our observations indicated that although there were many migratory species, the four major nesting species were *Anastomus oscitans, Phalacrocorax niger, Nycticorax nycticorax* and *Egretta garzetta*. According to one estimate, Raiganj is one of the largest heronries of India as this place once had more than 10000 nests of certain bird species, of which the openbill nests were a vast majority (personal communications).

According to Pramanik *et al.* (2009), the sanctuary had a thick canopy of broadleaved deciduous trees, which provided the nesting place for the openbill. The sanctuary had a network of artificial canals connected with the river Kulik. The number of avian species recorded by Pramanik *et al.* (2010) was far less than the earlier records and the authors attributed this to destruction of habitat. Over the years, many anthropogenic activities have happened in the vicinity of the heronry that could effect the habitat requirements and thus minimize preference for the area among the migratory species. This review amalgamates findings of earlier papers with some observations made by the present authors throughout several years with the aim of considering the changing birds' habitat in the Raiganj (Kulik) bird sanctuary that could provide effective cues to future conservation initiatives.

Openbill Activities at Kulik Bird Sanctuary

At the Kulik bird sanctuary, the openbill bred together with five other members of the order Ciconiiformes (Datta and Pal, 1993). Pramanik *et al.* (2010) observed large numbers of openbill from the last week of June to mid of January of the following year. The mating behaviour had been studied in detail by Pramanik *et al.* (2009). Pairing started during the last week of June, when prospective mates flied close to each other. Many times the openbill tried to fly one above the other for a short distance. Pairs often occupied a branch of a tree and rested side by side for a few hours (Pramanik *et al.*, 2009). The birds also pecked each other's face or head. Some aggressive behaviour of both the male and female openbill was also observed by Pramanik *et al.* (2009). Copulation attempts were observed mainly on the branch of the trees. Successful cloacal contact was observed and during this time, the male stood on the back of the female and caressed the neck of the female with his beak (Pramanik *et al.*, 2009). Our observations revealed that a typical "khat khat" sound was produced due to rubbing/clasping of bills of the partners during courtship (Figure 21.1). The final stage was

Figure 21.1: Rubbing of Bills of Male and Female Openbill during Courtship

kissing of the anal pore. The mating duration was about 48 to 63 seconds. Immediately after mating, the pair got busy to improve the condition of their nest.

Each female laid 2 to 5 eggs in one reproductive cycle (Pramanik *et al.,* 2009). Initially the colour of the egg was white and after 7 days, it changed to brown (Pramanik *et al.,* 2009). Constant care of the eggs as well as the newly hatched chicks could be noticed. The most critical period of the reproductive cycle was from September and October when newly hatched individuals grew to matured form. This time, the parent birds foraged alternatively. At the age of 4-5 weeks onwards, the chicks were able to move around the nest. At about 60 days, the young ones could fly short distances. Both parents help to brood, feed and water the young (Pramanik *et al.,* 2009). The parents also supported in the training of flight (personal observations). The young ones were able to fly freely during November-December, when they could collect their own food. Another important observation was that the birds appeared shiny white when they arrived in the sanctuary (Figure 21.2), however, after the breeding phase, their colour became grayish (Figure 21.3). The openbill population left the sanctuary during the end of December (personal observations).

Nesting Requirements

The openbill used small branches of at least 12 plant species (*Ficus bengalensis, Anthocephalus indicus, Barringtonia acurangula, Ficus glomerata, Terminalia arjuna, Dalbergia sisoo, Bambusa tulda, Bombax ceiba, Alstonia scholaris, Streblus asper, Lagerstoemia*

Figure 21.2: Shiny White Coloured Openbill Collecting their Nesting Material

Figure 21.3: Grayish Coloured Openbill after Breeding Period

speciosa, Trewia nudiflora) to build their nests (Pramanik *et al.,* 2009). However, the maximum numbers of nests were built on *Lagerstoemia speciosa.* Perhaps, since these trees are branched, hard and bushy type, it helped in nest construction (Pramanik *et al.,* 2009; 2010). The openbill sometimes built their nest together with other birds like egrets, herons and cormorants in the core region (about 0.14 km^2) of the sanctuary (Pramanik *et al.,* 2010). Pal and Das (2008) reported intra and interspecific competition for roosting space and suitable nesting sites.

Although the storks occupied and defended their prospective nest sites as early as one month prior to egg laying they initiated nest building about only 15 days before. Males contributed more significantly to the finding and fetching the materials for nest building, whereas the females took charge of manipulation at the nest. Nest materials consisted of dry or wet and green branches, sticks and dry or green grasses (Pramanik *et al.,* 2009). Fresh leaves of trees such as *Ficus glomerata, Eucalyptus citriodora, Lagerstoemia speciosa, Anthocephalus indicus, Cynodon dactylon* and *Cyperus rotundus* were used more commonly in nest formation (Pramanik *et al.,* 2009). Our observations revealed that the nesting material consisted mainly of green flexible twigs as the birds bent them with their beaks to make a circle form, *i.e.,* to give a bowl shape to the nest.

Nest repairing was common throughout the season, particularly after the hatching of the young ones (personal observations). If the nests fell down due to lack of improper nest site selection or due to storm and heavy rain, the birds who did not lay eggs at that time started to rebuild a new nest (Pramanik *et al.,* 2009). An interesting

Figure 21.4: Openbill Nest Placed on a Fork of a Tree

observation made by the present authors was the changing structure of the forks of trees upon which the openbill built the nests. The nests were generally placed in the forks, *i.e.,* junction of one or more branch/small branch from a major shoot/branch of the trees (Figure 21.4). According to Pramanik *et al.* (2009), openbill always chose the top most branches of the tree for building their nests. About 98 per cent nests were situated at the junction of two or three branches and only 2 per cent nests were on the single thick strong branch of the trees. We have also collected some photographs that testify the same (Figures 21.5 and 21.6).

Due to the heavy weight of the nest material, along with the eggs/chicks and the parents, these fork areas often got deshaped in such a manner that the leaner branches around the fork bent and the area was not suitable for nest construction in the subsequent years (Figures 21.7 and 21.8). The polygamous habit of the openbill could also result in the deshaping of forks. In a study by Datta and Pal (1995), openbill usually started nesting from the first week of July. The polygamous nesting attempts were not seen until August, when almost all of the nesting trees were invaded by monogamous pairs. Nesting by monogamous pairs also continued until the last week of September. After the third week of July, most of the new settlers had no option other than to choose a preoccupied nesting tree for nesting, which often resulted in intraspecific fighting. In some cases, fighting continued for 5 to 18 minutes. The fight sometimes attracted other unattached birds of the colony, which rushed to that spot

Figures 21.5 and 21.6: Location of Openbill Nests on a Tree

Figures 21.7 and 21.8: Deshaped Fork Areas

and got involved in the disturbance. Perhaps, this congregation of large number of birds in a nest and their fighting add pressure on the branches and could result in deshaping of the forks. The birds too seem to be aware of such development since they have been observed to abandon the deshaped fork areas and select a neighbouring tree or other areas of the same tree for building their nests in the following years. If not managed properly, more and more trees would be unsuitable for nest construction. Figure 21.9 shows a tree abandoned by the birds. It is uncertain whether such deshaped trees have to be uprooted and replanted or whether effective pruning could solve the purpose.

Feeding Requirements

According to (Pramanik *et al.,* 2010), the migratory species utilized the water bodies in Raiganj bird sanctuary to feed on fish and mollusca. Openbill were chiefly mollusca eaters while herons, egrets and cormorants predominantly took fish (Pramanik *et al.,* 2010). Sharma (2007) also reported *Pila globosa* as the chief diet of the openbill in Raiganj bird sanctuary. Openbill are known to be "specialist mollusca eaters" (Sundar, 2006) and their feeding habit on snails particularly from marshy grounds has been reported by Kahl (1971). Usually, the openbill searches for prey while wading in shallow waters up to a maximum depth of about 20 cm (personal observations).

Sundar (2006) recorded the largest flock sizes of openbill in an agricultural landscape in Uttar Pradesh, India during summer and winter. The author reasoned that in summer, wetlands diminished in the landscape possibly forcing the birds to flock in the remaining (agricultural) lands. In winter, very few wetlands remained shallow after the monsoon and this again prompted flocking to the few shallow and suitable areas such as the agricultural fields (Sundar, 2006). Seasonal variations in water levels in natural habitats are known to cause cyclic variations in abundance (Powell, 1987), and in foraging behavior and diet of large, wading waterbirds due to changes in availability of resources (Kushlan, 1979, 1981; González, 1997). It thus appears that if the water level increases, it may be difficult for the openbill to get adequate food. The man made artificial canals inside the Raiganj bird sanctuary are generally quite deep, and in future, instead of creating more deep drains within the sanctuary, it would perhaps be more fruitful to create a network of shallow watered canals.

An interesting event in the areas surrounding the Raiganj bird sanctuary has been the changes in the agricultural practices adopted by the locals, *i.e.,* from rice paddies to cash crops like mustard and areca nut (personal observations). Though rice is the staple food in the region and this is unlikely to change much, however, the new dwarf rice varieties may have some impact as these dwarf varieties may not provide proper cover and shelter. Another issue is the effect of pesticides. Considerable chemical accumulation has been documented in the tissue and eggs of waterbirds and raptors that use rice fields, with evidence that birds obtained these chemicals through their prey (Kaphalia *et al.,* 1981; Ramesh *et al.,* 1992; Sanpera *et al.,* 2003; Sundar and Subramania, 2010). The accumulated levels of persistent organochlorines in south Indian birds were found to be the highest in the world with clear evidence of

Figure 21.9: A Tree Abandoned by the Birds

migratory birds acquiring considerable quantities of these chemicals from their wintering grounds in the Indian subcontinent (Ramesh *et al.,* 1992; Tanabe *et al.,* 1998). A direct result of pesticide use in agricultural fields has been the declining number of mollusks available for the openbill (personal observations). What is more alarming is the tendency among locals to cultivate rice in the Kulik river bank in areas adjacent to the sanctuary (Figure 21.10); the chemicals used thus find their way directly into the river.

Human Interference

Laying of permanent foot tracks to facilitate tourists inside the sanctuary which cut across the nesting habitats of colonial nesters may have altered the habitat to a certain extent. This development is about a decade old, when the forest department constructed concrete foot tracks (in 2001) for movement of the tourists across the sanctuary. At present, the total length of foot tracks and patrolling paths open for tourists is 2.7 km. Movement of people below nesting trees during nest- building and mating season can affect the breeding behaviour of openbill.

According to Pramanik *et al.* (2009), the majority of nest building materials were collected from the vicinity of the colony at distances ranging from a few meters up to 1 to1.5 km from the nests. During nest-building season, if tourists are allowed to venture below the nesting trees, it will prevent birds from collecting nest materials. Even though openbill are used to human interference in the Raiganj bird sanctuary

Figure 21.10: Rice Cultivation in the Kulik River Bank Adjacent to the Sanctuary

(Datta and Pal, 1993), movement of people just below the nests during peak breeding season may have its effect and this has to be studied properly.

According to Datta and Pal (1995), until 1984, people rarely ventured in the sanctuary during the openbill breeding season. After the declaration of sanctuary status in 1984/1985, there was an increasing presence of tourists along with an increase of local population (people building their residence in areas surrounding the sanctuary). Movement of tourist vehicles and forest personnel in the nest building areas did not affect the birdlife as with experience, the birds probably came to know that this kind of intervention was harmless (Datta and Pal, 1995). Rare incidents of stealing of cormorant or egret eggs and nestlings were reported by Datta and Pal (1995). Some incidents of collection of egret feathers were also reported by Datta and Pal (1995). The most direct and deleterious affect of human interference on the birdlife including openbill population within Raiganj bird sanctuary comes from people collecting firewood. Datta and Pal (1995) reported the use long bamboo poles fitted with a hook at the tip for cutting of branches located at high elevations. These activities frightened the birds which moved away to even higher branches or nearby trees leaving the nest unguarded for several hours. This disturbance also caused many well grown nestlings to fall out of their nest to the ground. Firewood collectors even collected the bird nests as these provided useful (about 0.3 to 0.5 kg) burning material (Datta and Pal, 1995). In another paper, the openbill were reported to nest in some unusual tree species at exceptionally greater heights to avoid human interference (Datta and Pal, 1990).

Whether the National Highway 34 that passes through the sanctuary creates any disturbance is not properly understood. Roads and other linear infrastructure are known to have important environmental impacts on natural habitats and ecosystems worldwide (Laurance *et al.*, 2009). However, it could be argued that the highway in mention has been in existence for many years, even before the area was declared as a bird sanctuary and the fact that the birds still keep coming year after year indicate that either highways do not severely affect the entire process of nesting, breeding or flight training of the young ones, or the cost is insignificant to the advantages offered. According to Birdlife International (2004), there are 9 recognized bird areas in India including the Raiganj Bird Sanctuary where openbill congregate (Table 21.1). Only in Raiganj, the National Highway passes through the sanctuary. Some of the other bird areas are not much affected by the highways (Table 21.1). It could be mentioned here that there are some other areas of openbill congregation that are not well documented. For example, in the Cooch Behar town of West Bengal, India, there is an area of openbill congregation located near busy roads (latitude: 26°19′34.21″ N longitude: 89°28′00.66″ E; personal observations). This site is not enlisted in Birdlife International (2004). It could be noted that the population of egrets, herons and cormorants were more in abundance in the core region than in the buffer region of Raiganj bird sanctuary but the openbill population was seen both in the core region as well as the buffer region (Pramanik *et al.*, 2010). The authors attributed this to the ability of the openbill to tolerate human interference (Pramanik *et al.*, 2010). Mukhopadhyay (1980) also reported openbill nesting colonies close to human settlements in the southern region of West Bengal.

Sharma (2007) reported certain events that could have serious implications, though. Openbill were reportedly killed by farmers by inserting poison (pesticides) in the shell of apple snail in agricultural fields to protect their newly sworn paddy saplings, because while foraging, these birds destroyed their crops (Sharma, 2007). Openbill were also trapped, poached and killed for their flesh (Sharma, 2007). The eggs were also found to be sold in the market. They were also killed for feathers for some ornamental purposes, *i.e.*, for fitting in a musical instrument known as Dhak (drum). The droppings of openbill have strong foul smell and for this reason, some people did not allow them to settle within the garden, backyard or agricultural trees. To get rid of them, some people even killed them (Sharma, 2007). These incidents, although not possible within the limits of the sanctuary implies that conservation initiatives in the catchments are not sufficient and more awareness needs to be created.

Conclusion

As referred throughout this paper, some workers have documented various aspects of the biology and conservation of the openbill in the Raiganj bird sanctuary. However, since the nature of bird habitat is ever changing, considerations for conservation should also focus on these ever-changing needs. The forest department has been trying to effectively conserve the bird habitat inside the sanctuary. Not all factors are under control of the forest management. The openbill as well as other

Table 21.1: Geographical Coordinates, Protection Status and Distance from Highways in Nine Recognized Bird Areas (According to Birdlife International, 2004) of India where Openbill Congregate

Name of Bird Area	*State*	*Coordinates*	*Altitude*	*Area*	*Habitat*	*Protected/ Unprotected*	*Distance from Highway*
Wild Ass Wildlife Sanctuary	Gujarat	23°43'N 71°1'E	3-75 m	495371 ha	Coastline, Desert, Forest, Wetlands	Protected	NH 15; 2.27 km away
Kaliveli Tank and Yeduyanthittu Estuary	Tamil Nadu	12°10'N 79°50'E	0-1 m	7500 ha	Wetlands	Unprotected	No NH
Veeranam Lake	Tamil Nadu	11°15'N 79°33'E	72 m	3885 ha	Wetlands	Unprotected	NH 45; C7.75 km away
Kolleru Lake Wildlife Sanctuary	Andhra Pradesh	16°48'N 81°23'E	0-5 m	67300 ha	Forest, Shrubland, Wetlands	Protected *	no NH
Chandka-Dampara Wildlife Sanctuary	Orissa	20°21'N 85°40'E	40-202 m	17579 ha	Forest	Protected	NH 5; 15 km away
Mangal Jodi	Orissa	20°10'N 85°37'E	74 m	unknown	Wetlands	Unprotected	NH 5; 14 km away
Danapur Cantonment area	Bihar	25°39'N 85°2'E	50 m	400 ha	Wetlands	Unprotected	Danapur city; 0.75 km away
Mokama Taal (Barah) Wetlands	Bihar	25°28'N 85°42'E	50 m	1000 ha	Wetlands	Unprotected	Danapur city; 0.8 km away
Kulik (Raiganj) Bird Sanctuary	West Bengal	25°58'N** 87°53'E**	35 m	130 ha	Forest, Wetlands	Protected	NH 34; through sanctuary

* Bird Area is partially or wholly a Ramsar Site; ** According to Birdlife International (2004).

NH: National Highway.

Figure 21.11: Construction Activities and Illegal Filling of a Roadside Water Body Near the Sanctuary

Figure 21.12: Encroached Areas Near the Sanctuary

migratory birds in the sanctuary are strongly dependent on areas outside the sanctuary for foraging, and there is considerable pressure on this land. In Raiganj, built-up areas have increased dramatically over the last two decades and most of the roadside burrow pits and a large number of ponds across the city have disappeared altogether (personal observations; Figure 21.11). In future, the sanctuary will become even more isolated by development activities and many of the prime foraging sites outside the sanctuary will most likely disappear. Besides, the incidents reported by Sharma (2007) as described earlier can hinder any conservation initiative.

Gathering of the nesting material can become a difficult task during monsoons since most of the area gets submerged. During heavy rains, Pramanik *et al.* (2010) observed the falling of many nestlings that were preyed upon by ground predators. During the floods of 1998, the present authors witnessed the drowning of several hundred young birds when they fell from their nests on the inundated grounds and were washed away by the water currents. It could thus be argued that some highlands within the sanctuary are to be made available for the birds.

Some private land patches are enclave within the sanctuary. The daily activities of the local populace along with increasing commercial activities in these areas to facilitate tourists frustrate conservation initiatives. About 25.99 acre of land adjuscent to the sanctuary owned by the zilla parishad and some private owners are used by the local people for pisciculture or paddy cultivation and could be acquired. A portion of land owned by the National Highway authority has been encroached over the years (Figure 21.12).

Climate change is definitely another issue. Conversation with locals revealed that the rainfall distribution pattern has been gradually shifting towards the autumn over the past decade. Rainfall is also irregular and this could affect birdlife since the migratory species generally arrive at the onset of monsoon (Pal and Das, 2008). That the frequency of spring rainfall could affect the timing of breeding in the Mauritius kestrel, *Falco punctatus* has been reported recently (Senapathi *et al.,* 2011). How the shifting rainfall pattern could affect birdlife in Raiganj is yet to be documented. Field studies aimed at determining important foraging sites for the heronry birds are also required. There is also an urgent need to closely monitor the birds and to implement conservation action plans.

References

Birdlife International, 2004. *Important Bird Areas in Asia: Key Sites for Conservation* (Birdlife Conservation Series No. 13). Birdlife International, Cambridge, UK.

Datta, T. and Pal, B.C., 1990. Space partitioning in relation to nesting in six bird species in the heronry of Kulik Bird Sanctuary, West Bengal. *Environment and Ecology*, 8: 86 – 91.

Datta, T. and Pal, B.C., 1993. The effect of human interference on the nesting of the openbill stork (*Anastomus oscitans*) at the Raiganj Wildlife Sanctuary, India. *Biological Conservation*, 64: 149 – 154.

Datta, T. and Pal, B.C., 1995. Polygyny in the Asian openbill (*Anastomus oscitans*). *The Auk*, 112: 257 – 260.

González, J.A., 1997. Seasonal variation in the foraging ecology of the wood stork in the southern Llanos of Venezuela. *Condor*, 99: 671–680.

Kahl, M.P., 1971. Food and feeding behaviour of the openbill stork. *Journal of Ornithology*, 112: 21–35.

Kaphalia, B.A., Husain, M.M., Seth, T.D., Kumar, A. and Murti, C.R.K., 1981. Organochlorine pesticide residues in some Indian wild birds. *Pesticides Monitoring Journal*, 15: 9–13.

Kushlan, J.A., 1979. Feeding ecology and prey selection in the White Ibis. *Condor*, 81: 376–389.

Kushlan, J.A., 1981. Resource use strategies of wading birds. *Wilson Bulletin*, 93: 145–163.

Laurance, W.F., Goosem, M. and Laurance, S.G.W., 2009. Impacts of roads and linear clearings on tropical forests. *Trends in Ecology and Evolution*, 24: 659 – 669.

Mukhopadhyay, A., 1980. Some observations on the biology of the openbill stork, *Anastomus oscitans* (Boddaert) in southern Bengal. *Journal of the Bombay Natural History Society*, 77: 133 – 137.

Pal, B.C. and Das, A.K., 2008. Nesting in night heron, *Nycticorax nycticorax* (Linnaeus) at Raiganj Wildlife Sanctuary, West Bengal, India. *Environment and Ecology*, 26: 91–99.

Powell, G.V.N., 1987. Habitat use by wading birds in a semitropical estuary: implications of hydrography. *The Auk*, 104: 740–749.

Pramanik, A.K., Santra, K.B. and Manna, C.K., 2009. Nest–building behaviour of the Asian open billed stork *Anastomus oscitans*, in the Kulik bird sanctuary, Raiganj, India. *Our Nature*, 7: 39–47.

Pramanik, A.K., Santra, K.B. and Manna, C.K., 2010. Abundance and diversity of plants and animals in the Kulik bird sanctuary, Raiganj, West Bengal, India. *Journal of Biodiversity*, 1: 13–17.

Ramesh, A., Tanabe, S., Kannan, K., Subramaniam, A.N., Kumaran, P.L. and Tatsukawa, R., 1992. Characteristic trend of persistent organochlorine contamination in wildlife from a tropical agricultural watershed, south India. *Archives of Environmental Contamination and Toxicology*, 23: 26–36.

Sanpera, C., Ruiz, X., Jover, L., Lorente, G., Jabeen, R., Muhammad, A., Boncompagni, E. and Fasola, M., 2003. Persistent organic pollutants in little egret eggs from selected wetlands in Pakistan. *Archives of Environmental Contamination and Toxicology*, 44: 360–368.

Senapathi, D., Nicoll, M.A.C., Teplitsky, C., Jones, C.G. and Norris, K., 2011. Climate change and the risks associated with delayed breeding in a tropical wild bird population. *Proceedings of the Royal Society of London. Series B: Biological Sciences*, 278: 3184–3190.

Sharma, A., 2007. A winged heritage site. *The Statesman* (*NB Extra*), August, 3, 2007.

Sundar, K.S.G., 2006. Flock size, density and habitat selection of four large waterbirds species in an agricultural landscape in Uttar Pradesh, India: implications for management. *Waterbirds*, 29: 365–374.

Sundar, K.S.G. and Subramanya, S., 2010. Bird use of rice fields in the Indian subcontinent. *Waterbirds*, 33 (Special Publication 1): 44–70.

Tanabe, S., Senthilkumar, K., Kannan, K. and Subramaniam, A.N., 1998. Accumulated features of polychlori nated biphenyls and organochlorine pesticides in resident and migratory birds from south India. *Archives of Environmental Contamination and Toxicology*, 34: 387–397.

2013, Animal Diversity, Natural History and Conservation, Vol. 3 *Pages* **417–427**
Editors: **V.K. Gupta and Anil K. Verma**
Published by: **DAYA PUBLISHING HOUSE, NEW DELHI**

Chapter 22

Human-Elephant Conflict: Case Studies from Karanjia and Suggested Measures for Mitigation

Sandeep Ranjan Mishra*[1] and Anup Kumar Nayak[2]

[1]*Research Fellow, Odisha Forest Department (Wildlife) 0/0-Regional Chief Conservator of Forest and field director, Bhanjapur, Baripada, Mayurbhanja, Odisha, India*

[2]*Regional Chief Conservator of Forest and Field director Bhanjapur, Baripada, Mayurbhanja, Odisha*

ABSTRACT

Man-elephant conflict is a major problem in Northeast and Eastern India. The present paper analysis the conflict in Karanjia Forest Division in Orissa, the worst affected by the conflict in the State, which could be symptomatic of the problem of the mining areas of adjoining forest division Keonjhar and state Jharkhand.

Conflict in Karanjia resulted in 05 human death, injury to 09 human and unnatural death/poaching of at least 13 elephant from 2005 to 2010-11. Elephant destroy at least 196 houses, partly 98 and fully 98 and damaged a minimum 522.10 of acres of paddy from 2005 to 2010-11.Peak depredation seasons were July to October and the most vulnerable hours were 1800-2400 hrs. Most human killings were by solitary male elephants and depredation of crop by family groups.

The main causes of conflicts were loss of habitat, conversion of forests to other use, increasing anthropogenic pressure on habitat, increasing elephant

* *Corresponding author.* E-mail: mishra_sandeepranjan@yahoo.co.in

population, irrigation projects and changing demographic patterns in and around forests and irrational allocation of land for mining, area affected by non-forestry land use and shifting cultivation. Most of the deaths of elephant's occurred on the unprotected movement paths of elephants, sometimes loosely called corridors. Mitigation measures presently adopted involved traditional drive-away techniques including making noise by shouting, drum beating, bursting fire crackers and fire gun shots into the air, and using torch light, pelting stones and throwing burning torches have been used in severe cases.

Keywords: *Man-animal conflict, Orissa, Karanjia, Mitigation measures.*

Introduction

The Asian Elephant *Elephas maximums* (Linn.)was once common all over the tropical south and south-east Asia, from India to Vietnam and Sumatra. Although its general range has remained almost the same, expansion of human habitation, destruction of habitat for agriculture and poaching have resulted in a sharp decline in wild populations and severe habitat fragmentation (Choudhury, 1999).Asian elephants are presently confined to 13 Asian range countries of which India holds over 50 per cent of the global population-approximately 24,000-28,000 distributed across 18 States (Menon, 2003; Sukumar 2003). In recent years, increasing human-elephant conflict has been a major issue for managing wild elephant populations in India (Sukumar, 1990; Dey, 1991; Johnsingh and Panwar, 1992; Daniel *et al.,* 1995; Nath and sukumar, 1998). Eastern Part of India particularly in Orissa elephants were abundant in dense *Shorea robusta* (Sal) forests. Orissa has about 57 per cent of the elephant habitat in central India with 1800-2000 elephant spread over 11000 km (Swain and Patnaik, 2002). Nearly 44 per cent elephant habitat falls within eleven protected areas and three elephant reserves, *viz.,* Mayurbhanj, Mahanadi and Sambalpur have been declared. With two proposed Elephant Reserves *viz.,* Brahmani-Baitarani and South Orissa Elephant reserve. This paper presents an analysis of past conflict records maintained by the Forest Department and Preliminary observations from two year survey.

Study Area

Karanjia division (maximum longitude 86°.2′-minimum longitude 85°.44′ and maximum latitude 22°.07′ to minimum latitude 21°.15′) located in Mayurbhanj district of Orissa. The reserve starts from Taradiha nala at its confluence with Baitarani river near Sarangi village. Then it proceeds along upstream of river Baitarani (enclosing within it the Santoshpur RF of Keonjhar division) up to Dalki Sahi village, *i.e.,* the boundary of Thakurmunda range of Karanjia division (which is also the district boundary between Mayurbhanj and Keonjhar), then follows the district boundary in a northerly direction upto West Deo river near Mandua village, and then proceeds upstream along West Deo river up to its crossing-point with Karanjia-Thakurmunda road. Then the boundary moves along the road up to Karanjia, then up to Singada chhak, touches NH-6, then along NH-6 up to the crossing of Nanjura nalla, from there it proceeds north-west direction along Nanjura nalla up to the confluence of

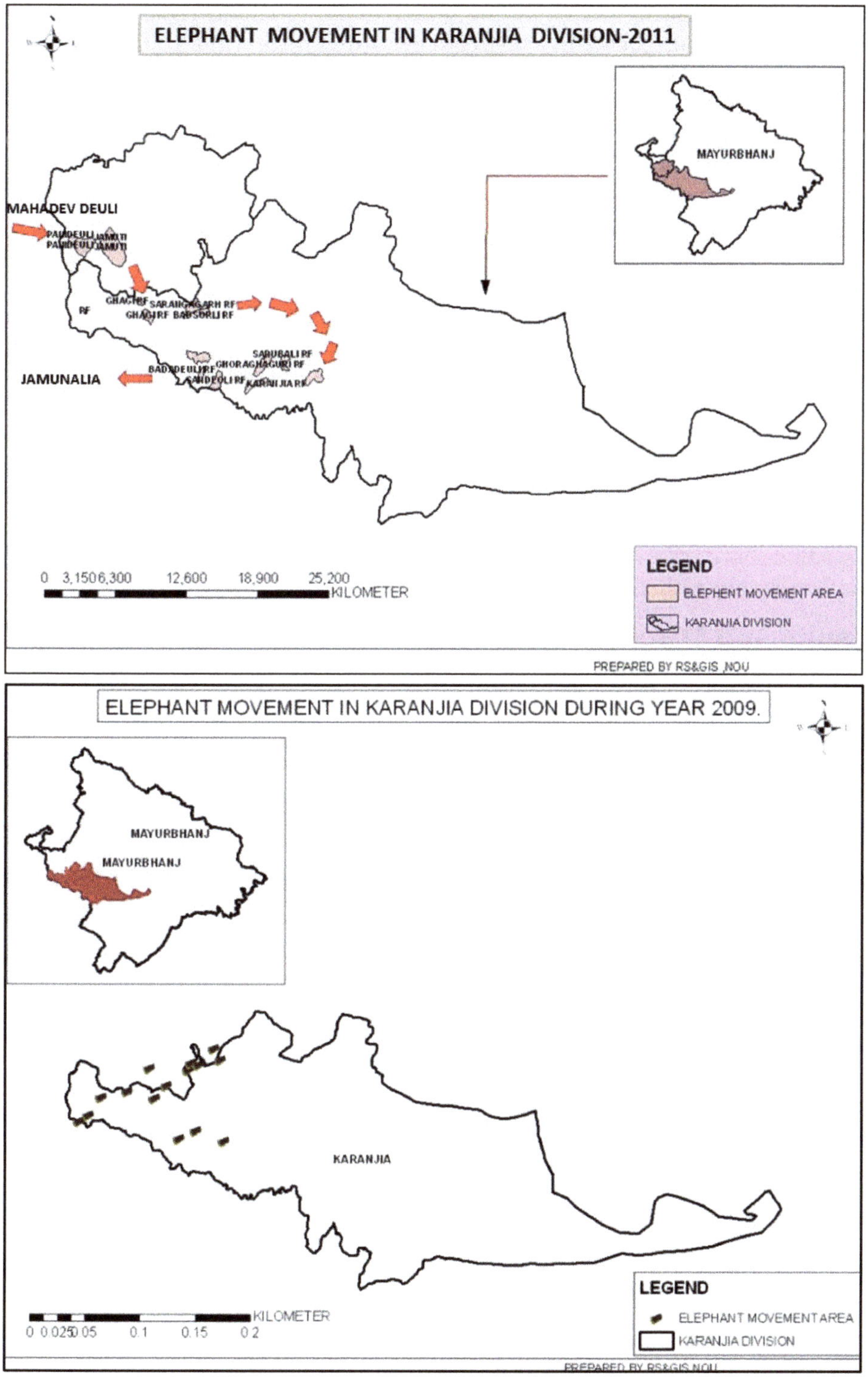

Figure 22.1: Map of the Study Area

Khairi nalla, then proceeds along the downstream of Khairi nalla up to Panposi village. Then the boundary runs upstream in north direction along Dubajora nala up to state boundary of Bihar-Orissa. Then the boundary of the elephant reserve proceeds along the Bihar-Orissa state boundary along eastern and northern direction upto north of Manbir reserved forest near village Manbir. It moves thereafter in eastern direction and encircles Manbir RF and Budheirana RF up to Edelbeda village. Then the boundary of the elephant reserve runs in an easterly direction along the boundary of the Badampahar range upto Kanhu river near Kaduani village and further runs upstream along the Kanhu river upto the confluence of Pokharia nala in the north of Pokharia village, and then runs upstream of Pokharia nalla up to boundary of Badampahar RF near Kasiabeda village. Then the boundary runs along the north of Badampahar RF in an easterly direction up to Badampahar Forest Range Hd.Qr., and then it proceeds upto Badampahar railway station and moves in a northerly direction along the Badmpahar-Tatanagar railway line up to crossing of Barhai nalla near Bahalda road railway station. Then the boundary proceeds westward along the downstream of Barhai nalla up to Dharamdihi village and then moves in north-west direction along the ridges of Churia hill, Sadam hill and Jerai RF, where it touches Bihar-Orissa State border. Then the boundary runs north-eastward along the Bihar-Orissa State border near Pandupani village up to the boundary of Baripada division near Satabankaka hill top.

Results and Discussion

Conflict in Karanjia division resulted in five human deaths, injury to nine human beings and unnatural death/poaching of at least 13 elephant from 2005 to 2010-11. Elephant destroyed at least 196 houses (Partly 98 and fully 98) and damaged a minimum 522.10 of acres of paddy from 2005 to 2010-11. Peak depredation seasons were July to October and the most vulnerable hours were 1800-2400 hrs. Most human killings were by solitary male elephants and depredation of crop by family groups.

Present Elephant Status, Movements and Habitat Utilization

Karanjia division is one of the major migration places for elephants From Keonjhar division. Elephants frequently migrated from Joda and Badbil of Keonjhar division to Karanjia division due to several reasons like developmental activities going on in Joda and Badbil, mining, habitat destruction, scarcity of food, water which are easily available in Karanjia division. Present herd consisted of male-4, female07 and calf-4 as the constituted members.

Mitigation Measures

Mitigation measures presently adopted involved traditional drive-away techniques including making noise by shouting, drum beating, bursting fire crackers and firing gun shots into the air and using torch light, pelting stones. Machans are generally used for guarding the crops.

Step Taken by Forest Department

The forest department staff of the concerned area undertake patrolling day and night and always try to drive away wild elephants from the agriculture/paddy fields and human settlements nearest reserve forest.

The police and local people jointly participate with the forest department to track and monitor the migratory elephants. Anti-depredation squads comprising of the forest official's daily wage elephant trackers.

Level of Anti-Depredation

The use of various instruments for the depredation, which has been categorized on the level of sophistication as:

Un Sophisticated

Shouting, lighted torches, noise making instruments and beating of empty tin cans

Sophisticated

Large drum, crackers, motor horns, spray of kerosene on the lighted fire torch to create fireballs, use of loud speakers and burning vehicles tyres.

Attitude and Expectation of the Local People

In India, traditional, cultural and religious attitudes towards wild animals make local people tolerant towards wild life, despite the damage to crops and livestock (Imam *et al.,* 20 02). The general reverence towards plants and animals in some Indian regions has often been reported to be the main reason for a positive attitude towards wild life and nature reserves (Sekhar 1998; Vijayan and Pati 2002; Madhusudan 2003; Mishra *et al.,* 2003) Though positive attitudes towards elephants still persists among the local people in Karanjia division. Surveys suggest that the locals are emotionally attached to elephants because of their religious beliefs. Since most of the villagers do not have any previous experience of dealing with wild elephants, they live in great fear.

Suggests Measure to Reduce Elephant-Human Conflict

The following measures may help to prevent elephant-human conflict:

1. Rigid protection of existing forests.
2. Enriching the forest with edible species of plants favoured by elephants.
3. Minimize the grazing pressure from domestic livestock and prevent forest fire.
4. Improving the water availability inside the forest by creating check dams and construction of tanks.
5. Provide corridors between large forest patches for safe movement of elephants.
6. Creation of awareness among the village and improving their skill in anti depredation measure.
7. Planting of crop not favoured by the elephants which are otherwise economically valuable.
8. Capacity building of the villagers to take up other economic activities for reducing their dependence on the forest for their sustenance

Anti-Depredation Measures

We have to locate the causal factors if it is due to migration of elephants we should not obstruct in the natural movement of elephants. Elephants are extremely intelligent animals if we chase them, obstruct their natural path, cause injury to them and even brust crackers, blank firing etc. they become violent and revengeful and create more damage. It is wise to allow their free movement from one habitat to another.

There are some wild elephant that enter villages in search of brewed rice (Handia) country liquor (Mahuli) and Paddy etc. to deal with such elephants we have to extremely cautious and Care full. Best thing is to identify such elephants to drive them out of the village we have to form a team consisting of experience villagers who have to deal such situation. These elephant generally operate in light. The team should have powerful search lights, vehicle, public address system, gun and cartridges etc. For such habitual problem elephants permanent solution would be to capture.

Dos and Don't in Elephant Depredation

- ☆ Do Not Obstruct natural movement of elephants.
- ☆ Do not allow the people to come near elephants.
- ☆ Do not tease the animal.
- ☆ Do not take part in the elephant driving operation taking alcohol or in a drunken condition. Do not allow preparation country liquor or handia inside the villages during elephant depredation.
- ☆ Never run uphill during elephant attack.
- ☆ Do not allow encroachers, illicit felling of timber and mining in the elephant habitat and in the elephant migration routes.

Do's

1. Allow the natural movement of elephant.
2. Keep the crowd away from elephants.
3. Ask the people in vicinity not to come out of their houses from sunset to sunrise.
4. Keep the people, district authority and press well informed about the movement of elephants.
5. Use powerful search light, public address system, experienced staff and local people in driving the elephants.

Reason for House Damage and Crop Damage

The house damage is only due to storage of paddy, mahua flower, handia, brewed rice and boiled rice.

Reasons for Crop Damage

Trampling and Unruly behaviour of the local people.

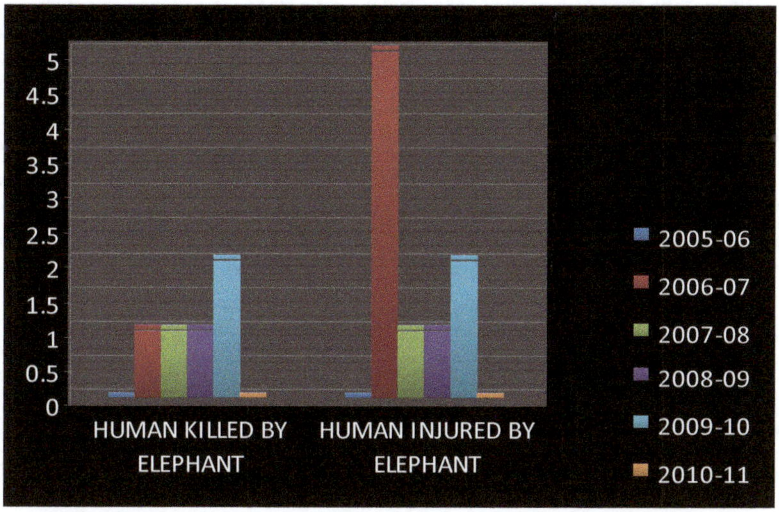

Figure 22.2: House Damaged and Crops Damaged by Elephant

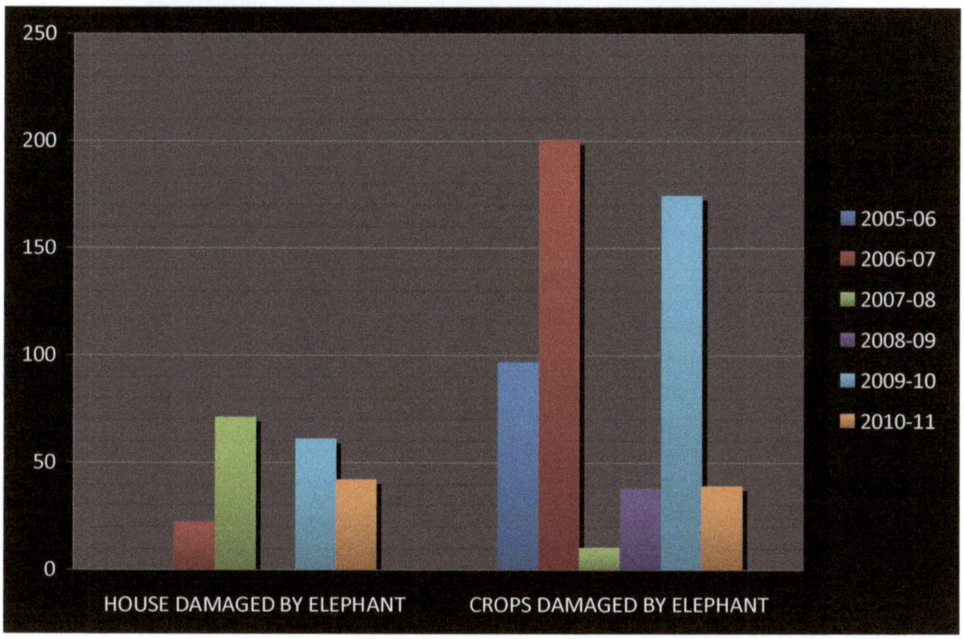

Figure 22.3: Human Killed and Injured by Elephant

Conclusions

In many parts of the world, people and animals are increasingly coming into conflict over living space and food. This is mainly due to expanding human populations and the continued loss of natural habitats. The impacts are often huge. People lose their crops, livestock, property and sometimes their lives. The animals

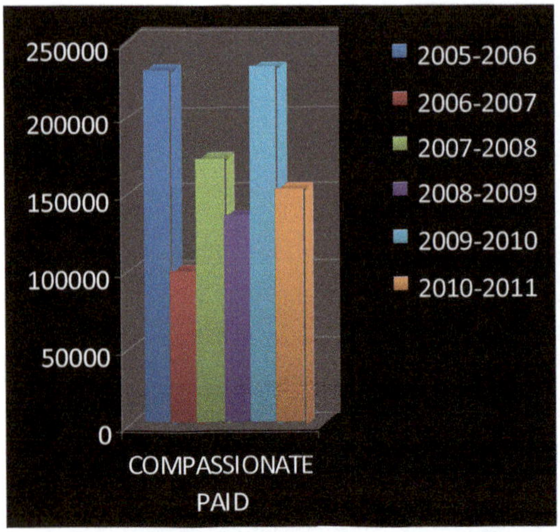

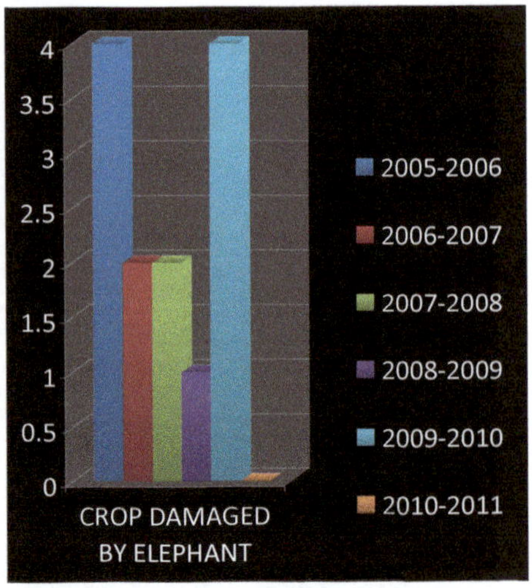

Figures 22.4 and 22.5: Crops Damaged by Elephant and Compassionate Paid

Figure 22.6: House Damaged by Elephant and Night Patrolling by Forest Department and Villagers

Figures 22.7 and 22.8: Tusker Inside the Reserve Forests and Crop Damaged by Elephant Herd

Figure 22.9: Elephant Herd Inside the Reserve Forest during Day Time

many of which are already threatened or endangered are often killed in retaliation or to prevent future conflicts.

Conflict is often between wildlife and the rural population and the poor and in many cases it is between large species and forest dwelling tribes. Most forest dwellers living inside our protected areas do not possess agricultural land or other means of livelihood. Even land is provided inside the forests the losses due to crop raiding animals will ensure that the farmer does not secure his sustenance.

The conflict between elephants and humans is not only pitching two intelligent and adaptable animals against each other but it is also face off between two complex societies competing for resources. Therefore conflict should be addressed as a whole rather than focusing attention on the reported dysfunctional behaviour of a few recalcitrant elephants. Individuals with in both communities are capable of learning although their behavior is rooted in culture. Elephants and humans are intelligent and resourceful mammals competing for resources and mitigation should not only involve minimizing conflict but also compensate the affected people while educating public and media.

"In case of elephant and man have one of the best examples known of two superficially dissimilar animals sharing common biological needs, and therefore, competing vigorously whenever they contact each other".— Alistair Graham (1973).

References

Choudhury, A.U., 1999. Status and Conservation of the Asian elephant Elephas maximums in north-eastern India. *Mammal Review*, 29(3): 141–173.

Dey, S.C., 1991. Depredation by Wildlife in fringe areas of North Bengal forest with special reference to elephant damage. *Indian Foresters*,117 (10): 901–908.

Daniel *et al.*, 1995. Ecology of the Asian elephant final report. *J.BNHS.*, 92: 1–10.

Imam *et al.*, 2002. A successful mass translocation of commercial rhesus monkeys *Macaca mulata* in Vrindaban, India. *Oryx*, 36(1): 87–93.

Johnsingh and Panwar, 1992. Elephant Conservation in India–problems and prospects. Mammals conservation in Developing countries: A New approach (P.Wegge,ed). Agriculture University of Norway, Aas, Norway, pp. 35–56.

Menon, V., 2003. A field Guide to Indian Mammals. Dorling Kindersley (India) Pvt Limited, p. 201.

Madhusudan, M.D., 2003. Living amidst large Wildlife: livestock and crop depredation by large mammals in the interior villages of Bhadra Tiger Reserve, South india. *Environmental management*, 31(4): 466–475.

Mishra, C., 2003. The role of incentive programmes in conserving the Snow leopard. *Conservation biology*, 17(6): 1512–1520

Nath and Sukumar 1998. Elephant–Human conflict in Kodagu, Southern India, distribution patterns, people's perception and mitigation methods. Asian Elephant Conservation Center, Center for Ecological Science, Indian institute of Science, Bangalore, India.

Sukumar, R., 2003. The Living Elephants: Evolutionary Ecology, Behavior and Conservation, Oxford University Press, p. 476.

Sukumar, R., 1990. Ecology of the Asian elephant in southern India. Feeding habits and Crop raiding patterns. *J. Trop. Ecol.*, 6: 33–53.

Swain and Patnaik, 2002. Elephants Of Orissa, Conservation Issues and Management Options. *Indian Foresters*, 128(2): 145–154.

Sekhar, N.U., 1998. Crop and livestock depredation caused by wild animals in protected areas: the case of Sariska Tiger Reserve, Rajasthan, India. *Environmental conservation*, 25(2): 160–171.

Vijayan, S. and Pati, B.P., 2002. Impact Of Changing Cropping Patterns on Man–Animal Conflicts Around Gir Protected Area with Specific Reference to Talala Sub–District, Gujarat, India. *Population and Environment*, 23(6): 541–559.

Previous Volumes

— Volume 1 —

2011, xv+480p., col. plts., figs., tabls., ind., 25 cm Rs. 2200

ISBN 978-81-7035-752-0

— Volume 2 —

2013, xiii+490p., col. plts., figs., tabls., ind., 25 cm **Rs. 3500**

ISBN 978-81-7035-831-2

Index

T

www.ingramcontent.com/pod-product-compliance
Ingram Content Group UK Ltd.
Pitfield, Milton Keynes, MK11 3LW, UK
UKHW022216051026
15624UKWH00001B/194

9 789351 301165